OXFORD GRADUATE TEXTS IN MATHEMATICS

An Introduction to Algebraic Geometry and Algebraic Groups

Meinolf Geck

Department of Mathematics, University of Stuttgart.

OXFORD

UNIVERSITY PRESS

OXFORD
UNIVERSITY PRESS

Great Clarendon Street, Oxford OX2 6DP
United Kingdom

Oxford University Press is a department of the University of Oxford.
It furthers the University's objective of excellence in research, scholarship,
and education by publishing worldwide. Oxford is a registered trade mark of
Oxford University Press in the UK and in certain other countries

First published 2003
First published in paperback 2013
Reprinted 2014

Published in the United States of America by Oxford University Press
198 Madison Avenue, New York, NY 10016, United States of America

British Library Cataloguing in Publication Data
Data available

Library of Congress Cataloging in Publication Data
Data available

ISBN 978-0-19-967616-3

Preface

Algebraic geometry, in its classical form, is the study of algebraic sets in affine or projective space. By definition, an algebraic set in k^n (where k is a field) is the set of all common zeros of a collection of polynomials in n variables. Algebraic groups are both groups and algebraic sets, where the group operations are given by polynomial functions. For example, the special linear group $\mathrm{SL}_n(k)$ consisting of all $n \times n$ matrices with determinant 1 is an algebraic group.

Historically, these groups were first studied in an analytic context, where the ground field k is $\mathbb{R}$ or $\mathbb{C}$. This is the classical theory of 'Lie groups' (see Chevalley (1946) or Rossmann (2002), for example), which plays an important role in various branches of mathematics. Through the fundamental work of Borel and Chevalley in the 1950s, it is known that this theory also makes sense over an arbitrary algebraically closed field. This book contains an introduction to the theory of 'groups of Lie type' over such a general ground field k; consequently, the main flavour of the exposition is purely algebraic. In the last chapter of this book, we will even exclusively study the case where k is an algebraic closure of a finite field of characteristic $p > 0$. Then the corresponding algebraic groups give rise to various families of *finite* groups. By the *classification of the finite simple groups*, every non-abelian finite simple group arises from an algebraic group over a field of characteristic $p > 0$, except for the alternating groups and 26 sporadic simple groups; see Gorenstein *et al.* (1994). This is one reason why algebraic groups over fields of positive characteristic also play an important role.

Although large parts of this book are developed in a general setting, the choice of the material and the orientation is towards groups over fields of positive characteristic and the corresponding 'finite groups of Lie type'. Many research articles in this area begin with a statement like

'Let G be a connected reductive algebraic group over $\mathbb{F}_q$; we assume that G is defined over $\mathbb{F}_q$, with corresponding Frobenius map F'

or variants thereof. Later on, it is very likely that the author will fix an F-stable Borel subgroup, an F-stable maximal torus therein, and so on. One of the aims of this book is to explain these notions carefully, to prove some fundamental results about them, and to illustrate the general theory with many examples which are worked out in great detail. In fact, this book arose from the desire to provide a general introduction into the background material from algebraic geometry which is needed to enter this web of ideas. This introduction should be at the same time as elementary as possible and yet lead to some substantial results.

One of the most striking advances in this area is the *Deligne–Lusztig* theory of representations of finite groups of Lie type. In the last three sections of this book, we will arrive at a point where we can give a first introduction to these ideas and study one family of finite groups of Lie type in detail. For those who wish to read more on this subject, there should then be no difficulty in continuing with Lusztig (1977), Carter (1985), or Digne and Michel (1991).

Each chapter of this book has an introduction which explains in more detail the content and the main results; at the end, the reader will find bibliographic remarks and exercises. Usually, there are rather explicit hints to solutions, at least for those exercises which are used in the text. As for the prerequisites, I assume a good knowledge of the material in a standard algebra course (the basics about groups, commutative rings, polynomials, and fields). In Chapter 1, we will make use of elementary properties of tensor products of vector spaces and separable field extensions. In Chapter 3, we shall need at one point the Schur–Zassenhaus theorem (on the existence and conjugacy of complements of abelian normal subgroups). In Chapter 4, we assume that the reader is willing to accept a very deep result (Grothendieck's trace formula) without proof. In the last parts of Chapter 4, we assume some familiarity with the character theory of finite groups. Otherwise, the exposition is completely self-contained and proceeds in a fairly elementary and concrete way, giving 'full proofs for everything'.

Among the potential readers I imagine a student who wishes to get a starting point for reading more advanced textbooks or more specialized research articles on groups of Lie type and their representations. I could also imagine a lecturer or researcher with a specific interest in finite simple groups, say, who wishes to have an easily accessible reference for basic *geometric* facts about groups of

Lie type. And last but not least, I imagine someone who just wishes to study some basic aspects in a beautiful area of mathematics, where algebraic geometry and finite groups (which are, a priori, quite unrelated areas!) get mixed.

In the elaboration of this book, I have used a number of sources. As far as algebraic geometry and commutative algebra are concerned, my favourite references are Atiyah and Macdonald (1969), Mumford (1988), and Shafarevich (1994). The standard textbooks on algebraic groups are Borel (1991), Humphreys (1991), and Springer (1998), which contain much more material than we can cover here. Our treatment owes a lot to the lecture notes of Steinberg (1974) and the article by Steinberg (1977), in which Steinberg gave new and, in many cases, more elementary proofs for standard results on algebraic groups. For the more specific theory of finite groups of Lie type and their characters (following Deligne–Lusztig), the standard reference is Carter (1985). Finally, the work of George Lusztig on characters of finite groups of Lie type has had the most important influence on the development of my own understanding of all these matters.

For the experts, let me try to mention some points which I consider to be the distinctive features of this book.

(1) The running examples in this book are the 'classical groups', that is, groups of matrices which leave a non-degenerate bilinear or quadratic form invariant. We will not discuss the classification of such forms, for which one can consult Dieudonné (1971) or Grove (2002), for example. Here, we introduce three families of groups (which are precisely the 'groups of Lie type' B_m, C_m, D_m) corresponding to three fixed choices of a symmetric or alternating form. We will give a complete treatment of these groups in the framework of so-called 'groups with a BN-pair', a most useful concept introduced by Tits in the 1960s. In this approach, for example, the Dickson pseudodeterminant for orthogonal groups in characteristic 2 is not needed at all; see Section 1.7.

(2) In this book, we exclusively work in the framework of affine varieties. There is no mention at all of general (quasi-projective) varieties, sheaves of functions, or such things. We do introduce projective space, but this will only play an auxiliary role, as a topological space on which an algebraic group acts. All the properties of projective varieties, grassmannian varieties, and flag varieties that are needed can be quite naturally obtained by working directly with

closed cones in the underlying affine spaces. This is the approach taken by Steinberg (1977), and it will be used to establish the main results on Borel subgroups in Section 3.4.

(3) The general structure theory of connected reductive groups shows that these groups always have a BN-pair in which B is a Borel subgroup. This important result is achieved as the culmination of a long series of arguments (see §14.15 of Borel (1991), for example), and we will not endeavour to prove it. In this book, we turn these matters upside down: We shall consider algebraic groups G with a BN-pair satisfying an additional condition which guarantees that G is reductive and that B is a Borel subgroup. As mentioned in (1), we establish the existence of such BN-pairs for all groups of classical type. In this way, we can avoid various classification results which form a tricky part in the general theory (for example, the classification of algebraic groups of dimension one).

(4) There is a general theory of fields of definition and rationality properties for algebraic varieties and groups. In this book, we only discuss this in the case of fields of positive characteristic where the rational structure is determined by a Frobenius map. As a highlight, we obtain the general order formula for finite groups of Lie type; see Section 4.2. Finally, following Carter (1985), we present some basic properties of the theory of ℓ-adic cohomology in the form of axioms, on which a number of concrete applications and explicit computations can be based. This will be illustrated in detail for the example of the finite Suzuki groups.

Most of the material in the first two chapters has been used as a basis for my undergraduate courses on elementary algebraic geometry at the University of Lyon 1 in 2000/2001/2002. The first five sections of Chapter 1 might actually be used as the core of a short course in algebraic geometry. For example, one could say more on algorithmic aspects using the excellent text of Cox *et al.* (1992); or treat in more detail algebraic curves in two-dimensional space, using Fischer (1994) or Reid (1988), for example; or one could extend the discussion of tangent spaces and Lie algebras in Section 1.4, by including more examples concerning classical groups, following Goodman and Wallach (1998). I used parts of Chapters 3 and 4 for a recent graduate course on algebraic groups over a field of characteristic $p > 0$ in Lyon and, much earlier, in 1995, in a two-semester course on algebraic geometry and algebraic groups at the RWTH Aachen.

Finally, I wish to thank David Hézard, Gerhard Hiss, and Gunter Malle, who carefully read various chapters and sent me detailed lists of corrections. For the typesetting of this text, the '*LATEX Companion*' by M. Goossens, F. Mittelbach, and A. Samarin (Addison–Wesley 1994) was an indispensible tool.

Lyon, February 2003

Contents

1

Algebraic sets and algebraic groups

In this first chapter, we introduce the basic objects: algebraic sets in affine space, the corresponding affine algebras, and algebraic groups. We proceed in a completely elementary way and assume only some very basic knowledge about groups, polynomial rings, quotient rings, and fields. The general theory will be mainly illustrated with examples concerning hypersurfaces and algebraic groups.

In Section 1.1, we give the basic definitions concerning algebraic sets and prove a version of Hilbert's nullstellensatz for hypersurfaces. As a tool for working out non-trivial examples, we introduce Groebner bases and some of their applications in Section 1.2. The dimension of an algebraic set will be defined using the Hilbert polynomial of an ideal. In Section 1.3, we introduce regular maps, direct products, and algebraic groups. As an important class of examples, we define symplectic and orthogonal groups. In Section 1.4, we prove the existence of non-singular points in irreducible algebraic sets. This theme will be continued in Section 1.5, where we introduce the Lie algebra of an algebraic group. As an application, we determine the dimensions of the orthogonal and symplectic groups.

The final two sections contain a self-contained introduction to the theory of groups with a *BN*-pair. This concept has been introduced by Tits, and it has turned out to be extremely useful. It does not only apply to algebraic groups, but also to the finite groups arising as the fixed-point set under a Frobenius map. (The latter aspect will be studied in more detail in Chapter 4.) We shall describe such *BN*-pairs for the symplectic and orthogonal groups. This will allow us to determine exactly which of these groups are connected.

1.1 The Zariski topology on affine space

In this section, we will define the basic objects of our study: algebraic sets in affine space k^n. Before we do this, we establish Hilbert's fundamental 'basis theorem'. Throughout, let k be a field and consider the polynomial ring $k[X_1, \ldots, X_n]$, where the X_i are independent variables. For any subset $S \subseteq k[X_1, \ldots, X_n]$, we denote by (S) the ideal generated by S. Thus, (S) consists of all finite sums of terms of the form hf where $h \in k[X_1, \ldots, X_n]$ and $f \in S$.

Recall that a commutative ring A with 1 is called a *noetherian* ring if every ideal of A is generated by a finite number of elements. (Note that, if I is an ideal in a noetherian ring A generated by $S \subseteq I$, then we can always choose a finite generating set for I from the elements of S.)

1.1.1 Theorem (Hilbert's basis theorem)   *If A is noetherian and X is an indeterminate over A, then $A[X]$ is also noetherian.*

In particular, if k is a field, then $k[X_1, \ldots, X_n]$ is noetherian.

Proof   The second statement follows from the first by induction on n, starting with the fact that a field k certainly is noetherian.

Now let $I \subseteq A[X]$ be a non-zero ideal and assume that I is not finitely generated. Let $d_1 \geqslant 0$ be the minimum of all $\deg(f)$ (where $\deg(f)$ denotes the degree of f) for $f \in I_0 := I \setminus \{0\}$; choose $f_1 \in I_0$ with $\deg(f_1) = d_1$. Since I is not finitely generated, $I_1 := I \setminus (f_1)$ is not empty. Let $d_2 \geqslant d_1$ be the minimum of all $\deg(f)$ for $f \in I_1$; choose $f_2 \in I_1$ with $\deg(f_2) = d_2$. Since I is not finitely generated, $I_2 := I \setminus (f_1, f_2)$ is not empty. Going on as above, we obtain a sequence of polynomials $f_1, f_2, \ldots \in I$, of degrees $d_1 \leqslant d_2 \leqslant \ldots$, such that $f_{i+1} \in I \setminus (f_1, \ldots, f_i)$ for all i. Now let $a_i \in A$ be the leading coefficient of f_i. Then the ideal $I' = (a_i \mid i \geqslant 1) \subseteq A$ is finitely generated by assumption, and so there exists some $h \geqslant 1$ such that $I' = (a_1, \ldots, a_h)$. Then $a_{h+1} \in I'$ means that $a_{h+1} = \sum_{i=1}^{h} x_i a_i$ with $x_i \in A$. We set

$$g := f_{h+1} - \sum_{i=1}^{h} x_i X^{d_{h+1} - d_i} f_i \in I \setminus (f_1, \ldots, f_h).$$

By definition, we have $\deg(g) \leqslant d_{h+1}$, but the coefficient of $X^{d_{h+1}}$ in g is $a_{h+1} - \sum_{i=1}^{h} x_i a_i = 0$. So we have $\deg(g) < d_{h+1}$, contradicting the choice of f_{h+1}. $\square$

We shall fix the following notation. Given $x = (x_1, \ldots, x_n) \in k^n$, there exists a unique k-algebra homomorphism $\varepsilon_x \colon k[X_1, \ldots, X_n] \to k$ such that $\varepsilon_x(X_i) = x_i$ for all i. To simplify notation, we shall write $f(x) = f(x_1, \ldots, x_n) = \varepsilon_x(f)$ for $f \in k[X_1, \ldots, X_n]$. In particular, any $f \in k[X_1, \ldots, X_n]$ defines a function $\dot{f} \colon k^n \to k$, $x \mapsto f(x)$. Note that, in general, f is not uniquely determined by $\dot{f}$. However, this is certainly the case if $|k| = \infty$; see Exercise 1.8.2.

1.1.2 Definition Let $S \subseteq k[X_1, \ldots, X_n]$ be any subset. Then

$$\mathbf{V}(S) := \{x \in k^n \mid f(x) = 0 \text{ for all } f \in S\}$$

is called the *algebraic set* defined by S. A subset of k^n is called *algebraic* if it is of the form $\mathbf{V}(S)$ for some subset $S \subseteq k[X_1, \ldots, X_n]$. It is readily checked that $\mathbf{V}(S) = \mathbf{V}(I)$, where $I = (S)$ is the ideal generated by S. Then, using Hilbert's basis theorem, we see that there exist $f_1, \ldots, f_r \in S$ such that

$$\mathbf{V}(S) = \mathbf{V}(\{f_1, \ldots, f_r\}).$$

Conversely, for any subset $V \subseteq k^n$, the ideal

$$\mathbf{I}(V) := \{f \in k[X_1, \ldots, X_n] \mid f(x) = 0 \text{ for all } x \in V\}$$

is called the *vanishing* ideal of V. Note that $\mathbf{I}(V)$ is an ideal of $k[X_1, \ldots, X_n]$. The quotient $A[V] := k[X_1, \ldots, X_n]/\mathbf{I}(V)$ is called the *affine algebra* of V.

1.1.3 Example (a) Consider the constant $1 \in k[X_1, \ldots, X_n]$. Then, clearly, $\mathbf{V}(\{1\}) = \varnothing$. On the other hand, $\mathbf{V}(\{0\}) = k^n$. Thus, $\varnothing$ and k^n are algebraic sets. We certainly have $\mathbf{I}(\varnothing) = k[X_1, \ldots, X_n]$. Similarly, one might expect that $\mathbf{I}(k^n) = \{0\}$, but this will only be true if $|k| = \infty$; see Exercise 1.8.2.

(b) Let $x = (x_1, \ldots, x_n) \in k^n$. Then $\{x\} = \mathbf{V}(\{X_1 - x_1, \ldots, X_n - x_n\}) \subseteq k^n$, and so the singleton set $\{x\}$ is an algebraic set in k^n.

(c) Consider the case $n = 1$. Then an algebraic set $V \subseteq k$ is the set of zeros of a collection of polynomials in $k[X_1]$. Since every non-zero polynomial in one variable has at most finitely many zeros, we conclude that $|V| < \infty$ or $V = k$. Conversely, assume that we have a finite subset $V = \{v_1, \ldots, v_m\} \subseteq k$. Then $V = \mathbf{V}(\{f\})$, where $f = (X_1 - v_1) \cdots (X_1 - v_m) \in k[X_1]$. Thus, we see that the algebraic sets in k are precisely the finite sets and k itself.

(d) Assume that k is infinite. Let $I \subseteq k[X_1, X_2, X_3]$ be the ideal generated by $f = X_2 - X_1^2$ and $g = X_3 - X_1^3$. The algebraic set $C = \mathbf{V}(I) \subseteq k^3$ is called the *twisted cubic*. Explicitly, we have

$$C = \{(x, x^2, x^3) \mid x \in k\} \subseteq k^3.$$

It is already a non-trivial task to find $\mathbf{I}(C)$. Using Exercise 1.8.4(a), we obtain that $\mathbf{I}(C) = (f, g)$.

1.1.4 Remark The following elementary properties are easily proved.

    (a) For subsets $S \subseteq S' \subseteq k[X_1, \ldots, X_n]$, we have $\mathbf{V}(S') \subseteq \mathbf{V}(S)$.

    (b) For any $S \subseteq k[X_1, \ldots, X_n]$, we have $S \subseteq \mathbf{I}(\mathbf{V}(S))$.

Similarly, we also have the following two further properties.

    (c) For subsets $V \subseteq V' \subseteq k^n$, we have $\mathbf{I}(V') \subseteq \mathbf{I}(V)$.

    (d) For any subset $V \subseteq k^n$, we have $V \subseteq \mathbf{V}(\mathbf{I}(V))$.

Let $\{V_\lambda\}_{\lambda \in \Lambda}$ be a family of subsets of k^n. Then it is clear that

$$\mathbf{I}\left(\bigcup_{\lambda \in \Lambda} V_\lambda\right) = \bigcap_{\lambda \in \Lambda} \mathbf{I}(V_\lambda).$$

Now let $\{S_\lambda\}_{\lambda \in \Lambda}$ be a family of subsets of $k[X_1, \ldots, X_n]$. Then it is clear that

$$\mathbf{V}\left(\bigcup_{\lambda \in \Lambda} S_\lambda\right) = \bigcap_{\lambda \in \Lambda} \mathbf{V}(S_\lambda).$$

In particular, the intersection of an arbitrary number of algebraic sets in k^n is again an algebraic set. Note that infinite unions of algebraic sets need not be algebraic. For example, assume that $|k| = \infty$ and let $Z \subsetneq k$ be a subset which is still infinite. Then the union of the algebraic sets $\{z\} \subseteq k$, for $z \in Z$, is not algebraic; see Example 1.1.3(c). However, we have the following result.

1.1.5 Lemma   *For any subsets $S, S' \subseteq k[X_1, \ldots, X_n]$, we have*

$$\mathbf{V}(\{f \cdot g \mid f \in S, g \in S'\}) = \mathbf{V}(S) \cup \mathbf{V}(S').$$

If S, S' are ideals in $k[X_1, \ldots, X_n]$, then we also have $\mathbf{V}(S \cap S') = \mathbf{V}(S) \cup \mathbf{V}(S')$.

Proof   Clearly, we have $\mathbf{V}(S) \cup \mathbf{V}(S') \subseteq \mathbf{V}(\{f \cdot g \mid f \in S, g \in S'\})$. To prove the reverse inclusion, let $x \in k^n$ be such that $f(x)g(x) = 0$ for all $f \in S$ and $g \in S'$. Assume, if possible, that $x \notin \mathbf{V}(S)$ and $x \notin \mathbf{V}(S')$. Then $f(x) \neq 0$ for some $f \in S$ and $g(x) \neq 0$ for some $g \in S'$. Hence, $f(x)g(x) \neq 0$, a contradiction.

    Finally, assume that S, S' are ideals. Then we have $\mathbf{V}(S) \subseteq \mathbf{V}(S \cap S')$ and $\mathbf{V}(S') \subseteq \mathbf{V}(S \cap S')$. Furthermore, for $f \in S$ and $g \in S'$, we have $f \cdot g \in S \cap S'$, and so $\mathbf{V}(S \cap S') \subseteq \mathbf{V}(\{f \cdot g \mid f \in S, g \in S')\} = \mathbf{V}(S) \cup \mathbf{V}(S')$. $\square$

1.1.6 The Zariski topology We have seen in Remark 1.1.4 that arbitrary intersections of algebraic sets in k^n are again algebraic. It follows from Lemma 1.1.5 that finite unions of algebraic sets in k^n are again algebraic. Finally, by Example 1.1.3, the empty set $\varnothing$ and k^n itself are algebraic. Thus, the algebraic sets in k^n form the closed sets of a topology on k^n, which is called the *Zariski* topology. A subset $X \subseteq k^n$ is open if $k^n \setminus X$ is closed (that is, algebraic).

Now every algebraic set $V \subseteq k^n$ is itself a topological space, with the induced topology. Explicitly, a subset $Z \subseteq V$ is closed (algebraic) if Z is the intersection of V and an algebraic set in k^n. Thus, by Remark 1.1.4, a closed subset of V is nothing but an algebraic set in k^n which is contained in V.

1.1.7 The closure of a set Given any subset $V \subseteq k^n$ (not necessarily algebraic), we denote by $\bar{V} \subseteq k^n$ its closure in the Zariski topology on k^n. Thus, $\bar{V}$ is the intersection of all algebraic subsets of k^n which contain V. We have

$$\mathbf{V}(\mathbf{I}(V)) = \bar{V}.$$

Indeed, by Remark 1.1.4, we have $V \subseteq \mathbf{V}(\mathbf{I}(V))$ and so $\bar{V} \subseteq \mathbf{V}(\mathbf{I}(V))$. Conversely, assume that $W \subseteq k^n$ is any algebraic set containing V. We write $W = \mathbf{V}(S)$ for some $S \subseteq k[X_1, \ldots, X_n]$. Then, again by Remark 1.1.4, we have $S \subseteq \mathbf{I}(\mathbf{V}(S)) = \mathbf{I}(W) \subseteq \mathbf{I}(V)$ and so $\mathbf{V}(\mathbf{I}(V)) \subseteq \mathbf{V}(S) = W$, as required.

The above result shows, in particular, that we have $\mathbf{V}(\mathbf{I}(V)) = V$ for any algebraic set $V \subseteq k^n$. Hence the operator

$$\mathbf{I} \colon \{V \subseteq k^n \mid V \text{ algebraic}\} \to \{I \subseteq k[X_1, \ldots, X_n] \mid I \text{ ideal}\}$$

is injective. The following example shows that $\mathbf{I}$ is not surjective.

1.1.8 Example Let $f \in k[X_1, \ldots, X_n]$ be non-constant. Then the algebraic set $H_f := \mathbf{V}(\{f\}) \subseteq k^n$ is called a *hypersurface*. If $n = 2$, then H_f is called a *plane curve*. Note that, with no further assumption, it may happen that $H_f = \varnothing$. (For example, if $k = \mathbb{R}$ and $f = X_1^2 + 1$.) However, if k is algebraically closed and $n \geqslant 2$, then $|H_f| = \infty$; see Exercise 1.8.2(b).

Now assume that $m \geqslant 2$. We claim that $\mathbf{I}(V) \neq (f^m)$ for any algebraic set $V \subseteq k^n$. Indeed, if we had $\mathbf{I}(V) = (f^m)$, then we would also have $V = \mathbf{V}(\mathbf{I}(V)) = \mathbf{V}(\{f^m\}) = \mathbf{V}(\{f\})$, and so $f \in \mathbf{I}(V) = (f^m)$, which is impossible.

The following result describes the vanishing ideal of a hypersurface.

1.1.9 Theorem (Hilbert's nullstellensatz for hypersurfaces)
Let k be an algebraically closed field. Let $f \in k[X_1, \ldots, X_n]$ be non-constant and $\varnothing \neq H_f \subseteq k^n$ the corresponding hypersurface; see Example 1.1.8. Write $f = f_1^{n_1} \cdots f_r^{n_r}$, where $f_1, \ldots, f_r$ are irreducible and pairwise coprime to each other. Then we have

$$H_f = H_{f_1} \cup \cdots \cup H_{f_r} \quad \text{and} \quad \mathbf{I}(H_f) = (f_1 \cdots f_r),$$

In particular, if f is irreducible (or, more generally, separable), then $\mathbf{I}(H_f) = (f)$.

Proof Assume first that f is irreducible. Relabelling the variables X_i if necessary, we may assume that X_n occurs in f, i.e. we have

$$f = a_l + a_{l-1}X_n + \cdots + a_0 X_n^l, \quad \text{where } a_i \in k[X_1, \ldots, X_{n-1}],$$

$$a_0 \neq 0, \ l \geqslant 1.$$

Now let $0 \neq g \in \mathbf{I}(H_f)$. Assume, if possible, that f does not divide g. Setting $A := k[X_1, \ldots, X_{n-1}]$, this means that $f \in A[X_n]$ is a non-constant irreducible polynomial which does not divide $g \in A[X_n]$. By Gauss' lemma, this statement remains true if we regard f and g as polynomials in $K[X_n]$, where K is the field of fractions of A. Hence, since $K[X_n]$ is a principal ideal domain, we can write $1 = \tilde{G}f + \tilde{F}g$, where $\tilde{F}, \tilde{G} \in K[X_n]$. So there exist $F, G \in A[X_n]$ such that

$$0 \neq d := Gf + Fg \in A = k[X_1, \ldots, X_{n-1}].$$

Since $|k| = \infty$, there exists some $x' = (x_1, \ldots, x_{n-1}) \in k^{n-1}$ with $a_0(x')d(x') \neq 0$; see Exercise 1.8.2. Now consider the polynomial

$$\tilde{f} := a_l(x') + a_{l-1}(x')X_n + \cdots + a_0(x')X_n^l \in k[X_n].$$

Since $a_0(x') \neq 0$ and $l \geqslant 1$, this is a non-constant polynomial. Hence, since k is algebraically closed, there exists some $x_n \in k$ such that $\tilde{f}(x_n) = 0$. Now, set $x = (x_1, \ldots, x_{n-1}, x_n) \in k^n$. Then we have $f(x) = \tilde{f}(x_n) = 0$ and so $x \in H_f$. Since $g \in \mathbf{I}(H_f)$, we also have $g(x) = 0$ and, consequently, $d(x') = d(x) = G(x)f(x) + F(x)g(x) = 0$, a contradiction. Thus, f divides g and so $g \in (f)$.

Now consider the general case and write $f = f_1^{n_1} \cdots f_r^{n_r}$, where the f_i are irreducible and pairwise coprime. Then, clearly, we have

$H_f = \mathbf{V}(\{f_1 \cdots f_r\})$ and, by Lemma 1.1.5, we have $H_f = H_{f_1} \cup \cdots \cup H_{f_r}$. This yields

$$\mathbf{I}(H_f) = \mathbf{I}(H_{f_1}) \cap \cdots \cap \mathbf{I}(H_{f_r}) = (f_1) \cap \cdots \cap (f_r),$$

where the first equality holds by Remark 1.1.4 and the second equality holds since each f_i is irreducible. Finally, we have $(f_1) \cap \cdots \cap (f_r) = (f_1 \cdots f_r)$ since $f_1 \cdots f_r$ is the least common multiple of $f_1, \ldots, f_r$. $\square$

We now discuss some topological notions concerning algebraic sets; in particular, we will discuss consequences of the fact that $k[X_1, \ldots, X_n]$ is noetherian.

1.1.10 Definition Let $Z \neq \varnothing$ be a topological space. We say that Z is *reducible* if we can write $Z = Z_1 \cup Z_2$, where $Z_1, Z_2 \subseteq Z$ are non-empty closed subsets with $Z_1 \neq Z$, $Z_2 \neq Z$. Otherwise, we say that Z is an *irreducible* topological space. A subset $Y \subseteq Z$ is called irreducible if Y is irreducible with the induced topology. (The empty set will not be considered as irreducible.)

Furthermore, we say that Z is a *noetherian* topological space if every chain of closed sets $Z_1 \supseteq Z_2 \supseteq \cdots$ in Z becomes stationary, i.e., if there exists some $m \geqslant 1$ such that $Z_m = Z_{m+i}$ for all $i \geqslant 1$.

1.1.11 Proposition *Let $Z \neq \varnothing$ be a noetherian topological space. Then there are only finitely many maximal closed irreducible subsets in Z; they are called the irreducible components of Z. If these are $Z_1, \ldots, Z_r$, we have $Z = Z_1 \cup \cdots \cup Z_r$.*

Proof First we show that Z can be written as a finite union of closed irreducible subsets. Assume, if possible, that this is not the case. Then, in particular, Z itself cannot be irreducible, and so we can write $Z = Z_1 \cup Z_1'$, where Z_1, Z_1' are proper closed subsets of Z. Now at least one of these two subsets, Z_1 say, is not irreducible, and so we can write $Z_1 = Z_2 \cup Z_2'$, where Z_2, Z_2' are proper closed subsets of Z_1. Going on this way, we find a strictly decreasing chain of closed subsets $Z_1 \supsetneq Z_2 \supsetneq Z_3 \supsetneq \cdots$ in Z, which is impossible since Z is assumed to be noetherian. Thus, our assumption was wrong and we can write $Z = Z_1 \cup \cdots \cup Z_r$, where $Z_1, \ldots, Z_r$ are closed irreducible subsets of Z. We may assume that Z_i is not contained in Z_j, for all $i \neq j$.

Next we show that each closed irreducible subset $Y \subseteq Z$ is contained in some Z_i. Indeed, we have $Y = Y \cap Z = Y \cap (Z_1 \cup \cdots \cup Z_r) = (Y \cap Z_1) \cup \cdots \cup (Y \cap Z_r)$. Now each $Y \cap Z_i$ is closed in Y. So, since Y is irreducible, we must have $Y \cap Z_i = Y$ for some i and, hence, $Y \subseteq Z_i$. Thus, the Z_i are precisely the maximal closed irreducible subsets of X. $\qquad\square$

1.1.12 Proposition   *Let $V \subseteq k^n$ be a non-empty algebraic set.*

(a) *V is noetherian with respect to the Zariski topology. Thus, we have $V = V_1 \cup \cdots \cup V_r$, where the V_i are the maximal closed irreducible subsets of V.*

(b) *V is irreducible (in the Zariski topology) if and only if $\mathbf{I}(V)$ is a prime ideal. (Note that $\mathbf{I}(V) \neq k[X_1, \ldots, X_n]$ since $V \neq \varnothing$.)*

Proof   (a) Let $V_1 \supseteq V_2 \supseteq V_3 \supseteq \cdots$ be a chain of closed (i.e algebraic) sets in V. Then we obtain a corresponding chain $\mathbf{I}(V_1) \subseteq \mathbf{I}(V_2) \subseteq \cdots$ of ideals in $k[X_1, \ldots, X_n]$. Let $I := \bigcup_{i \geqslant 1} \mathbf{I}(V_i)$. Then I is an ideal in $k[X_1, \ldots, X_n]$ and so, by Hilbert's basis theorem 1.1.1, I is finitely generated. A finite generating set for I will lie in $\mathbf{I}(V_m)$ for some $m \geqslant 1$. Then we have $\mathbf{I}(V_m) = \mathbf{I}(V_{m+1}) = \cdots$. By §1.1.7, we have $V_i = \mathbf{V}(\mathbf{I}(V_i))$ for all i, and so $V_m = V_{m+1} = \cdots$.

(b) Assume first that V is irreducible, and let $f, g \in k[X_1, \ldots, X_n]$ be such that $fg \in \mathbf{I}(V)$. Then $V \subseteq H_{fg} = H_f \cup H_g$ (see Lemma 1.1.5 for the last equality). So we obtain $V = V \cap H_{fg} = (V \cap H_f) \cup (V \cap H_g)$. Since V is irreducible, we must have $V \cap H_f = V$ or $V \cap H_g = V$, which means that $f \in \mathbf{I}(V)$ or $g \in \mathbf{I}(V)$, as desired. Conversely, assume that $\mathbf{I}(V)$ is a prime ideal and consider a decomposition $V = V_1 \cup V_2$, where V_1, V_2 are closed subsets. Assume, if possible, that $V_1 \neq V$ and $V_2 \neq V$. Using 1.1.7, this implies $\mathbf{I}(V) \subsetneqq \mathbf{I}(V_1)$ and $\mathbf{I}(V) \subsetneqq \mathbf{I}(V_2)$. So there exist some $f \in \mathbf{I}(V_1) \setminus \mathbf{I}(V)$ and some $g \in \mathbf{I}(V_2) \setminus \mathbf{I}(V)$. But, since $V = V_1 \cup V_2$, we also have $fg \in \mathbf{I}(V_1) \cap \mathbf{I}(V_2) = \mathbf{I}(V)$, which is impossible. $\qquad\square$

1.1.13 Example   (a) A singleton set is obviously irreducible. Thus, if k is finite, then any algebraic set $V \subseteq k^n$ is the finite union of its points, which are all closed and irreducible. Now assume that k is infinite. Then, by Exercise 1.8.2, we have $\mathbf{I}(k^n) = \{0\}$, and this is a prime ideal. Thus, k^n is irreducible.

(b) Assume that $|k| = \infty$ and consider the twisted cubic $C \subseteq k^3$ introduced in Example 1.1.3(d). We have $\mathbf{I}(C) = (X_2 - X_1^2, X_3 - X_1^3)$, and this is a prime ideal by Exercise 1.8.4. Thus, C is an irreducible algebraic set.

(c) Assume that k is algebraically closed and consider the hypersurface $H_f \subseteq k^n$ defined by a non-constant polynomial $f \in k[X_1, \ldots, X_n]$. By Theorem 1.1.9, we have $H_f = H_{f_1} \cup \cdots \cup H_{f_r}$, where the f_i are irreducible and pairwise coprime. We have $\mathbf{I}(H_{f_i}) = (f_i)$ for each i, and so H_{f_i} is irreducible by Proposition 1.1.12(b). Furthermore, we have $H_{f_i} \not\subseteq H_{f_j}$ for $i \neq j$, since the f_i are coprime. Thus, $H_{f_1}, \ldots, H_{f_r}$ are the irreducible components of H_f.

1.1.14 Principal open subsets Let $V \subseteq k^n$ be a non-empty algebraic set. Let $f \in k[X_1, \ldots, X_n]$ be such that $f \notin \mathbf{I}(V)$, and set

$$V_f := \{v \in V \mid f(v) \neq 0\} \subseteq V.$$

Then V_f is a non-empty open set in V and is called a *principal open set*. Any open set in V is a finite union of principal open sets. Indeed, let $U \subseteq V$ be open. Then $V \setminus U$ is closed and so there exists an algebraic set $W \subseteq k^n$ such that $V \setminus U = W \cap V$. Now, by Hilbert's basis theorem, $W \cap V$ is defined by a finite collection of polynomials, $f_1, \ldots, f_r$ say. Then we have $U = V_{f_1} \cup \cdots \cup V_{f_r}$.

There is another reason why principal open sets are important: in a suitable sense, they can be regarded as algebraic sets in their own right. (This will play a role, for example, in Section 2.2). To make this more precise, consider the polynomial ring $k[X_1, \ldots, X_n, Y]$. We identify $k^n \times k = k^{n+1}$. Then we define

$$\tilde{V}_f := \{(v, y) \in V \times k \mid f(v)y = 1\} \subseteq k^{n+1}.$$

The set $\tilde{V}_f$ is algebraic in k^{n+1} and the projection map $\pi_f \colon \tilde{V}_f \to V$, $(v, y) \mapsto v$, defines a bijection between $\tilde{V}_f$ and V_f.

1.1.15 Lemma *In the above setting, we have*

$$\mathbf{I}(\tilde{V}_f) = (I, fY - 1) \quad and \quad A[\tilde{V}_f] \cong (A[V])[Y]/(\bar{f}Y - 1),$$

where $I := \mathbf{I}(V) \subseteq k[X_1, \ldots, X_n]$ and $\bar{f}$ denotes the image of f in $A[V]$. Thus, in the notation of Exercise 1.8.11, $A[\tilde{V}_f] \cong A[V]_{\bar{f}}$ is the localization of $A[V]$ in $\bar{f}$.

Proof We certainly have $(I, fY - 1) \subseteq \mathbf{I}(\tilde{V}_f)$. To prove the reverse inclusion, let $g \in k[X_1, \ldots, X_n, Y]$ be such that $g(v, y) = 0$ for all $(v, y) \in \tilde{V}_f$. We must show that $g \in (I, fY - 1)$. To see this, let us write $g = \sum_{i=0}^{r} g_i Y^i$, where $g_i \in k[X_1, \ldots, X_n]$; we set $g' := \sum_{i=0}^{r} f^{r-i} g_i \in k[X_1, \ldots, X_n]$. Then we have $f^{r+1} g \equiv f g' \bmod (fY - 1)$ and so $fg' \in \mathbf{I}(\tilde{V}_f)$. Now, since $g' \in k[X_1, \ldots, X_n]$, we have $g'(v) = g'(v, y) = 0$ for all $v \in V_f$ (where $y \in k$ is such that $f(v)y = 1$). On the other hand, $f(v) = 0$ for all $v \in V \setminus V_f$, and so $fg' \in I = \mathbf{I}(V)$. Thus, we have shown that $f^{r+1} g \in (I, fY - 1)$. Writing $g = (1 - (fY)^{r+1})g + Y^{r+1}(f^{r+1}g)$ and noting that $1 - (fY)^{r+1}$ is divisible by $1 - fY$ in $k[X_1, \ldots, X_n, Y]$, we conclude that $g \in (I, fY - 1)$. The statement about $A[\tilde{V}_f]$ is then clear. $\qquad\square$

1.2   Groebner bases and the Hilbert polynomial

The aim of this section is to develop some basic tools for working with ideals in $k[X_1, \ldots, X_n]$. More precisely, we introduce Groebner basis and the Hilbert polynomial of an ideal. For this purpose, let us fix some notation. Let

$$\mathbb{Z}_{\geqslant 0}^n := \{\alpha = (\alpha_1, \ldots, \alpha_n) \in \mathbb{Z}^n \mid \alpha_i \geqslant 0 \text{ for all } i\}.$$

For $\alpha = (\alpha_1, \ldots, \alpha_n) \in \mathbb{Z}_{\geqslant 0}^n$, we denote the corresponding monomial by $X^\alpha := X_1^{\alpha_1} \cdots X_n^{\alpha_n}$. Then any non-zero $f \in k[X_1, \ldots, X_n]$ is expressed uniquely as

$$f = \sum_{\alpha} a_\alpha X^\alpha, \quad \text{where } \alpha \in \mathbb{Z}_{\geqslant 0}^n \text{ and } a_\alpha \in k,$$

with $a_\alpha = 0$ for all but finitely many α. We write $|\alpha| = \alpha_1 + \cdots + \alpha_n$. The maximum of all $|\alpha|$ such that $a_\alpha \neq 0$ is called the degree of f and denoted by $\deg(f)$. (If $f = 0$, we set $\deg(f) = -\infty$.)

1.2.1 Definition   A total order $\preceq$ on $\mathbb{Z}_{\geqslant 0}^n$ is called a *monomial order* if the following conditions hold.

    (a) We have $(0, \ldots, 0) \preceq \alpha$ for all $\alpha \in \mathbb{Z}_{\geqslant 0}^n$.
    (b) For any $\alpha, \beta, \gamma \in \mathbb{Z}_{\geqslant 0}^n$ we have: $\alpha \preceq \beta \Rightarrow \alpha + \gamma \preceq \beta + \gamma$.

Now let $0 \neq f = \sum_{\alpha} a_\alpha X^\alpha \in k[X_1, \ldots, X_n]$. The expressions $a_\alpha X^\alpha$, where $a_\alpha \neq 0$, are called the *terms* of f. Let α_0 be maximal (with respect to a given monomial order $\preceq$) such that $a_{\alpha_0} \neq 0$. Then $\mathrm{LT}(f) := a_{\alpha_0} X^{\alpha_0}$ is called the *leading* term and $\mathrm{LM}(f) := X^{\alpha_0}$ is called the *leading* monomial of f.

Given $\alpha, \beta \in \mathbb{Z}_{\geqslant 0}^n$, we also write $X^\alpha \preceq X^\beta$ if $\alpha \preceq \beta$. Note that

$$X^\alpha \mid X^\beta \Rightarrow \alpha \preceq \beta.$$

1.2.2 Example   (a) The lexicographic order **LEX**: Let $\alpha, \beta \in \mathbb{Z}_{\geqslant 0}^n$, $\alpha \neq \beta$. Let $i \in \{1, \dots, n\}$ be minimal such that $\alpha_i \neq \beta_i$. Then we write $\alpha \preceq \beta$ if $\alpha_i < \beta_i$.

(b) The graded lexicographic order **GR-LEX**: Let $\alpha, \beta \in \mathbb{Z}_{\geqslant 0}^n$, $\alpha \neq \beta$. If $|\alpha| \neq |\beta|$, we write $\alpha \preceq \beta$ if $|\alpha| < |\beta|$. On the other hand, if $|\alpha| = |\beta|$, we write $\alpha \preceq \beta$ if $\alpha_i < \beta_i$ where, as before, i is minimal such that $\alpha_i \neq \beta_i$.

It is easily verified that these two order relations indeed are monomial orders.

1.2.3 Lemma   *Let $I \subseteq k[X_1, \dots, X_n]$ be an ideal which is generated by a set G of monomials. Then a polynomial $f \in k[X_1, \dots, X_n]$ lies in I if and only if every term of f is divisible by some $g \in G$.*

Proof   If every term of f is divisible by some $g \in G$, then f certainly lies in I. Now assume that $f \in I$ and write $f = \sum_{i=1}^{r} h_i g_i$, where $h_i \in k[X_1, \dots, X_n]$ and $g_i \in G$. Expressing each h_i as a k-linear combination of monomials and expanding the products, we see that f is a k-linear combinations of terms of form $X^\alpha g_i$, for various i and $\alpha \in \mathbb{Z}_{\geqslant 0}^n$, as required. $\square$

1.2.4 Lemma   *Let $(g_i)_{i \geqslant 1}$ be a sequence of monomials in $k[X_1, \dots, X_n]$ such that $g_1 \succeq g_2 \succeq g_3 \succeq \cdots$ for some monomial order $\preceq$. Then there exists some $r \geqslant 1$ such that $g_r = g_{r+1} = g_{r+2} = \cdots$.*

Proof   Let I be the ideal generated by all g_i ($i \geqslant 1$). By Hilbert's basis theorem, I is generated by the first, r say, elements $g_1, \dots, g_r$. Now let $i > r$; then $g_r \succeq g_i$. On the other hand, since $g_i \in I$, there exists some $j \in \{1, \dots, r\}$ such that g_j divides g_i; see Lemma 1.2.3. Then $g_i \succeq g_j \succeq g_r$ and so $g_i = g_r$, as desired. $\square$

1.2.5 Proposition (division algorithm)   *Let us fix a monomial order $\preceq$ and let $f, f_1, \dots, f_s \in k[X_1, \dots, X_n]$ be non-zero. Then we have $f = h_1 f_1 + \cdots + h_s f_s + r$, where $r, h_1, \dots, h_s \in k[X_1, \dots, X_n]$ are such that $\mathrm{LT}(h_i f_i) \preceq \mathrm{LT}(f)$ for all i with $h_i \neq 0$, and where $r = 0$ or no term of r is divisible by $\mathrm{LT}(f_i)$ for any $1 \leqslant i \leqslant s$.*

Proof Since every non-empty collection of monomials in $k[X_1, \ldots, X_n]$ has a minimal element by Lemma 1.2.4, we can proceed by downward induction on $\mathrm{LT}(f)$ with respect to $\preceq$. If f is constant, we can take $h_i = 0$ and $r = f$. Now assume that f is non-constant. If some $\mathrm{LT}(f_i)$ divides $\mathrm{LT}(f)$, we set

$$\tilde{f} := f - \frac{\mathrm{LT}(f)}{\mathrm{LT}(f_i)}\, f_i.$$

If $\tilde{f} = 0$, we are done. Otherwise, we have $\mathrm{LT}(\tilde{f}) \prec \mathrm{LT}(f)$ and we can apply induction to $\tilde{f}$. Thus, we have an expression $\tilde{f} = \tilde{h}_1 f_1 + \cdots + \tilde{h}_s f_s + r$ satisfying the above conditions. Then $f = h_1 f_1 + \cdots + h_s f_s + r$, where $h_j = \tilde{h}_j$ for $j \neq i$ and $h_i = \tilde{h}_i + \mathrm{LT}(f)/\mathrm{LT}(f_i)$, and all the required conditions are satisfied. On the other hand, if no such i exists, we set $\tilde{f} := f - \mathrm{LT}(f)$. If $\tilde{f} = 0$, we are done. Otherwise, we can apply induction to $\tilde{f}$ and obtain an expression $\tilde{f} = h_1 f_1 + \cdots + h_s f_s + \tilde{r}$ satisfying the above conditions. Then we have $f = h_1 f_1 + \cdots + h_s f_s + r$, where $r = \tilde{r} + \mathrm{LT}(f)$, and all the required conditions are satisfied. $\qquad\square$

1.2.6 Example Let us consider the order **LEX** in $k[X, Y]$ with $X \succeq Y$. Let $f = XY^2 - X$ and $f_1 = XY + 1$, $f_2 = Y^2 - 1$. Then we have

$$f = Y \cdot f_1 + 0 \cdot f_2 + (-X - Y) \quad \text{and} \quad f = 0 \cdot f_1 + X \cdot f_2 + 0.$$

Both expressions satisfy all the conditions in Proposition 1.2.5. So the 'remainder' r is not uniquely determined. Even worse, the second expression shows that $f \in (f_1, f_2)$, while this is not obvious from the first expression.

1.2.7 Definition (Buchberger) Let $I \subseteq k[X_1, \ldots, X_n]$ be a non-zero ideal and $\preceq$ be a monomial order on $\mathbb{Z}^n_{\geqslant 0}$. A finite subset $G \subseteq I \setminus \{0\}$ is called a *Groebner basis* of I if the monomials $\mathrm{LT}(g)$ $(g \in G)$ generate the ideal

$$(\mathrm{LT}(I)) := (\mathrm{LT}(f) \mid 0 \neq f \in I) \subseteq k[X_1, \ldots, X_n].$$

By Hilbert's basis theorem, every non-zero ideal has a Groebner basis.

1.2.8 Theorem *Let $I \subseteq k[X_1, \ldots, X_n]$ be a non-zero ideal and $\preceq$ be a monomial order on $\mathbb{Z}^n_{\geqslant 0}$. Let G be a Groebner basis of I. Then* *we have $I = (G)$, and a k-basis of $k[X_1, \ldots, X_n]/I$ is given by the residue classes of X^α, where α runs over the elements in*

$$C(I) := \{\alpha \in \mathbb{Z}_{\geqslant 0}^n \mid X^\alpha \text{ is not divisible by } \mathrm{LT}(g) \text{ for any } g \in G\}.$$

Proof Let $G = \{f_1, \ldots, f_s\}$ and $0 \neq f \in k[X_1, \ldots, X_n]$. By Proposition 1.2.5, we can write $f = h_1 f_1 + \cdots + h_s f_s + r$, where $r = 0$ or no term of r is divisible by $\mathrm{LT}(f_i)$ for any i. Note that this means that r is a k-linear combination of monomials X^α, where $\alpha \in C(I)$. This already shows that the residue classes of the monomials X^α $(\alpha \in C(I))$ span $k[X_1, \ldots, X_n]/(G)$.

Now let $f \in I$ and assume, if possible, that $r \neq 0$. Then $0 \neq r = f - \sum_{i=1}^s h_i f_i \in I$ and so $\mathrm{LT}(r) \in (\mathrm{LT}(f_1), \ldots, \mathrm{LT}(f_s))$. By Lemma 1.2.3, some $\mathrm{LT}(f_i)$ must divide $\mathrm{LT}(r)$, contradicting the conditions on r in Proposition 1.2.5. Thus, we have $r = 0$. This shows, in particular, that $I = (G)$ and that the residue classes of X^α $(\alpha \in C(I))$ are linearly independent in $k[X_1, \ldots, X_n]/I$. $\square$

1.2.9 Example (a) Let $0 \neq f \in k[X_1, \ldots, X_n]$ and $I = (f)$. Choose any monomial order on $\mathbb{Z}_{\geqslant 0}^n$. Then $\mathrm{LT}(f)$ divides $\mathrm{LT}(g)$ for any non-zero $g \in I$. (Indeed, write $g = hf$ with $h \in k[X_1, \ldots, X_n]$. Using Definition 1.2.1(b), we see that $\mathrm{LT}(g) = \mathrm{LT}(h)\,\mathrm{LT}(f)$.) Hence, $\{f\}$ is a Groebner basis of I.

(b) Consider the ideal $I = (X_2 - X_1^2, X_3 - X_1^3) \subseteq k[X_1, X_2, X_3]$; see Example 1.1.3(d). If we choose a lexicographic order with $X_1 \preceq X_2 \preceq X_3$, then Exercise 1.8.4 shows that $\{X_2 - X_1^2, X_3 - X_1^3\}$ is a Groebner basis of I. On the other hand, choosing a graded lexicographic order with $X_3 \preceq X_2 \preceq X_1$, it turns out that $\{X_1^2 - X_2, X_1 X_2 - X_3, X_1 X_3 - X_2^2\}$ is a Groebner basis of I.

(c) Let $\{0\} \neq I \subseteq k[X_1, \ldots, X_n]$ be an ideal which is generated by a finite set of monomials, G say. Then G is a Groebner basis for any monomial order $\preceq$. Indeed, we clearly have $G \subseteq (\mathrm{LT}(I))$. Conversely, let $0 \neq f \in I$. Then $\mathrm{LT}(f)$ is divisible by some $g \in G$ (see Lemma 1.2.3) and so $\mathrm{LT}(f) \in (G)$.

1.2.10 Remark There are algorithms for computing Groebner bases of any non-zero ideal in $k[X_1, \ldots, X_n]$. Without giving the proof, let us just describe a simple version of *Buchberger's* algorithm, following §2.7 of Cox *et al.* (1992). First, we shall need the following definition. Let $f, g \in k[X_1, \ldots, X_n]$ be non-zero. We write $\mathrm{LM}(f) = X^\alpha$ and $\mathrm{LM}(g) = X^\beta$ and set $\gamma = (\gamma_1, \ldots, \gamma_n)$ where

$\gamma_i = \max\{\alpha_i, \beta_i\}$ for all i. Then the *S-polynomial* of f and g is defined by

$$S(f, g) := \frac{X^\gamma}{\mathrm{LT}(f)} \cdot f - \frac{X^\gamma}{\mathrm{LT}(g)} \cdot g,$$

Now let $I = (f_1, \ldots, f_s)$ be an ideal in $k[X_1, \ldots, X_n]$ where $f_i \neq 0$ for all i. Then a Groebner basis G is constructed by the following steps:

Initialize $G' := \{f_1, \ldots, f_s\}$. For each pair $p \neq q$ in G', compute $S(p, q)$ and divide $S(p, q)$ by G', leaving a remainder. Add all non-zero remainders that you obtain in this way to G' and repeat the above procedure. After a finite number of iterations, you will find that, for any pair $p \neq q$ in G', the remainder of $S(p, q)$ under division by G' is zero. Then $G = G'$ is a Groebner basis for I.

For practical issues (computer implementation etc.) see, for example, the **SINGULAR** system described by Greuel and Pfister (2002).

We also remark that, in general, Groebner bases are not unique. A Groebner basis G is called *reduced* if all $g \in G$ satisfy the following conditions.

(a) The coefficient of $\mathrm{LT}(g)$ is 1.
(b) No term of g is divisible by $\mathrm{LT}(g')$ for any $g' \in G$, $g' \neq g$.

Given a Groebner basis G, it is easy to construct a new Groebner basis which is reduced; furthermore, such a reduced Groebner basis is uniquely determined. For more details on all of this, see Cox *et al.* (1992).

Our next aim is to define an invariant which measures the size of an ideal $I \subseteq k[X_1, \ldots, X_n]$. For this purpose, we study the 'growth' of $k[X_1, \ldots, X_n]/I$.

1.2.11 Definition Let $I \subseteq k[X_1, \ldots, X_n]$ be an ideal. For any integer $s \geqslant 0$, we set $k[X_1, \ldots, X_n]_{\leqslant s} := \{f \in k[X_1, \ldots, X_n] \mid \deg(f) \leqslant s\}$ and $I_{\leqslant s} := I \cap k[X_1, \ldots, X_n]_{\leqslant s}$. Then $k[X_1, \ldots, X_n]_{\leqslant s}$ is a finite-dimensional vector space over k and $I_{\leqslant s}$ is a subspace of it. Thus, we can define a function

$$^{\mathrm{a}}\mathrm{HF}_I \colon \mathbb{Z}_{\geqslant 0} \to \mathbb{Z}_{\geqslant 0}, \quad s \mapsto \dim_k(k[X_1, \ldots, X_n]_{\leqslant s}/I_{\leqslant s}).$$

This function is called the (*affine*) *Hilbert function* of I.

1.2.12 Example   (a) Let $I = k[X_1, \ldots, X_n]$. Then ${}^{\mathrm{a}}\mathrm{HF}_I(s) = 0$ for all $s \geqslant 0$.

(b) Let $I = \{0\}$. Then ${}^{\mathrm{a}}\mathrm{HF}_I(s) = \dim_k(k[X_1, \ldots, X_n]_{\leqslant s})$, and this is the number of all monomials $X_1^{\alpha_1} \cdots X_n^{\alpha_n}$ such that $|\alpha| \leqslant s$. Thus, we have to count all n-tuples of non-negative integers whose sum is $\leqslant s$. It is well-known that the number of such tuples is given by the binomial coefficient $\binom{s+n}{s}$. So we have

$$
{}^{\mathrm{a}}\mathrm{HF}_{\{0\}}(s) = \binom{s+n}{s} = \binom{s+n}{n} = \frac{1}{n!} \underbrace{(s+n)(s+n-1) \cdots (s+1)}_{n \text{ factors}}
$$

$$
= \frac{1}{n!} s^n + \frac{1}{n!} \binom{n+1}{2} s^{n-1} + \cdots + \left(\sum_{i=1}^{n} \frac{1}{i} \right) s + 1.
$$

(c) Let $I = (f)$, where $0 \neq f \in k[X_1, \ldots, X_n]$. Then a k-vector-space basis of I is given by all products $X_1^{\alpha_1} \cdots X_n^{\alpha_n} f$, where $\alpha = (\alpha_1, \ldots, \alpha_n) \in \mathbb{Z}_{\geqslant 0}^n$; the degree of such a product is $|\alpha| + d$, where $d = \deg(f)$. Now assume that $s \geqslant d$. Then $\dim_k(I_{\leqslant s})$ is the number of all α such that $|\alpha| + d \leqslant s$ or, equivalently, $|\alpha| \leqslant s - d$. Using the formulas in (b), we find that

$$
{}^{\mathrm{a}}\mathrm{HF}_{(f)}(s) = \binom{s+n}{s} - \binom{s-d+n}{s-d} = \binom{s+n}{n} - \binom{s-d+n}{n}
$$

$$
= \frac{d}{(n-1)!} s^{n-1} + \text{linear combination of lower powers of } s.
$$

Note that the above formula is only valid for $s \geqslant d$; if $s < d$, we have ${}^{\mathrm{a}}\mathrm{HF}_{(f)}(s) = {}^{\mathrm{a}}\mathrm{HF}_{\{0\}}(s) = \binom{s+n}{n}$.

In the above examples, the Hilbert function is given by a polynomial in s. Our aim is to show that this is true for any ideal in $k[X_1, \ldots, X_n]$.

1.2.13 Lemma (Macaulay)   *Let $\preceq$ be a graded lexicographic monomial order. For an ideal $I \subseteq k[X_1, \ldots, X_n]$, we have ${}^{\mathrm{a}}\mathrm{HF}_I(s) = {}^{\mathrm{a}}\mathrm{HF}_{(\mathrm{LT}(I))}(s)$ for all $s \geqslant 0$.*

Proof   If $I = \{0\}$, there is nothing to prove. So let us assume now that $I \neq \{0\}$ and let $s \geqslant 0$. Then, clearly, there are only finitely many monomials occurring in the elements of $I_{\leqslant s}$. In particular, we can write

$$
\{\mathrm{LM}(f) \mid 0 \neq f \in I_{\leqslant s}\} = \{\mathrm{LM}(f_1), \ldots, \mathrm{LM}(f_m)\}, \qquad (\dagger)
$$

where $f_i \in I_{\leqslant s}$ and $\mathrm{LM}(f_m) \not\succeq \cdots \not\succeq \mathrm{LM}(f_1)$. We shall show that

   (a) $\{f_1, \ldots, f_m\}$ is a k-basis of $I_{\leqslant s}$;
   (b) $\{\mathrm{LM}(f_1), \ldots, \mathrm{LM}(f_m)\}$ is a k-basis of $(\mathrm{LT}(I))_{\leqslant s}$.

(Note that, once (a) and (b) are established, we are done since then $I_{\leqslant s}$ and $(\mathrm{LT}(I))_{\leqslant s}$ have the same dimension.)

First, we check that the sets in (a) and (b) are linearly independent. This is clear in (b) since any set of distinct monomials is linearly independent. Now assume that we have a relation $f = \sum_{i=1}^{m} a_i f_i$, where not all $a_i \in k$ are 0. We must show that $f \neq 0$. Let i_0 be minimal such that $a_{i_0} \neq 0$. Then, since $\mathrm{LM}(f_i) \not\succeq \mathrm{LM}(f_{i_0})$ for all $i > i_0$, we have $\mathrm{LM}(f) = a_{i_0} \mathrm{LM}(f_{i_0}) \neq 0$, as desired.

Next, we check that the sets in (a) and (b) are generating sets. We begin with (b). Let $g \in (\mathrm{LT}(I))_{\leqslant s}$. This means that we can write g in the form $g = \sum_j h_j \mathrm{LM}(g_j)$, where $h_j \in k[X_1, \ldots, X_n]$ and $g_j \in I$. Writing each h_j as a sum of its terms and expanding the product, we see that we may assume without loss of generality that each h_j lies in k. Then, since we have chosen a graded lexicographic order, this implies that $g_j \in I_{\leqslant s}$ for all j. Using ($\dagger$), we conclude that g is a k-linear combination of $\{\mathrm{LM}(f_1), \ldots, \mathrm{LM}(f_m)\}$.

Finally, consider (a). Let $0 \neq f \in I_{\leqslant s}$. Then, by ($\dagger$), we have $\mathrm{LM}(f) = \mathrm{LM}(f_i)$ for some i, and so $\mathrm{LT}(f) = a\,\mathrm{LT}(f_i)$, where $0 \neq a \in k$. Now set $f' := f - af_i$. If $f' = 0$, we are done. If $f' \neq 0$, we repeat the whole argument. Since $\mathrm{LT}(f') \not\succeq \mathrm{LT}(f)$, this process stops after a finite number of iterations, by Lemma 1.2.4.   □

1.2.14 Theorem   *For any ideal $I \subseteq k[X_1, \ldots, X_n]$, there exists a unique polynomial $^{\mathrm{a}}\mathrm{HP}_I(t) \in \mathbb{Q}[t]$ (where t is an indeterminate) and $s_0 \geqslant 0$ such that*

$$^{\mathrm{a}}\mathrm{HP}_I(s) = {}^{\mathrm{a}}\mathrm{HF}_I(s) = \dim_k(k[X_1, \ldots, X_n]_{\leqslant s}/I_{\leqslant s}) \quad \textit{for all } s \geqslant s_0.$$

If $I = k[X_1, \ldots, X_n]$, we have $^{\mathrm{a}}\mathrm{HP}_I(t) = 0$. If $I \neq k[X_1, \ldots, X_n]$, then $^{\mathrm{a}}\mathrm{HP}_I(t)$ is a non-zero polynomial with the following properties.

   (a) *The degree of $^{\mathrm{a}}\mathrm{HP}_I(t)$ is the largest d with the property that there exist indices $1 \leqslant i_1 < \cdots < i_d \leqslant n$ such that $I \cap k[X_{i_1}, \ldots, X_{i_d}] = \{0\}$.*

   (b) *Let $d = \deg {}^{\mathrm{a}}\mathrm{HP}_I(t)$. Then we have $^{\mathrm{a}}\mathrm{HP}_I(t) = a_d t^d + \cdots + a_1 t + a_0$ with $d!\,a_i \in \mathbb{Z}$ for all i and $d!\,a_d > 0$.*

The polynomial $^{\mathrm{a}}\mathrm{HP}_I(t)$ is called the (*affine*) *Hilbert polynomial* of I.

Proof First note that the polynomial $^{\mathrm{a}}\mathrm{HP}_I(t)$ is unique, once we know that it exists. If $I = \{0\}$ or $I = k[X_1, \ldots, X_n]$, then everything follows from the explicit computations in Example 1.2.12. So let us now assume that $\{0\} \neq I \neq k[X_1, \ldots, X_n]$. Let G be a Groebner basis of I (with respect to a graded lexicographic order) and write $\{\mathrm{LM}(g) \mid g \in G\} = \{X^\beta \mid \beta \in M\}$. We set

$$C(I) = \{\alpha \in \mathbb{Z}^n_{\geqslant 0} \mid X^\alpha \text{ is not divisible by } X^\beta \text{ for any } \beta \in M\}.$$

For any $s \geqslant 0$, let $C(I)_{\leqslant s}$ be the set of all $\alpha \in C(I)$ such that $|\alpha| \leqslant s$. We claim that $^{\mathrm{a}}\mathrm{HF}_I(s) = |C(I)_{\leqslant s}|$ for all $s \geqslant 0$. Indeed, by Lemma 1.2.13, we have $^{\mathrm{a}}\mathrm{HF}_I(s) = {}^{\mathrm{a}}\mathrm{HF}_{(\mathrm{LT}(I))}(s)$. Furthermore, by Example 1.2.9(c), $\{X^\beta \mid \beta \in M\}$ is a Groebner basis of $(\mathrm{LT}(I))$. Hence, by Lemma 1.2.3, we have that $\dim \mathrm{LT}(I)_{\leqslant s}$ equals the number of all α such that $|\alpha| \leqslant s$ and $X^\beta \mid X^\alpha$ for some $\beta \in M$. Thus, we have $^{\mathrm{a}}\mathrm{HF}_I(s) = |C(I)_{\leqslant s}|$ as required. Now we continue in four steps.

Step 1. Given a subset $J \subseteq \{1, \ldots, n\}$ and a function $\tau \colon J \to \mathbb{Z}_{\geqslant 0}$, we define $C(J, \tau) := \{(\alpha_1, \ldots, \alpha_n) \in \mathbb{Z}^n_{\geqslant 0} \mid \alpha_j = \tau(j) \text{ for all } j \in J\}$. Now we claim that there exists a finite collection $\mathcal{J}$ of pairs (J, τ) as above such that

$$C(I) = \bigcup_{(J,\tau) \in \mathcal{J}} C(J, \tau). \tag{$*$}$$

This is seen as follows. For $\beta = (\beta_1, \ldots, \beta_n) \in \mathbb{Z}^n_{\geqslant 0}$, we define $C(\beta)$ to be the set of all $\alpha \in \mathbb{Z}^n_{\geqslant 0}$ such that X^β does not divide X^α. Then we have $C(I) = \bigcap_{\beta \in M} C(\beta)$. Now note that if (J, τ) and (J', τ') are two pairs as above, then

$$C(J, \tau) \cap C(J', \tau') = \begin{cases} \varnothing & \text{if } \tau(j) \neq \tau'(j) \text{ for some} \\ & j \in J \cap J', \\ C(J \cup J', \tau_0) & \text{otherwise,} \end{cases}$$

where τ_0 is defined by $\tau_0(j) = \tau(j)$ if $j \in J$ and $\tau_0(j) = \tau'(j)$ if $j \in J'$.

Thus, it remains to show that $C(\beta)$ (for any $\beta \in \mathbb{Z}^n_{\geqslant 0}$) is a finite union of sets of the form $C(J, \tau)$. But, we have $(\alpha_1, \ldots, \alpha_n) \in C(\beta)$ if and only if $\alpha_i < \beta_i$ for some $1 \leqslant i \leqslant n$. Thus, as required, we have

$$C(\beta) = \bigcup_{i=1}^{n} \bigcup_{t_i=0}^{\beta_i-1} \{(\alpha_1, \ldots, \alpha_n) \in \mathbb{Z}_{\geqslant 0}^n \mid \alpha_i = t_i\}$$

$$= \bigcup_{i=1}^{n} \bigcup_{t_i=0}^{\beta_i-1} C(\{i\}, \tau \colon i \mapsto t_i).$$

Step 2. We claim that, for any pair (J, τ) as in Step 1, there exists a polynomial $F_{J,\tau}(t) \in \mathbb{Q}[t]$ such that

$$\deg F_{J,\tau}(t) = n - |J| \quad \text{and} \quad F_{J,\tau}(s) = |C(J,\tau)_{\leqslant s}|$$

for all $s \geqslant |\tau| := \sum_{j \in J} \tau(j)$. Indeed, let $\gamma = (\gamma_1, \ldots, \gamma_n) \in \mathbb{Z}_{\geqslant 0}^n$ be such that $\gamma_j = 0$ for $j \notin J$ and $\gamma_j = \tau(j)$ for $j \in J$. Then $C(J,\tau)$ is the set of all $\alpha + \gamma$, where $\alpha = (\alpha_1, \ldots, \alpha_n) \in \mathbb{Z}_{\geqslant 0}^n$ is such that $\alpha_j = 0$ for $j \in J$. Thus, $|C(J,\tau)_{\leqslant s}|$ is the number of monomials in the variables X_j $(j \notin J)$ such that, when multiplied with X^γ, the total degree is at most s. This is the same as the number of monomials in $d := n - |J|$ variables of total degree at most $s - |\gamma|$ (note that $|\tau| = |\gamma|$). Hence, as in Example 1.2.12(b), we find that

$$|C(J,\tau)_{\leqslant s}| = \binom{d + s - |\tau|}{d} = \frac{1}{d!}(s^d + \text{combination of lower}$$
$$\text{powers of } s).$$

Step 3. If $A_1, \ldots, A_m$ are finite subsets of some set, then we have

$$|A_1 \cup \cdots \cup A_m| = \sum_{r=1}^{m} \sum_{1 \leqslant i_1 < \cdots < i_r \leqslant m} (-1)^{r-1} |A_{i_1} \cap \cdots \cap A_{i_r}|;$$

this is easily proved by induction on m and is called the *inclusion–exclusion* principle. We apply this principle to the computation of $\mathrm{^aHF}_I(s)| = |C(I)_{\leqslant s}|$. By (*), we can write $C(I)_{\leqslant s}$ as a finite union of sets of the form $C(J,\tau)_{\leqslant s}$. Now let s_0 be an integer which is bigger that $|\tau_0|$ for any pair (J_0, τ_0) occurring in that union or in any intersection of sets in that union. Let $s \geqslant s_0$. Then, by Step 2, $|C(J,\tau)_{\leqslant s}|$ is given by a polynomial in s (whose leading coefficient is $1/(n - |J|)!$). Furthermore, we have seen in Step 1 that the intersection of sets of the form $C(J,\tau)$ is either empty or of the form $C(J_0, \tau_0)$, where $J_0 \supseteq J$ (and $J_0 \neq J$ if the pairs in the intersection are not all equal). Thus, the inclusion–exclusion principle yields that $|C(I)_{\leqslant s}|$ is also given by a polynomial in s. This proves the existence of ${}^{\mathrm{a}}\mathrm{HP}_I(t)$. Let d_0 be the largest $d \geqslant 0$ such that there exists a pair (J, τ) with $C(J, \tau) \subseteq C(I)$ and $d = n - |J|$. Then we have

$$
{}^{\mathrm{a}}\mathrm{HP}_I(t) = \sum_{\substack{(J,\tau) \in \mathcal{J} \\ d_0 + |J| = n}} F_{J,\tau}(t) + \text{combination of polynomials of degree} < d_0.
$$

Using the expression for $F_{J,\tau}(t)$ in Step 2, we now see that the terms of degree d_0 in the above sum do not cancel. Hence ${}^{\mathrm{a}}\mathrm{HP}_I(t)$ has degree d_0, and the leading coefficient is of the form $a/d_0!$ for an integer $a > 0$. All the remaining coefficients are also seen to be integer multiples of $d_0!$.

Step 4. It remains to prove the characterization of the degree of ${}^{\mathrm{a}}\mathrm{HP}_I(t)$ in (a). By the discussion in Step 3, $\deg {}^{\mathrm{a}}\mathrm{HP}_I(t)$ is the largest d such that there exists a subset $J \subseteq \{1, \ldots, n\}$ of size $n - d$ and a function $\tau \colon J \to \mathbb{Z}_{\geqslant 0}$ such that $C(J, \tau) \subseteq C(I)$. First we show that $I \cap k[X_j \mid j \notin J] = \{0\}$ for any such pair (J, τ). Assume, if possible, that this is not the case and let $0 \neq f \in I \cap k[X_j \mid j \notin J]$. But then we also have $0 \neq \mathrm{LT}(f) \in (\mathrm{LT}(I)) \cap k[X_j \mid j \notin J]$. Writing $\mathrm{LT}(f) = aX^\alpha$, where $0 \neq a \in k$ and $\alpha = (\alpha_1, \ldots, \alpha_n) \in \mathbb{Z}_{\geqslant 0}^n$, we have $\alpha_j = 0$ for $j \in J$, and so $\gamma + \alpha \in C(J, \tau) \subseteq C(I)$ for any $\gamma \in C(J, \tau)$. On the other hand, we have $aX^{\gamma+\alpha} = X^\gamma \mathrm{LT}(f) \in (\mathrm{LT}(I))$, and so $X^{\gamma+\alpha}$ is divisible by X^β for some $\beta \in M$, contradicting $\gamma + \alpha \in C(I)$. Thus, we have

$$
\deg {}^{\mathrm{a}}\mathrm{HP}_I(t) \leqslant \max\left\{ 0 \leqslant d \leqslant n \,\middle|\, \begin{array}{l} \text{there exist } 1 \leqslant i_1 < \cdots < i_d \leqslant n \\ \text{such that } I \cap k[X_{i_1}, \ldots, X_{i_d}] = \{0\} \end{array} \right\}.
$$

Conversely, if $1 \leqslant i_1 < \cdots < i_d \leqslant n$ are such that $I \cap k[X_{i_1}, \ldots, X_{i_d}] = \{0\}$, then the natural map $\pi \colon k[X_{i_1}, \ldots, X_{i_d}] \to k[X_1, \ldots, X_n]/I$ induced by the inclusion $k[X_{i_1}, \ldots, X_{i_d}] \subseteq k[X_1, \ldots, X_n]$ is injective. Consequently, by the same computation as in Example 1.2.12(b), we have ${}^{\mathrm{a}}\mathrm{HP}_I(s) = {}^{\mathrm{a}}\mathrm{HF}_I(s) \geqslant \binom{d+s}{d}$ for all $s \geqslant 0$. Since the right-hand side is given by a polynomial of degree d, this implies that $\deg {}^{\mathrm{a}}\mathrm{HP}_I(t) \geqslant d$, as desired. $\qquad\square$

Note that the above proof provides in fact an algorithm for computing the polynomial ${}^{\mathrm{a}}\mathrm{HP}_I(t)$, once we know a Groebner basis for I with respect to a graded lexicographic order $\preceq$. See Chapter 9 of Cox *et al.* (1992) for more details.

We now apply the above results to algebraic sets in k^n.

1.2.15 Definition Let $V \subseteq k^n$ be an algebraic set and $^{\mathrm{a}}\mathrm{HP}_{\mathbf{I}(V)}(t)$ be the Hilbert polynomial of $\mathbf{I}(V) \subseteq k[X_1, \ldots, X_n]$; see Theorem 1.2.14. Assume that $V \neq \varnothing$ so that $\mathbf{I}(V) \neq k[X_1, \ldots, X_n]$. Then the *dimension* of V is defined as

$$\dim V = \deg {}^{\mathrm{a}}\mathrm{HP}_{\mathbf{I}(V)}(t).$$

By Theorem 1.2.14, $\dim V$ is the largest d with the property that there exist indices $1 \leqslant i_1 < \cdots < i_d \leqslant n$ such that $\mathbf{I}(V) \cap k[X_{i_1}, \ldots, X_{i_d}] = \{0\}$.

1.2.16 Example (a) Let $V = k^n$ and assume that $|k| = \infty$. Then $\mathbf{I}(V) = \{0\}$ and the formula in Example 1.2.12(b) shows that $\dim V = n$.

(b) Assume that k is algebraically closed and let $H_f \subseteq k^n$ be a hypersurface (where f is non-constant). By Theorem 1.1.9, we may assume that $\mathbf{I}(H_f) = (f)$. Then $\dim H_f = n - 1$; indeed, by Example 1.2.12(c), we have

$$^{\mathrm{a}}\mathrm{HP}_{(f)}(t) = \frac{\deg(f)}{(n-1)!}t^{n-1} + \text{linear combination of lower powers of } t.$$

(c) Let $V \subseteq k^n$ be an algebraic set and $W \subseteq V$ be a closed subset. Then we have $\dim W \leqslant \dim V$. Indeed, consider the ideals $J = \mathbf{I}(V)$ and $I = \mathbf{I}(W)$ in $k[X_1, \ldots, X_n]$. Since $J \subseteq I$, we have $J_{\leqslant s} \subseteq I_{\leqslant s}$ and so $^{\mathrm{a}}\mathrm{HF}_I(s) \leqslant {}^{\mathrm{a}}\mathrm{HF}_J(s)$ for all $s \geqslant 0$. It follows that $\dim W = \deg {}^{\mathrm{a}}\mathrm{HP}_I(t) \leqslant \deg {}^{\mathrm{a}}\mathrm{HP}_J(t) = \dim V$.

1.2.17 Proposition *Let $V \subseteq k^n$ be a non-empty algebraic set and $V = V_1 \cup \cdots \cup V_r$ be its decomposition into irreducible components (see Proposition 1.1.12). Then*

$$\dim V = \max\{\dim V_1, \ldots, \dim V_r\}.$$

Proof We have $\dim V_j \leqslant \dim V$ by Example 1.2.16(c). Now let $d = \dim V$. Then, by Theorem 1.2.14, there exist $1 \leqslant i_1 < \cdots < i_d \leqslant n$ such that $\mathbf{I}(V) \cap k[X_{i_1}, \ldots, X_{i_d}] = \{0\}$. Now assume, if possible, that $\dim V_j < d$ for all j. Then there exist some $0 \neq F_j \in \mathbf{I}(V_j) \cap k[X_{i_1}, \ldots, X_{i_d}]$ (again by Theorem 1.2.14). Now set $F := F_1 \cdots F_r \in k[X_{i_1}, \ldots, X_{i_d}]$. Then we have $F \neq 0$ and

$$F \in \mathbf{I}(V_1) \cap \cdots \cap \mathbf{I}(V_r) = \mathbf{I}(V_1 \cup \cdots \cup V_r) = \mathbf{I}(V),$$

a contradiction. So there exists some j such that $\dim V = \dim V_j$. $\qquad\square$

The following result provides an alternative characterization of $\dim V$. First, we recall some general notions. Let A be a k-*algebra*. (Here, and throughout this book, this means that A is a commutative associative k-algebra with 1.) We say that a set of elements $a_1, \ldots, a_m \in A$ is *algebraically independent* if there exists no non-zero polynomial $F \in k[X_1, \ldots, X_m]$ such that $F(a_1, \ldots, a_m) = 0$. We set

$$\partial_k(A) := \sup \left\{ m \geqslant 0 \;\middle|\; \begin{array}{l} \text{there exist } m \text{ algebraically} \\ \text{independent elements in } A \end{array} \right\}.$$

If A is a field, then $\partial_k(A)$ is called the *transcendence degree* of A over k. See Exercise 1.8.12 for some properties of $\partial_k(A)$.

1.2.18 Proposition *Let* $A = k[X_1, \ldots, X_n]/I$ *where* $I \subseteq k[X_1, \ldots, X_n]$ *is a proper ideal. Then* $\deg {}^{\mathrm{a}}\mathrm{HP}_I(t) = \partial_k(A)$. *If, moreover,* A *is an integral domain and* K *is the field of fractions of* A, *then* $\deg {}^{\mathrm{a}}\mathrm{HP}_I(t) = \partial_k(A) = \partial_k(K)$.

In particular, we have $\dim V = \partial_k(A[V])$ for any non-empty algebraic set $V \subseteq k^n$.

Proof We denote by $[f]$ the image of a polynomial $f \in k[X_1, \ldots, X_n]$ in A. Let $d = \deg {}^{\mathrm{a}}\mathrm{HP}_I(t)$. Then, by Theorem 1.2.14(a), there exist $1 \leqslant i_1 < \cdots < i_d \leqslant n$ such that $I \cap k[X_{i_1}, \ldots, X_{i_d}] = \{0\}$. This means that $[X_{i_1}], \ldots, [X_{i_d}]$ are algebraically independent in A and so $d \leqslant \partial_k(A)$. Now we set $R = K$ if A is an integral domain, and $R = A$ otherwise. Then, clearly, $\partial_k(A) \leqslant \partial_k(R)$. Hence it remains to prove that, if $\phi_1, \ldots, \phi_r \in R$ are algebraically independent, then we must have $r \leqslant d$. Let us write $\phi_i = [f_i]/[f]$, where $f_i \in k[X_1, \ldots, X_n]$ and $f \in k[X_1, \ldots, X_n]$ is such that $[f] \neq 0$ (if $R = K$) or $f = 1$ (if $R = A$). Let $N := \max \{\deg(f), \deg(f_i)\}$ and consider the polynomial ring $k[Y_1, \ldots, Y_r]$. For any $s \geqslant 0$, we construct an injective map

$$\beta \colon k[Y_1, \ldots, Y_r]_{\leqslant s} \to k[X_1, \ldots, X_n]_{\leqslant Ns}/I_{\leqslant Ns}.$$

This is done as follows. Let $g \in k[Y_1, \ldots, Y_r]_{\leqslant s}$. It is readily checked that we have $f^s g(f_1/f, \ldots, f_r/f) \in k[X_1, \ldots, X_n]_{\leqslant Ns}$. Then we define $\beta(g)$ to be the class of $f^s g(f_1/f, \ldots, f_r/f)$ modulo $I_{\leqslant Ns}$.

To show that β is injective, suppose g is such that $f^s g(f_1/f, \ldots, f_r/f) \in I_{\leqslant Ns} \subseteq I$. Working in R, we can write this as $[f]^s g(\phi_1, \ldots, \phi_r) = 0$. Since $[f] \neq 0$, we deduce that $g(\phi_1, \ldots, \phi_r) = 0$

and so $g = 0$. Thus, we have an injective map as desired. It follows that

$$^{\mathrm{a}}\mathrm{HF}_I(Ns) \geqslant \dim_k k[Y_1, \ldots, Y_r]_{\leqslant s} = \binom{s+r}{r} \quad \text{by Example 1.2.12(b).}$$

Since the above inequality holds for all $s \geqslant 0$ and since, for large enough s, the left and the right-hand side are given by polynomials of degree d and r, respectively, we conclude that $d \geqslant r$, as desired. $\square$

1.2.19 Example (a) Assume that $V = \{v\} \subseteq k^n$ is a singleton set. Let $v = (v_1, \ldots, v_n)$. Then we have $V = \mathbf{V}(I)$, where $I = (X_1 - v_1, \ldots, X_n - v_n)$. Since $k[X_1, \ldots, X_n]/I \cong k$, we conclude that I is maximal and that $\mathbf{I}(V) = I$. So we have $k[X_1, \ldots, X_n]/\mathbf{I}(V) \cong k$ and, hence, $^{\mathrm{a}}\mathrm{HF}_{\mathbf{I}(V)}(s) = 1$ for all $s \geqslant 1$. Thus, we have $\dim\{v\} = 0$. Using Proposition 1.2.17, it follows that $\dim V = 0$ whenever $|V| < \infty$. One can in fact show that $^{\mathrm{a}}\mathrm{HP}_I(t) = |V|$ in this case; see Exercise 11 in §9.4 of Cox *et al.* (1992).

(b) Consider the twisted cubic $C = \mathbf{V}(X_2 - X_1^2, X_3 - X_1^3) \subseteq k^3$ and assume that $|k| = \infty$. Then, by Example 1.1.8(b), we have $\mathbf{I}(C) = (X_2 - X_1^2, X_3 - X_1^3)$ and Exercise 1.8.4 shows that $k[X_1, X_2, X_3]/\mathbf{I}(C) \cong k[Y_1]$, where Y_1 is an indeterminate. Thus, we conclude that $\dim C = 1$.

An irreducible algebraic set of dimension 1 is called an *affine curve*; we have just seen that the twisted cubic is an affine curve. In Exercise 1.8.10(b) it is shown that the affine curves in k^2 (where k is algebraically closed) are precisely the algebraic sets defined by one non-constant irreducible polynomial.

1.2.20 Proposition *Let $V \subseteq k^n$ be an irreducible algebraic set and $W \subseteq V$ be a closed subset. Then we have $\dim W < \dim V$ if $W \subsetneq V$.*

Proof By Example 1.2.16(c), we have $\dim W \leqslant \dim V$. Assume, if possible, that $d := \dim W = \dim V$ and $W \subsetneq V$. Then $\mathbf{I}(V) \subsetneq \mathbf{I}(W)$ by §1.1.7. Let $f \in \mathbf{I}(W) \setminus \mathbf{I}(V)$. By Theorem 1.2.14, there exist indices $1 \leqslant i_1 < \cdots < i_d \leqslant n$ such that $\mathbf{I}(W) \cap k[X_{i_1}, \ldots, X_{i_d}] = \{0\}$. Now $f, X_{i_1}, \ldots, X_{i_d}$ cannot be algebraically independent modulo $\mathbf{I}(V)$ (see Proposition 1.2.18), and so there exists a polynomial $0 \neq F \in k[Y, Y_1, \ldots, Y_d]$ such that $F(f, X_{i_1}, \ldots, X_{i_d}) \in \mathbf{I}(V)$. Since V is irreducible, $\mathbf{I}(V)$ is a prime ideal and so we can assume that F is irreducible. Let us write $F = \sum_{i=0}^{r} F_i Y^i$, where $F_i \in k[Y_1, \ldots, Y_d]$. Since $f \notin \mathbf{I}(V)$ and F is irreducible, F cannot be a multiple of Y and so $F_0 \neq 0$. But we have $F_0(X_{i_1}, \ldots, X_{i_d}) \equiv F(f, X_{i_1}, \ldots, X_{i_d}) \equiv 0 \bmod \mathbf{I}(W)$, and so $X_{i_1}, \ldots, X_{i_d}$ are not algebraically independent modulo $\mathbf{I}(W)$, a contradiction. $\square$

1.3 Regular maps, direct products, and algebraic groups

This section introduces two new concepts: regular maps between algebraic sets and direct products of algebraic sets. These two concepts are needed to define algebraic groups; these are algebraic sets on which a group multiplication is defined which is given by regular maps.

1.3.1 Definition Let $V \subseteq k^n$ and $W \subseteq k^m$ be non-empty algebraic sets. We say that $\varphi \colon V \to W$ is a *regular map*, if there exist $f_1, \ldots, f_m \in k[X_1, \ldots, X_n]$ (where the X_i are indeterminates) such that

$$\varphi(x) = (f_1(x), \ldots, f_m(x)) \quad \text{for all } x \in V.$$

Such a map φ is continuous in the Zariski topology. Indeed, let $Z \subseteq W$ be a closed set. Then $Z = \mathbf{V}(R) \subseteq k^m$ for some $R \subseteq k[Y_1, \ldots, Y_m]$ (where the Y_j are indeterminates) and so

$$\varphi^{-1}(Z) = \mathbf{V}(\{g(f_1, \ldots, f_m) \mid g \in R\}) \subseteq k^n.$$

Here, $g(f_1, \ldots, f_m)$ means that each variable Y_i is substituted by f_i. Thus, $g(f_1, \ldots, f_m) \in k[X_1, \ldots, X_n]$ for all $g \in R$, and so $\varphi^{-1}(Z)$ is algebraic in k^n.

1.3.2 Remark Let $\varphi \colon V \to W$ be a regular map as in Definition 1.3.1. Assume that V is irreducible. Then the Zariski closure $\overline{\varphi(V)} \subseteq W$ is also irreducible.

Indeed, let $f, g \in k[Y_1, \ldots, Y_m]$ be such that $fg \in \mathbf{I}(\overline{\varphi(V)})$. This means that $f(\varphi(v))g(\varphi(v)) = 0$ for all $v \in V$, and so $f(f_1, \ldots, f_m)g(f_1, \ldots, f_m) \in \mathbf{I}(V)$. Since V is irreducible, $\mathbf{I}(V)$ is a prime ideal, and so $f(f_1, \ldots, f_m) \in \mathbf{I}(V)$ or $g(f_1, \ldots, f_m) \in \mathbf{I}(V)$. Assume, for example, that $f(f_1, \ldots, f_m) \in \mathbf{I}(V)$. Then we have $f(\varphi(v)) = 0$ for all $v \in V$, and so $\varphi(V) \subseteq \mathbf{V}(\{f\})$. But $\mathbf{V}(\{f\}) \subseteq k^n$

is a closed set and, hence, f also vanishes on $\overline{\varphi(V)}$; that is, we have $f \in \mathbf{I}(\overline{\varphi(V)})$.

1.3.3 Regular functions Let $V \subseteq k^n$ be a non-empty algebraic set and consider the special case where $W = k$. Then a regular map $\varphi\colon V \to k$ is given by a single polynomial $f \in k[X_1, \ldots, X_n]$ such that $\varphi(x) = f(x)$ for all $x \in V$. Now note that, for any $g \in k[X_1, \ldots, X_n]$, we have $g(x) = f(x)$ for all $x \in V$ if and only if $g - f \in \mathbf{I}(V)$. Thus, φ uniquely determines an element of $A[V] = k[X_1, \ldots, X_n]/\mathbf{I}(V)$. Conversely, for any residue class $\bar{f} = f + \mathbf{I}(V) \in A[V]$, we obtain a well-defined regular map $\varphi\colon V \to k$ such that $\varphi(x) = f(x)$ for all $x \in V$.

This discussion shows that $A[V]$ may also be regarded as the set of all regular maps $V \to k$. For $f \in k[X_1, \ldots, X_n]$, we will usually identify the function $f\colon V \to k$ with $\bar{f} \in A[V]$ and call $A[V]$ the *algebra of regular functions* on V. Using this interpretation of the affine algebra, we see that a regular map $\varphi\colon V \to W$ as in Definition 1.3.1 induces a k-algebra homomorphism

$$\varphi^*\colon A[W] \to A[V], \quad \bar{g} \mapsto \bar{g} \circ \varphi.$$

Indeed, if $g \in k[Y_1, \ldots, Y_m]$ and φ is given by the polynomials $f_1, \ldots, f_m \in k[X_1, \ldots, X_n]$, then $\bar{g} \circ \varphi$ is given by the polynomial $g(f_1, \ldots, f_m)$, obtained by substituting $Y_i \mapsto f_i$ for $1 \leqslant i \leqslant m$.

The assignment $\varphi \mapsto \varphi^*$ is (contravariant) functorial, in the following sense. If $\varphi\colon V \to W$ and $\psi\colon W \to Z$ are regular (where $Z \subseteq k^l$ is algebraic), then we have $(\psi \circ \varphi)^* = \varphi^* \circ \psi^*$. Furthermore, for $V = W$, we have $\mathrm{id}_V^* = \mathrm{id}_{A[V]}$.

1.3.4 Proposition *Let $V \subseteq k^n$ and $W \subseteq k^m$ be non-empty algebraic sets. Then the assignment $\varphi \mapsto \varphi^*$ defines a bijection*

$$\{\textit{regular maps } V \to W\} \xrightarrow{\sim}$$
$$\{k\textit{-algebra homomorphisms } A[W] \to A[V]\}.$$

In particular, $\varphi\colon V \to W$ is an isomorphism of algebraic sets (that is, φ is bijective and its inverse is regular) if and only if φ^* is a k-algebra isomorphism.

Proof Let $\varphi\colon V \to W$ be a regular map, given by $f_1, \ldots, f_m \in k[X_1, \ldots, X_n]$. Then we have $\varphi^*(\bar{Y}_i) = \bar{Y}_i \circ \varphi = \bar{f}_i$ for all i. This shows that $\varphi \mapsto \varphi^*$ is injective. Conversely, let $\alpha\colon A[W] \to A[V]$

be a k-algebra homomorphism. For any $i \in \{1, \ldots, m\}$, let $f_i \in k[X_1, \ldots, X_n]$ be such that $\bar{f_i} = \alpha(\bar{Y_i})$. We consider the regular map $\varphi \colon V \to k^m$ given by the polynomials $f_1, \ldots, f_m$. Writing an arbitrary polynomial in $k[Y_1, \ldots, Y_m]$ as a linear combination of monomials and using the fact that α is a k-algebra homomorphism, we conclude that

$$\alpha(\bar{g})(v) = g(f_1(v), \ldots, f_m(v)) = g(\varphi(v)) \qquad (*)$$

for any $g \in k[Y_1, \ldots, Y_m]$ and $v \in V$. We will check that $\varphi(V) \subseteq W$. To see this, let $v \in V$ and $g \in \mathbf{I}(W)$. Then $\bar{g} = 0$ and $(*)$ shows that $g(\varphi(v)) = 0$. Thus, we have $\varphi(v) \in \mathbf{V}(\mathbf{I}(W)) = W$. Finally, $(*)$ now also shows that $\alpha = \varphi^*$.

The statement about isomorphisms is now clear. Indeed, if φ is an isomorphism, then the functoriality properties in §1.3.3 show that φ^* is an isomorphism. Conversely, if φ^* is an isomorphism, then $(\varphi^*)^{-1} \colon A[V] \to A[W]$ is a k-algebra homomorphism, and so there exists some regular map $\psi \colon W \to V$ such that $\psi^* = (\varphi^*)^{-1}$. Then the above properties show that ψ is inverse to φ. $\qquad\square$

1.3.5 Example (a) Consider the map $\varphi \colon k \to k^3$, $x \mapsto (x, x^2, x^3)$. Then φ is certainly regular and we have $\varphi(k) = C$, the twisted cubic. Furthermore, the description of $A[C]$ in Example 1.2.19(b) shows that $\varphi^* \colon A[C] \to A[k]$ is an isomorphism. Thus, k and C are isomorphic algebraic sets.

(b) Let $\varphi \colon k^2 \to k^2$ be the regular map defined by $\varphi(x, y) = (xy, y)$. We have $\varphi(k^2) = \{(0,0)\} \cup \{(x, y) \in k^2 \mid x, y \in k, y \neq 0\}$. This shows that, in general, the image of a regular map need not be open or closed. We will study this problem in more detail in Section 2.2.

(c) Assume that k is algebraically closed of characteristic $p > 0$, and consider the regular map $\varphi \colon k \to k$, $x \mapsto x^p$. Then φ is bijective but $\varphi^* \colon k[X] \to k[X]$, $X \mapsto X^p$, is not surjective. Thus, φ is not an isomorphism!

1.3.6 Example Let $f_1, \ldots, f_m \in k[X_1, \ldots, X_n]$ and define $\varphi \colon k^n \to k^m$ by $\varphi(x) = (f_1(x), \ldots, f_m(x))$. As we have seen in the above example, the image of φ need not be closed. So let us set $V_\varphi := \overline{\varphi(k^n)} \subseteq k^m$. Can we find a set of polynomials $S \subseteq k[Y_1, \ldots, Y_m]$ such that $V_\varphi = \mathbf{V}(S)$? The theory of Groebner bases provides a constructive answer to this question. To see this, we regard $f_1, \ldots, f_m$ as elements of $R := k[X_1, \ldots, X_n, Y_1, \ldots, Y_m]$ and consider the ideal

$$I = \hat{I} \cap k[Y_1, \ldots, Y_m] \quad \text{where} \quad \hat{I} := (Y_1 - f_1, \ldots, Y_m - f_m) \subseteq R.$$

Then consider the algebraic set $\hat{V} = \mathbf{V}(\hat{I}) \subseteq k^n \times k^m = k^{n+m}$ and the projection $\pi_m \colon k^n \times k^m \to k^m$. We claim that, if $|k| = \infty$, then

$$\pi_m(\hat{V}) = \varphi(k^n) \quad \text{and} \quad V_\varphi = \mathbf{V}(I) \subseteq k^m.$$

Proof For any $(x, y) \in k^n \times k^m$, we have $(x, y) \in \hat{V}$ if and only if $y = \varphi(x)$. Thus, we see that $\pi_m(\hat{V}) = \varphi(k^n)$, and it remains to show that $\mathbf{V}(I) = \overline{\pi_m(\hat{V})}$.

Now, if $(x, y) \in \hat{V}$ and $f \in I$, then $f(y) = f(x, y) = 0$, where the first equality holds since f only involves the variables $Y_1, \ldots, Y_m$ and the second equality holds since $f \in \hat{I}$ and $(x, y) \in \mathbf{V}(\hat{I})$. Thus, we have $\pi_m(\hat{V}) \subseteq \mathbf{V}(I)$.

Conversely, let $f \in \mathbf{I}(\pi_m(\hat{V}))$. Since f only involves the variables $Y_1, \ldots, Y_m$, we have $f \in \mathbf{I}(\hat{V})$ and so $f \in \hat{I}$, by Exercise 1.8.4. Thus, $f \in \hat{I} \cap k[Y_1, \ldots, Y_m] = I$, and it follows that $V_\varphi = \mathbf{V}(I) \subseteq \mathbf{V}(\mathbf{I}(\pi_m(\hat{V}))) = \overline{\pi_m(\hat{V})}$ using §1.1.7. $\square$

For example, consider the map $\varphi \colon k \to k^3$, $x \mapsto (x, x^2, x^3)$. Then the ideal $\hat{I}$ is generated by $\{Y_1 - X, Y_2 - X^2, Y_3 - X^3\} \subseteq k[X, Y_1, Y_2, Y_3]$. A Groebner basis with respect to **LEX** with $Y_1 \preceq Y_2 \preceq Y_3 \preceq X$ is given by $\{Y_2 - Y_1^2, Y_3 - Y_1^3, X - Y_1\}$. By Exercise 1.8.5, $I = \hat{I} \cap k[Y_1, Y_2, Y_3]$ is generated by $Y_2 - Y_1^2$ and $Y_3 - Y_1^3$. Thus, we have recovered the defining equations of the twisted cubic.

In the above setting, there are examples which show that the image of a closed subset $\hat{V} \subseteq k^n \times k^m$ under the projection map π_m need not be closed; see Exercise 1.8.9. We will see later in Section 3.2 that this phenomenon can not occur when we consider algebraic sets in *projective space*.

1.3.7 Direct products of algebraic sets Let $V \subseteq k^n$ and $W \subseteq k^m$ be (non-empty) algebraic sets, defined by sets $S \subseteq k[X_1, \ldots, X_n]$ and $T \subseteq k[Y_1, \ldots, Y_m]$, respectively. Identifying $k^n \times k^m$ with k^{n+m} and $k[X_1, \ldots, X_n]$, $k[Y_1, \ldots, Y_m]$ with subrings of

$k[X_1, \ldots, X_n, Y_1, \ldots, Y_m]$, we see that

$$V \times W = \left\{ (v, w) \in k^{n+m} \;\middle|\; \begin{array}{l} f(v) = 0 \text{ for all } f \in S, \\ g(w) = 0 \text{ for all } g \in T \end{array} \right\}$$

$$= \mathbf{V}(S \cup T) \subseteq k^{n+m},$$

where we use the identities $f(v) = f(v, w)$ for all $f \in S$ and $g(w) = g(v, w) = 0$ for all $g \in T$. Thus, the direct product $V \times W$ is an algebraic set in k^{n+m}. Since $(\mathbf{I}(V), \mathbf{I}(W)) \subseteq \mathbf{I}(V \times W)$, we have a well-defined k-bilinear map

$$A[V] \times A[W] \to A[V \times W], \quad (\bar{f}, \bar{g}) \mapsto \bar{f} \times \bar{g} := fg + \mathbf{I}(V \times W).$$

So we get an induced k-linear map $A[V] \otimes_k A[W] \to A[V \times W]$.

1.3.8 Proposition *Let $V \subseteq k^n$ and $W \subseteq k^m$ be non-empty algebraic sets.*

(a) *If V and W are irreducible, then so is $V \times W$.*

(b) *We have $\mathbf{I}(V \times W) = (\mathbf{I}(V), \mathbf{I}(W)) \subseteq k[X_1, \ldots, X_n, Y_1, \ldots, Y_m]$; the map $A[V] \otimes_k A[W] \to A[V \times W]$, $\bar{f} \otimes \bar{g} \mapsto \bar{f} \times \bar{g}$, is an isomorphism.*

(c) *We have $\dim(V \times W) = \dim V + \dim W$.*

Proof (a) We must show that $\mathbf{I}(V \times W) \subseteq k[X_1, \ldots, X_n, Y_1, \ldots, Y_m]$ is a prime ideal; see Proposition 1.1.12. So let $f_1, f_2 \in k[X_1, \ldots, X_n, Y_1, \ldots, Y_m]$ be such that $f_1 f_2 \in \mathbf{I}(V \times W)$. For a fixed $w \in W$ and $i = 1, 2$, we define an algebraic set by $V_i(w) := \{v \in V \mid f_i(v, w) = 0\} \subseteq V$. Since $f_1 f_2 \in \mathbf{I}(V \times W)$, we have $V = V_1(w) \cup V_2(w)$. Since V is irreducible, we conclude that $V = V_1(w)$ or $V = V_2(w)$. Consequently, we have $W = W_1 \cup W_2$, where $W_i := \{w \in W \mid V_i(w) = V\}$ for $i = 1, 2$. Now note that $W_i = \{w \in W \mid f_i(v, w) = 0 \text{ for all } v \in V\}$ and so W_i is algebraic. Since W is irreducible, we may therefore conclude that $W = W_1$ or $W = W_2$. In the first case, we have $f_1(v, w) = 0$ for all $v \in V$ and $w \in W$, and so $f_1 \in \mathbf{I}(V \times W)$. Similarly, in the second case, we have $f_2 \in \mathbf{I}(V \times W)$. Thus, we have shown that $\mathbf{I}(V \times W)$ is a prime ideal, as desired.

(b) Since $(\mathbf{I}(V), \mathbf{I}(W)) \subseteq \mathbf{I}(V \times W)$, we have a canonical surjection

$$B := k[X_1, \ldots, X_n, Y_1, \ldots, Y_m]/(\mathbf{I}(V), \mathbf{I}(W)) \longrightarrow A[V \times W]. \quad (*)$$

We have to show that it is also injective. For this purpose, let $\{\bar{f}_\alpha \mid \alpha \in I\}$ be a k-basis of $A[V]$ (where $f_\alpha \in k[X_1, \ldots, X_n]$) and

$\{\bar{g}_\beta \mid \beta \in J\}$ be a k-basis of $A[W]$ (where $g_\beta \in k[Y_1, \ldots, Y_m]$). For example, we could take the bases provided by Theorem 1.2.8. Then the residue classes in B of the products $f_\alpha g_\beta$ span B over k. Hence it will be enough to show that all the products $f_\alpha g_\beta$ are linearly independent modulo $\mathbf{I}(V \times W)$. So let $x_{\alpha,\beta} \in k$ be such that $\sum_{\alpha \in I} \sum_{\beta \in J} x_{\alpha,\beta} f_\alpha g_\beta \in \mathbf{I}(V \times W)$, that is, we have

$$\sum_{\alpha \in I} \sum_{\beta \in J} x_{\alpha,\beta}\, f_\alpha(v) g_\beta(w) = 0 \quad \text{for all } v \in V,\, w \in W.$$

We must show that $x_{\alpha,\beta} = 0$ for all α, β. To see this, fix $w \in W$ and set

$$y_\alpha(w) := \sum_{\beta \in J} x_{\alpha,\beta}\, g_\beta(w) \in k.$$

Then we have $\sum_{\alpha \in I} y_\alpha(w) f_\alpha(v) = 0$ for all $v \in V$. Since the polynomials $\{f_\alpha\}$ are linearly independent modulo $\mathbf{I}(V)$, we conclude that $y_\alpha(w) = 0$ for all α. This holds for all $w \in W$ and so, since the polynomials $\{g_\beta\}$ are linearly independent modulo $\mathbf{I}(W)$, we conclude that $x_{\alpha,\beta} = 0$ for all α, β, as desired.

Finally, consider the map $A[V] \otimes_k A[W] \to A[V \times W]$, which certainly is surjective. To prove injectivity, consider the substitution homomorphism from $k[X_1, \ldots, X_n, Y_1, \ldots, Y_m]$ into $A[V] \otimes_k A[W]$ defined by sending X_i to $\bar{X}_i \otimes 1$ and Y_j to $1 \otimes \bar{Y}_j$ for all i and j. The kernel of that map contains $\mathbf{I}(V)$ and $\mathbf{I}(W)$. Hence we get an induced map $B \to A[V] \otimes_k A[W]$. The composition of that map with $A[V] \otimes_k A[W] \to A[V \times W]$ is easily seen to be the map $(*)$ and, hence, is an isomorphism. Consequently, $A[V] \otimes_k A[W] \to A[V \times W]$ also is an isomorphism.

(c) By Theorem 1.2.14, there are subsets $I \subseteq \{1, \ldots, n\}$ and $J \subseteq \{1, \ldots, m\}$ such that $\dim(V \times W) = |I| + |J|$ and $k[X_i, Y_j \mid i \in I, j \in J] \cap \mathbf{I}(V \times W) = \{0\}$. Using (b), this implies $k[X_i \mid i \in I] \cap \mathbf{I}(V) = \{0\}$ and $k[Y_j \mid j \in J] \cap \mathbf{I}(W) = \{0\}$ and so $\dim(V \times W) \leqslant \dim V + \dim W$. Conversely, let $I \subseteq \{1, \ldots, n\}$ be such that $\dim V = |I|$ and $k[X_i \mid i \in I] \cap \mathbf{I}(V) = \{0\}$. Furthermore, let $J \subseteq \{1, \ldots, m\}$ be such that $\dim W = |J|$ and $k[Y_j \mid j \in J] \cap \mathbf{I}(W) = \{0\}$. It is easily seen (for example, using an argument analogous to that in the proof of (b)) that then we also have $k[X_i, Y_j \mid i \in I, j \in J] \cap \mathbf{I}(V \times W) = \{0\}$. Hence, by Theorem 1.2.14, we have $\dim(V \times W) \geqslant |I| + |J|$, as required. $\qquad\square$

Now we are ready to define algebraic monoids and algebraic groups.

1.3.9 Definition Consider $M_n(k) = k^{n \times n}$ as an algebraic set (under the identification $k^{n \times n} = k^{n^2}$). Let $\mu \colon M_n(k) \times M_n(k) \to M_n(k)$ be the usual matrix multiplication. Thus, for $A = (a_{ij})$ and $B = (b_{ij})$ in $M_n(k)$, we have $\mu(A, B) = (c_{ij})$ where $c_{ij} = \sum_{l=1}^{n} a_{il} b_{lj}$. This formula shows that μ is a regular map. Then $M_n(k)$ together with μ is called the *general linear algebraic monoid* of degree n; the identity element is the $n \times n$ identity matrix I_n. Setting $A[M_n(k)] = k[X_{ij} \mid 1 \leqslant i, j \leqslant n]/\mathbf{I}(M_n(k))$, the algebra homomorphism $\mu^* \colon A[M_n(k)] \to A[M_n(k)] \otimes_k A[M_n(k)]$ is given by

$$\mu^*(\bar{X}_{ij}) = \sum_{l=1}^{n} \bar{X}_{il} \otimes \bar{X}_{lj} \qquad \text{for all } i, j \in \{1, \ldots, n\}.$$

(Of course, we have $\mathbf{I}(M_n(k)) = \{0\}$ if k is an infinite field.) A *linear algebraic monoid* is an algebraic subset $G \subseteq M_n(k)$ such that $I_n \in G$ and $\mu(A, B) \in G$ for all $A, B \in G$. A homomorphism of algebraic monoids is a regular map $\varphi \colon G \to H$ between two linear algebraic monoids $G \subseteq M_n(k)$ and $H \subseteq M_m(k)$ such that $\varphi(AB) = \varphi(A)\varphi(B)$ for all $A, B \in G$ and $\varphi(I_n) = I_m$.

If, moreover, every element $A \in G$ has an inverse and the map $\iota \colon G \to G$, $A \mapsto A^{-1}$, is regular, then G is called a *linear algebraic group*.

1.3.10 Example Let $\mathrm{SL}_n(k) = \{A \in M_n(k) \mid \det(A) = 1\}$. Then $\mathrm{SL}_n(k)$ is closed under multiplication. Furthermore, let

$$\det := \sum_{\sigma \in \mathfrak{S}_n} \mathrm{sgn}(\sigma) X_{1\sigma(1)} \cdots X_{n\sigma(n)} \in k[X_{ij} \mid 1 \leqslant i, j \leqslant n],$$

where $\mathfrak{S}_n$ is the symmetric group of degree n. Then $\mathrm{SL}_n(k) = \mathbf{V}(\{\det -1\}) \subseteq M_n(k)$ is algebraic, and so $\mathrm{SL}_n(k)$ is a linear algebraic monoid. Furthermore, for any invertible matrix $A \in M_n(k)$, we have $A^{-1} = \det(A)^{-1} \tilde{A}^{\mathrm{tr}}$, where $\tilde{A}$ is the matrix of cofactors of A; the entries of $\tilde{A}$ are given by the determinants of various submatrices of A of size $n - 1$. Thus, $\iota \colon \mathrm{SL}_n(k) \to \mathrm{SL}_n(k)$, $A \mapsto A^{-1}$, is regular and so $\mathrm{SL}_n(k)$ is a linear algebraic group, which is called the *special linear group*.

Finally, if k is algebraically closed, then $\mathrm{SL}_n(k)$ is an irreducible hypersurface with $\mathbf{I}(\mathrm{SL}_n(k)) = (\det -1)$. For this purpose, by Theorem 1.1.9, one has to check that $\det -1 \in k[X_{ij} \mid 1 \leqslant i, j \leqslant n]$ is irreducible. (We leave this an exercise; a different proof that $\mathrm{SL}_n(k)$

is irreducible will be given in Remark 1.6.11.) In particular, we see that $\dim \mathrm{SL}_n(k) = n^2 - 1$ by Example 1.2.16(b).

More generally, let $e \geqslant 1$ and set $\mathrm{SL}_n^{(e)}(k) = \{A \in M_n(k) \mid \det(A)^e = 1\}$. Then, as above, we see that $\mathrm{SL}_n^{(e)}(k)$ is a linear algebraic monoid. Furthermore, for any $A \in \mathrm{SL}_n^{(e)}(k)$, we have $\det(A)^e = 1$ and so $\det(A)^{-1} = \det(A)^{e-1}$. Thus, A^{-1} can also be expressed by polynomials in the entries of A. Consequently, $\mathrm{SL}_n^{(e)}(k)$ is a linear algebraic group.

1.3.11 Example   Let $G \subseteq M_n(k)$ be an algebraic set which is a subgroup of $\mathrm{SL}_n^{(e)}(k)$ for some $e \geqslant 1$. Then G is a linear algebraic group.

(a) Let $U_n(k)$ be the set of all upper unitriangular matrices in $M_n(k)$, that is, we have $A = (a_{ij}) \in U_n(k)$ if and only if $a_{ii} = 1$ for all i and $a_{ij} = 0$ for all $i > j$. Then $U_n(k)$ is an algebraic subset of $M_n(k)$ and a subgroup of $\mathrm{SL}_n(k)$. Thus, $U_n(k)$ is a linear algebraic group. As an algebraic set, we have $U_n(k) \cong k^{n(n-1)/2}$ and so $U_n(k)$ is irreducible of dimension $n(n-1)/2$ (if $|k| = \infty$). If $n = 2$, then

$$k \to U_2(k), \quad a \mapsto \begin{bmatrix} 1 & a \\ 0 & 1 \end{bmatrix},$$

is an isomorphism of algebraic sets. Furthermore, the multiplication in $U_2(k)$ simply corresponds to the addition in k. Thus, the *additive group* of k may be regarded as a linear algebraic group, which we denote by $\mathbb{G}_a(k)$.

(b) Let G be any finite group. Then, by Cayley's theorem, G may be regarded as a subgroup of $\mathfrak{S}_n$ for some $n \geqslant 1$. Using the representation of $\mathfrak{S}_n$ by permutation matrices, we obtain an embedding of G into the linear algebraic group $\mathrm{SL}_n^{(2)}(k)$. Thus, being a finite set, G itself is an algebraic group.

1.3.12 Remark   Let $G \subseteq M_n(k)$ be a linear algebraic group. For a fixed $x \in G$, we define a map $\lambda_x \colon G \to G$ by $\lambda_x(g) := \mu(x, g)$. Thus, λ_x is the composition of μ with $\delta \colon G \to G \times G$, $g \mapsto (x, g)$. The map δ is certainly regular; the same holds for μ by definition. Hence λ_x is regular. Furthermore, the associativity of the multiplication in G implies that

$$\lambda_x \circ \lambda_y = \lambda_{xy} \qquad \text{for all } x, y \in G.$$

Since every element of G has an inverse, we conclude that λ_x is an isomorphism, the inverse being given by $\lambda_{x^{-1}}$. Similarly, we define a map $\rho_x\colon G \to G$ by $\rho_x(g) := \mu(g, x)$. As above, we see that ρ_x is an isomorphism. Furthermore, we have $\rho_x \circ \rho_y = \rho_{yx}$ for all $x, y \in G$.

1.3.13 Proposition   *Let $G \subseteq M_n(k)$ be a linear algebraic group. Let $G^\circ \subseteq G$ be an irreducible component containing 1. Then the following hold.*

(a) G° is a closed normal subgroup of G of finite index, and the cosets of G° are precisely the irreducible components of G. In particular, the irreducible components of G are disjoint, and G° is uniquely determined.

(b) Every closed subgroup of G of finite index contains G°.

(c) G is irreducible if and only if G is connected as a topological space (i.e. G cannot be expressed as the disjoint union of two non-empty open subsets).

Proof   (a) Consider the multiplication map $\mu\colon G \times G \to G$. Let $X \subseteq G$ be an irreducible component with $1 \in X$. Then $\overline{X \cdot G^\circ} = \mu(X \times G^\circ) \subseteq G$ is a closed irreducible subset of G (see Remark 1.3.2 and Proposition 1.3.8). Since $1 \in X$ and $1 \in G^\circ$, we have $X, G^\circ \subseteq X \cdot G^\circ \subseteq \overline{X \cdot G^\circ}$. Since X, G° are maximal closed irreducible subsets, we must have $X = \overline{X \cdot G^\circ}$ and $G^\circ = \overline{X \cdot G^\circ}$. In particular, we see that $X = G^\circ$ and so G° is uniquely determined. The above argument also shows that $G^\circ \cdot G^\circ \subseteq \overline{G^\circ \cdot G^\circ} = G^\circ$ and so G° is closed under multiplication. Now consider the inversion map $\iota\colon G \to G$. Since this is an isomorphism of algebraic sets, $(G^\circ)^{-1} = \iota(G^\circ)$ is also an irreducible component containing 1, and hence must be equal to G°. Thus, G° is closed under inversion and so is a subgroup of G.

Now let $x \in G$. We have seen in Remark 1.3.12 that ρ_x is an isomorphism of algebraic sets and so it maps irreducible components to irreducible components. Thus, the coset $G^\circ x = \rho_x(G^\circ)$ is an irreducible component of G. Conversely, let $X \subseteq G$ be any irreducible component and let $x \in G$ be such that $x^{-1} \in X$. Then $\rho_x(X) = Xx$ is an irreducible component containing 1 and so $Xx = G^\circ$. Hence X is equal to the coset $G^\circ x^{-1}$. Thus, the cosets of G° are precisely the irreducible components of G. Consequently, G° has finite index in G. Furthermore, for any $x \in G$, the two cosets $G^\circ x = \rho_x(G^\circ)$ and $xG^\circ = \lambda_x(G^\circ)$ are irreducible components with non-empty intersection; hence they must be equal. This shows that G° is a normal subgroup.

(b) Let $H \subseteq G$ be any closed subgroup of finite index. Let $g_1, \ldots, g_r \in G$ (with $g_1 = 1$) be such that G is the disjoint union

$\coprod_{i=1}^{r} Hg_i$. As above we see that each coset Hg_i is a closed subset of G. It follows that $G^\circ = G^\circ \cap G = \bigcup_{i=1}^{r}(G^\circ \cap Hg_i)$. Since G° is irreducible, we must have $G^\circ \cap Hg_i = G^\circ$ and so $G^\circ \subseteq Hg_i$ for some i. Since $1 \in G^\circ$, it follows that $i = 1$ and so $G^\circ \subseteq H$, as desired.

(c) It is clear that if G is a disjoint union of two open subsets, then each of these open subsets is also closed and so G is not irreducible. Conversely, if G is not irreducible, we write G as the union of its irreducible components. By (a), this union is disjoint and so each irreducible component is also open. $\qquad\square$

1.3.14 Corollary *All irreducible components of a linear algebraic group $G \subseteq M_n(k)$ have the same dimension. In particular, we have $\dim G = \dim G^\circ$.*

Proof We have seen above that the irreducible components of G are the cosets of G°. Such a coset is of the form $G^\circ x = \rho_x(G^\circ)$ for some $x \in G$, where $\rho_x \colon G \to G$ is an isomorphism of algebraic sets. Hence we have $\dim(G^\circ x) = \dim G^\circ$. The fact that $\dim G = \dim G^\circ$ follows from Proposition 1.2.17. $\qquad\square$

1.3.15 Example We shall now introduce an important class of groups: the so-called classical groups. Let $Q \in M_n(k)$ be an invertible matrix, and set

$$\Gamma_n(Q, k) := \{A \in M_n(k) \mid A^{\mathrm{tr}} Q A = Q\}.$$

First note that taking the determinant of $A^{\mathrm{tr}}QA = Q$ and using that $\det(Q) \neq 0$, we obtain $\det(A) = \pm 1$. Next, if $A, B \in \Gamma_n(Q, k)$, then we also have AB and $A^{-1} \in \Gamma_n(Q, k)$. Finally, writing out the equation $A^{\mathrm{tr}}QA = Q$ for all matrix entries, we see that $\Gamma_n(Q, k)$ is a closed subset of $M_n(k)$. Thus, $\Gamma_n(Q, k) \subseteq \mathrm{SL}_n^{(2)}(k)$ is a linear algebraic group, called a *classical* group.

If $Q' \in M_n(k)$ is another invertible matrix, we say that Q, Q' are equivalent if there exists some invertible matrix $R \in M_n(k)$ such that $Q' = R^{\mathrm{tr}}QR$. In this case, we have $A \in \Gamma_n(Q', k)$ if and only if $RAR^{-1} \in \Gamma_n(Q, k)$. Thus, the map

$$\varphi_R \colon \Gamma_n(Q', k) \to \Gamma_n(Q, k), \quad A \mapsto RAR^{-1},$$

is an isomorphism of algebraic groups. Now Q defines a bilinear form $\beta_Q \colon k^n \times k^n \to k$ by $\beta_Q(v, w) = v^{\mathrm{tr}}Qw$ (where v, w are regarded as column vectors).

(Sym) We say that Q is *symmetric* if $\mathrm{char}(k) \neq 2$ and $\beta_Q(v, w) = \beta_Q(w, v)$ for all $v, w \in k^n$. The latter condition certainly is equivalent to $Q = Q^{\mathrm{tr}}$.

(Alt) We say that Q is *alternating* if $\beta_Q(v, v) = 0$ for all $v \in k^n$. This certainly implies that $Q = -Q^{\mathrm{tr}}$. (If k has characteristic $\neq 2$, the latter condition is also necessary.)

Now, without any further assumption on k, there may be many pairwise inequivalent matrices Q as above. However, if k is algebraically closed, the picture simplifies drastically. In this case, there is only one symmetric Q and one alternating Q up to equivalence; see §2.10 and §4.4 of Grove (2002). Hence, since we will mainly be interested in the case where k is algebraically closed, it will be enough to study the groups $\Gamma_n(Q, k)$ for the following special choices of the matrix Q. (The motivation for these choices will be become clear in Section 1.7 where we construct a BN-pair for these groups.)

Assume first that we are in the case (Sym) where $\mathrm{char}(k) \neq 2$. Then

$$
\mathrm{O}_n(k) := \Gamma_n(Q_n, k), \quad \text{where} \quad Q_n := \begin{bmatrix} 0 & \cdots & 0 & 1 \\ \vdots & & \iddots & 0 \\ 0 & 1 & \iddots & \vdots \\ 1 & 0 & \cdots & 0 \end{bmatrix} \in M_n(k),
$$

is called the *orthogonal group* of degree n. Note that $\mathrm{O}_n(k)$ is not irreducible: indeed, the determinant defines a group homomorphism $\varepsilon \colon \mathrm{O}_n(k) \to \{\pm 1\}$, and there exist matrices $A \in \mathrm{O}_n(k)$ with determinant -1. (Check this!) Thus, $\mathrm{O}_n(k)$ is the union of two cosets of the closed normal subgroup $\mathrm{SO}_n(k) := \mathrm{O}_n(k) \cap \mathrm{SL}_n(k)$, which is called the *special orthogonal* group of dimension n. For $n = 1$, $\mathrm{SO}_1(k)$ is the trivial group and $\mathrm{O}_1(k) = \{\pm 1\}$. For $n = 2$, we obtain

$$
\mathrm{SO}_2(k) = \left\{ \begin{bmatrix} a & 0 \\ 0 & a^{-1} \end{bmatrix} \,\middle|\, 0 \neq a \in k \right\} \quad \text{and} \quad \begin{bmatrix} 0 & 1 \\ 1 & 0 \end{bmatrix} \in \mathrm{O}_2(k) \backslash \mathrm{SO}_2(k);
$$

see Exercise 1.8.19 for the case $n = 3$. Now assume that we are in the case (Alt) where k may have any characteristic. Then $n = 2m$ must be even, and

$$
\mathrm{Sp}_{2m}(k) := \Gamma_{2m}(Q_{2m}^-, k), \quad \text{where } Q_{2m}^- := \left[\begin{array}{c|c} 0 & Q_m \\ \hline -Q_m & 0 \end{array} \right] \in M_{2m}(k),
$$

is called the *symplectic* group of degree $2m$. Note that, for $m = 1$, we obtain

$$\mathrm{Sp}_2(k) = \mathrm{SL}_2(k).$$

We shall see in Theorems 1.7.4 and 1.7.8 that $\mathrm{SO}_n(k)$, $\mathrm{Sp}_{2m}(k)$ are connected.

1.3.16 Orthogonal groups and quadratic forms   When dealing with even-dimensional orthogonal groups, it is actually more convenient to work with a slightly different description, using quadratic forms. In particular, this will also give rise to a definition of orthogonal groups in characteristic 2. Let $n = 2m$ for some $m \geqslant 1$ and consider the matrix Q_{2m} defined above. Now let $P = (p_{ij}) \in M_{2m}(k)$ be such that $Q_{2m} = P + P^{\mathrm{tr}}$. Then the corresponding polynomial

$$f_P := \sum_{i,j=1}^{2m} p_{ij} X_i X_j \in k[X_1, \ldots, X_n]$$

defines in the usual way a function on k^{2m}. We set

$$\Gamma_{2m}(f_P, k) := \{ A \in M_{2m}(k) \mid f_P(Ax) = f_P(x) \text{ for all } x \in k^{2m} \},$$

where $x \in k^{2m}$ is regarded as a column vector. We claim that $\Gamma_{2m}(f_P, k)$ is a linear algebraic group. Indeed, first note that $\Gamma_{2m}(f_P, k)$ is closed in $M_{2m}(k)$ since the condition $f_P(Ax) = f_P(x)$ yields a system of polynomial equations in the coefficients of A. Furthermore, $\Gamma_{2m}(f_P, k)$ certainly is closed under multiplication. To see that every element in $\Gamma_{2m}(f_P, k)$ is invertible, we note that

$$f_P(x+y) - f_P(x) - f_P(y) = \sum_{i,j=1}^{2m} (p_{ij} + p_{ji}) x_i y_j = \beta_{Q_{2m}}(x, y), \qquad (*)$$

where $x = (x_1, \ldots, x_{2m})$, $y = (y_1, \ldots, y_{2m})$, and $\beta_{Q_{2m}}$ is defined as in §1.3.15. Consequently, any matrix $A \in \Gamma_{2m}(f_P, k)$ also satisfies the equation $A^{\mathrm{tr}} Q_{2m} A = Q_{2m}$, and so $\det(A) = \pm 1$. Hence, we have $\Gamma_{2m}(f_P, k) \subseteq \Gamma_{2m}(Q_{2m}, k)$. We remark that $\Gamma_{2m}(f_P, k)$ is defined by the polynomial equations defining $\Gamma_{2m}(Q_{2m}, k)$ plus the $2m$ equations $f_P(Ae_r) = f_P(e_r)$, where $\{ e_r \mid 1 \leqslant r \leqslant 2m \}$ is the standard basis of k^{2m}. (This easily follows using the relation $(*)$.) These $2m$

equations yield the relations $(A^{\mathrm{tr}}PA)_{rr} = P_{rr}$. Thus, we have

$$\Gamma_{2m}(f_P, k) = \{A \in \Gamma_{2m}(Q_{2m}, k) \mid (A^{\mathrm{tr}}PA)_{rr} = P_{rr} \text{ for } 1 \leqslant r \leqslant 2m\}.$$

If $f_{P'}$ is another polynomial as above, we say that $f_P, f_{P'}$ are *equivalent* if there exists an invertible $R \in M_{2m}(k)$ such that $Q_{2m} = R^{\mathrm{tr}}Q_{2m}R$ and $f_P(Rv) = f_{P'}(v)$ for all $v \in k^{2m}$. Hence, we have $A \in \Gamma_{2m}(f_{P'}, k)$ if and only if $RAR^{-1} \in \Gamma_{2m}(f_P, k)$. Thus, as before, the map $A \mapsto RAR^{-1}$ defines an isomorphism of algebraic groups $\Gamma_{2m}(f_{P'}, k) \to \Gamma_{2m}(f_P, k)$. Now, if k is algebraically closed, then there is (up to equivalence) a unique quadratic polynomial f_P as above; see Theorems 4.4 and 12.9 of Grove (2002). We make the following definite choice:

$$f_{2m} := \sum_{i=1}^{m} X_i X_{2m+1-i} \quad \text{with matrix} \quad P_{2m} := \left[\begin{array}{c|c} 0 & Q_m \\ \hline 0 & 0 \end{array}\right].$$

The corresponding group $\Gamma_{2m}(f_{2m}, k)$ will be denoted by $\mathrm{O}_{2m}^+(k)$.

Now, if $\mathrm{char}(k) \neq 2$, then $2f_{2m}(x) = \beta_{Q_{2m}}(x, x)$ for $x \in k^{2m}$. Thus, we have

$$\mathrm{O}_{2m}^+(k) = \mathrm{O}_{2m}(k) \quad (\mathrm{char}(k) \neq 2);$$

so we don't obtain anything new here. In order to have a uniform notation, we shall also set $\mathrm{SO}_{2m}^+(k) := \mathrm{SO}_{2m}(k)$ in this case. Now assume that $\mathrm{char}(k) = 2$. Then we have $\Gamma_{2m}(Q_{2m}, k) = \mathrm{Sp}_{2m}(k)$. Writing out the relations $(A^{\mathrm{tr}}PA)_{rr} = P_{rr}$ $(1 \leqslant r \leqslant 2m)$ for the above special choice of $P = P_{2m}$ yields

$$\mathrm{O}_{2m}^+(k) = \left\{(a_{ij}) \in \mathrm{Sp}_{2m}(k) \,\Big|\, \sum_{i=1}^{m} a_{ij}\, a_{2m+1-i,j} = 0 \text{ for } 1 \leqslant j \leqslant 2m\right\};$$

this will be called the *orthogonal group in characteristic 2*. For $m = 1$, we obtain

$$\mathrm{O}_2^+(k) = \left\{\left[\begin{array}{cc} a & 0 \\ 0 & a^{-1} \end{array}\right] \,\Big|\, 0 \neq a \in k\right\} \cup \left\{\left[\begin{array}{cc} 0 & a \\ a^{-1} & 0 \end{array}\right] \,\Big|\, 0 \neq a \in k\right\}.$$

Thus, $\mathrm{O}_2^+(k) \subsetneq \mathrm{Sp}_2(k)$ certainly is not connected. We shall set

$$\mathrm{SO}_{2m}^+(k) := \mathrm{O}_{2m}^+(k)^{\circ} \quad (\mathrm{char}(k) = 2).$$

In Theorem 1.7.8 it will be shown that $\mathrm{SO}_{2m}^+(k)$ has index 2 in $\mathrm{O}_{2m}^+(k)$.

Table 1.1 Series of groups of classical type

Type	Group	Remarks
A_{m-1}	$\mathrm{SL}_m(k)$	any $m \geqslant 1$, any characteristic
B_m	$\mathrm{SO}_{2m+1}(k)$	any $m \geqslant 1$, $\mathrm{char}(k) \neq 2$
C_m	$\mathrm{Sp}_{2m}(k)$	any $m \geqslant 1$, any characteristic
D_m	$\mathrm{SO}_{2m}^{+}(k)$	any $m \geqslant 1$, any characteristic

In Lie notation, $\mathrm{SL}_n(k)$, $\mathrm{SO}_n(k)$, $\mathrm{Sp}_{2m}(k)$, $\mathrm{SO}_{2m}^{+}(k)$ are groups of classical type A, B, C, D. The precise correspondence is given in Table 1.1; see Chapter 1 and §11.3 of Carter (1972) for more details. Note that we could have also defined $\mathrm{SO}_{2m+1}(k)$ in characteristic 2, as in §1.3.15. However, if $\mathrm{char}(k) = 2$, then there exists a bijective homomorphism of algebraic groups $\mathrm{SO}_{2m+1}(k) \xrightarrow{\sim} \mathrm{Sp}_{2m}(k)$; see Theorem 14.2 of Grove (2002).

1.4 The tangent space and non-singular points

The aim of this section is to introduce the tangent space of an algebraic set at a point and to prove the existence of non-singular points. This involves a process of 'linearization' of the polynomials defining an algebraic set. On a more formal level, this requires a further basic notion, that of derivations.

1.4.1 Definition Let A be a k-algebra and M be an A-module; we denote the action of A on M by $(a, m) \mapsto a.m$ ($a \in A$, $m \in M$). A k-linear map $D \colon A \to M$ is called a *derivation* if we have the product rule

$$D(ab) = a.D(b) + b.D(a) \quad \text{for all } a, b \in A.$$

The set of all derivations of A into M is denoted by $\mathrm{Der}_k(A, M)$. This set naturally is an A-module. Indeed, if $c \in A$ and $D \in \mathrm{Der}_k(A, M)$, then we define $c.D \colon A \to M$ by $(c.D)(a) := c.D(a)$ for all $a \in A$. It is readily checked that $c.D \in \mathrm{Der}_k(A, M)$.

1.4.2 Example (a) Let us take $M = A$, where the action is given by multiplication. Then $\mathrm{Der}_k(A, A)$ carries an additional structure, that of a *Lie algebra*. Recall that a Lie algebra is a vector space L

equipped with a bilinear product $[\ ,\ ]\colon L \times L \to L$ which satisfies $[x, x] = 0$, for all $x \in L$, and the Jacobi identity

$$[x, [y, z]] + [y, [z, x]] + [z, [x, y]] = 0 \quad \text{for all } x, y, z \in L.$$

For example, $M_n(k)$ is a Lie algebra with product $[A, B] := AB - BA$.

Now, $\mathrm{Der}_k(A, A)$ is a Lie algebra with product $[D, D'] := D \circ D' - D' \circ D$. Indeed, it is clear that $[D, D] = 0$; the Jacobi identity is checked by a straightforward verification. Moreover, we have the following *Leibniz rule*:

$$D^n(ab) = \sum_{i=0}^{n} \binom{n}{i} D^i(a) D^{n-i}(b) \quad \text{for all } a, b \in A.$$

If the characteristic of k is a prime $p > 0$, this shows that D^p is also a derivation.

(b) Let $R = k[X_1, \ldots, X_n]$ where X_i are indeterminates. We define $D_i \colon R \to R$ by $D_i(f) = \partial f / \partial X_i$ (the usual partial derivative with respect to X_i). Then $D_i \in \mathrm{Der}_k(R, R)$. Furthermore, we have

$$D = \sum_{i=1}^{n} D(X_i)\, D_i \quad \text{for any } D \in \mathrm{Der}_k(R, R).$$

To check this identity, we note that D is uniquely determined by its values on a set of algebra generators of R. (This follows from the product rule for derivations.) Hence it is enough to verify the identify on X_j for $1 \leqslant j \leqslant n$, in which case it is trivial. Since $D_1, \ldots, D_n$ certainly are linearly independent, we conclude that $\mathrm{Der}_k(R, R)$ is a free R-module with basis $D_1, \ldots, D_n$.

1.4.3 Lemma *Let $A = k[X_1, \ldots, X_n]/I$ where $I \subseteq k[X_1, \ldots, X_n]$ is an ideal. Let M be an A-module, and set*

$$\mathfrak{T}_{A,M} := \left\{ (v_1, \ldots, v_n) \in M^n \,\middle|\, \sum_{i=1}^{n} \overline{D_i(f)}.v_i = 0 \text{ for all } f \in I \right\},$$

where D_i denotes partial derivative with respect to X_i and the bar denotes the canonical map $k[X_1, \ldots, X_n] \to A$. Then, for any $v = (v_1, \ldots, v_n) \in \mathfrak{T}_{A,M}$, we have a well-defined derivation

$D_v \in \operatorname{Der}_k(A, M)$ *given by*

$$D_v(\bar{f}) = \sum_{i=1}^{n} \overline{D_i(f)}.v_i \quad \text{for } f \in k[X_1, \ldots, X_n].$$

The map $\Phi \colon \mathfrak{T}_{A,M} \to \operatorname{Der}_k(A, M)$, $v \mapsto D_v$, *is an A-module isomorphism.*

Proof Let $R := k[X_1, \ldots, X_n]$ and consider the canonical map $\pi \colon R \to A$, $f \mapsto \bar{f}$, with kernel I. Then we can also regard M as an R-module via π.

First we must check that D_v is well-defined. For this purpose we must check that $\sum_{i=1}^{n} \overline{D_i(f)}.v_i = 0$ for any $f \in I$, which is true by the definition of $\mathfrak{T}_{A,M}$. Next we show that $D_v \in \operatorname{Der}_k(A, M)$. Indeed, for any $f, g \in R$, we have

$$D_v(\bar{f}\,\bar{g}) = \sum_{i=1}^{n} \overline{D_i(fg)}.v_i = \sum_{i=1}^{n} \left(\overline{gD_i(f)}.v_i + \overline{fD_i(g)}.v_i \right)$$
$$= \bar{g}\, D_v(\bar{f}) + \bar{f}\, D_v(\bar{g}),$$

as required. Thus, we have a well-defined A-linear map $\Phi \colon \mathfrak{T}_{A,M} \to \operatorname{Der}_k(A, M)$. We have $D_v(\bar{X}_j) = v_j$ for $1 \leqslant j \leqslant n$, and so Φ is injective. It remains to show that Φ is also surjective. So let $D \in \operatorname{Der}_k(A, M)$. Then the composition $\tilde{D} = D \circ \pi$ is a derivation in $\operatorname{Der}_k(R, M)$. As in Example 1.4.2(b), one sees that

$$\tilde{D}(f) = \sum_{i=1}^{n} D_i(f)\, \tilde{D}(X_i) \quad \text{for all } f \in R.$$

Now set $v_i := \tilde{D}(X_i) \in M$ for $1 \leqslant i \leqslant n$. We claim that $v = (v_1, \ldots, v_n) \in \mathfrak{T}_{A,M}$. Indeed, we have

$$D(\bar{f}) = \tilde{D}(f) = \sum_{i=1}^{n} D_i(f).v_i = \sum_{i=1}^{n} \overline{D_i(f)}.v_i \quad \text{for all } f \in R.$$

Thus, if $f \in I$, then $\bar{f} = 0$ and so $v \in \mathfrak{T}_{A,M}$ as claimed. Since $D_v(\bar{X}_j) = v_j = D(\bar{X}_j)$, we see that $D_v = D$. Thus, Φ is surjective. $\qquad\square$

1.4.4 Remark In the setting of Lemma 1.4.3, assume that the ideal I is generated by $f_1, \ldots, f_m \in k[X_1, \ldots, X_n]$. Then we claim that

$$\mathfrak{T}_{A,M} = \left\{ (v_1, \ldots, v_n) \in M^n \,\Big|\, \sum_{i=1}^{n} \overline{D_i(f_j)}.v_i = 0 \text{ for } 1 \leqslant j \leqslant m \right\}.$$

Indeed, the inclusion '$\subseteq$' is clear. To prove the reverse inclusion, assume that $(v_1, \ldots, v_n) \in M^n$ lies in the right-hand side of the above identity. Now let $f \in I$. Then we can write $f = \sum_{j=1}^{m} h_j f_j$ with $h_j \in k[X_1, \ldots, X_n]$, and it follows that

$$\overline{D_i(f)} = \sum_{j=1}^{m} \overline{D_i(h_j f_j)} = \sum_{j=1}^{m} \left(\overline{h_j D_i(f_j)} + \overline{f_j D_i(h_j)} \right) = \sum_{j=1}^{m} \overline{h_j D_i(f_j)}.$$

Hence we also have $\sum_{i=1}^{n} \overline{D_i(f)}.v_i = 0$, as required.

We now obtain a new characterization of the transcendence degree of a finitely generated field extension $K \supseteq k$. Recall that k is called a *perfect* field if either k has characteristic 0 or k has characteristic $p > 0$ and $k = \{x^p \mid x \in k\}$. For example, finite fields or algebraically closed fields are perfect.

1.4.5 Proposition *Assume that k is a perfect field, and let A be a finitely generated k-algebra which is an integral domain.*

(a) Let K be the field of fractions of A. Then $\partial_k(A) = \dim_K \mathrm{Der}_k(K, K)$.

(b) If $A = k[X_1, \ldots, X_n]/(f_1, \ldots, f_m)$, where $f_i \in k[X_1, \ldots, X_n]$, then

$$\partial_k(A) = n - \mathrm{rank}_K \left(\overline{D_i(f_j)} \right)_{\substack{1 \leqslant i \leqslant n \\ 1 \leqslant j \leqslant m}}.$$

Proof We begin with the following observations. Let $R \subseteq K$ be any k-subalgebra such that K is the field of fractions of R. We regard K as an R-module (by left multiplication). Now consider the restriction map

$$\rho \colon \mathrm{Der}_k(K, K) \to \mathrm{Der}_k(R, K), \quad D \mapsto D|_R.$$

Since R generates K, it is clear that ρ is injective. On the other hand, if $D \in \mathrm{Der}_k(R, K)$, then we can extend D to a derivation

$\tilde{D} \in \mathrm{Der}_k(K, K)$ by setting

$$\tilde{D}(f/g) = \frac{1}{g^2}(gD(f) - fD(g)) \quad \text{for } f, g \in R,\ g \neq 0.$$

It is straightforward to check that $\tilde{D}$ is well-defined and lies in $\mathrm{Der}_k(K, K)$. (This is just similar to the rule for the derivative of a quotient of two functions.) Since $\rho(\tilde{D}) = D$, we see that ρ is surjective. Also note that $\mathrm{Der}_k(R, K)$ still is a K-vector space and ρ is K-linear. Thus, we have

$$\dim_K \mathrm{Der}_k(K, K) = \dim_K \mathrm{Der}_k(R, K).$$

Furthermore, let us assume that R can be written as a quotient of a polynomial ring. Then the R-module $\mathfrak{T}_{R,K}$ defined in Lemma 1.4.3 certainly is a K-vector space and the map Φ is seen to be K-linear. Thus, we have in fact

$$\dim_K \mathrm{Der}_k(K, K) = \dim_K \mathfrak{T}_{R,K}. \qquad (*)$$

To prove (a), let $d = \partial_k(A)$. Then we also have $d = \partial_k(K)$ by Proposition 1.2.18. So, since k is perfect, we can apply Exercise 1.8.15 and conclude that there exist $z_1, \ldots, z_{d+1} \in K$ such that $K = k(z_1, \ldots, z_{d+1})$, $\{z_1, \ldots, z_d\}$ are algebraically independent over k and z_{d+1} is separable algebraic over $k(z_1, \ldots, z_d)$. We now apply the above discussion to the subalgebra $R := k[z_1, \ldots, z_{d+1}] \subseteq K$. Since $\{z_1, \ldots, z_{d+1}\}$ are not algebraically independent, there exists some $0 \neq f \in k[Y_1, \ldots, Y_{d+1}]$ such that $f(z_1, \ldots, z_{d+1}) = 0$. Since K is a field, we may assume that f is irreducible. Now consider the map

$$\alpha \colon k[Y_1, \ldots, Y_{d+1}] \to K, \quad g \mapsto g(z_1, \ldots, z_{d+1}).$$

We claim that $\ker(\alpha) = (f)$. Indeed, it is clear that $\alpha(f) = 0$. Conversely, let $g \in k[Y_1, \ldots, Y_{d+1}]$ be such that $\alpha(g) = 0$, and assume that f does not divide g. As in the proof of Theorem 1.1.9, there exist $F, G \in k[Y_1, \ldots, Y_{d+1}]$ such that $0 \neq d := Gf + Fg \in k[Y_1, \ldots, Y_d]$; note that Y_{d+1} occurs in some term of f. Then we obtain $d(z_1, \ldots, z_d) = \alpha(Gf + Fg) = 0$, contradicting the fact that $z_1, \ldots, z_d$ are algebraically independent. So we have $\ker(\alpha) = (f)$ as claimed. On the other hand, the image of α is just R. Hence we have $R \cong k[Y_1, \ldots, Y_{d+1}]/(f)$. It follows that $\mathfrak{T}_{R,K}$ is a subspace of K^{d+1} defined by one linear equation; see Remark 1.4.4. That linear equation is non-zero since some partial derivative of f will be non-zero; see Exercise 1.8.14. So we have $\dim_K \mathfrak{T}_{R,K} = d$ and $(*)$ yields (a). Finally, (b) follows from (a) and by applying $(*)$ to $R = A$. $\qquad \square$

1.4.6 Corollary   *Assume that k is a perfect field. Let $V \subseteq k^n$ be an irreducible algebraic set and assume that $\mathbf{I}(V) \subseteq k[X_1, \ldots, X_n]$ is generated by $f_1, \ldots, f_m$. Then we have*

$$\dim V = n - \operatorname{rank}\left(\overline{D_i(f_j)}\right)_{\substack{1 \leqslant i \leqslant n \\ 1 \leqslant j \leqslant m}},$$

where D_i denotes the partial derivative with respect to X_i and the bar denotes the canonical map $k[X_1, \ldots, X_n] \to A[V]$; the rank is taken over the field of fractions of $A[V]$.

Proof   See Proposition 1.2.18 and Proposition 1.4.5. $\square$

The above results were developed into a purely algebraic setting. We now translate them into geometric properties of algebraic sets.

1.4.7 Taylor expansion   Let $V \subseteq k^n$ be a non-empty algebraic set with vanishing ideal $\mathbf{I}(V) \subseteq k[X_1, \ldots, X_n]$. Let us fix a point $p = (p_1, \ldots, p_n) \in V$. For $f \in k[X_1, \ldots, X_n]$, we define a linear polynomial

$$d_p(f) := \sum_{i=1}^{n} D_i(f)(p)\, X_i \in k[X_1, \ldots, X_n],$$

where D_i denotes the usual partial derivative with respect to X_i. We have the usual rules for computing derivates: $d_p(f+g) = d_p(f) + d_p(g)$ and $d_p(fg) = d_p(f)\, g(p) + f(p)\, d_p(g)$ for all $f, g \in k[X_1, \ldots, X_n]$. Furthermore, we have $d_p(f) = f - f(0)$ if f is linear.

For any $v = (v_1, \ldots, v_n) \in k^n$, we now consider the 'line through p in direction v' defined by $L_v := \{p + tv \mid t \in k\} \subseteq k^n$. Then, given any $f \in k[X_1, \ldots, X_n]$, we consider the Taylor expansion of $f(p+tv)$ (as a function in t) around $t = 0$:

$$f(p + tv) = f(p) + \underbrace{\left(\sum_{i=1}^{n} D_i(f)(p)\, v_i\right)}_{=d_p(f)(v)} t + \text{combination of higher powers of } t.$$

We say that L_v is a *tangent line* at $p \in V$ if $d_p(f)(v) = 0$ for all $f \in \mathbf{I}(V)$.

1.4.8 Definition   Let $V \subseteq k^n$ be a non-empty algebraic set. For fixed $p \in V$, let us regard k as an $A[V]$-module where $\bar{f} \in A[V]$

acts by multiplication with $f(p)$. We denote this $A[V]$-module by k_p. Then we define

$$T_p(V) := \mathfrak{T}_{A[V],k_p} = \{v \in k^n \mid d_p(f)(v) = 0 \text{ for all } f \in \mathbf{I}(V)\}$$
$$= \{v \in k^n \mid L_v \text{ is a tangent line at } p \in V\}$$

to be the *tangent space* at $p \in V$. Thus, $T_p(V)$ is at the same time a linear subspace and an algebraic subset of k^n. Assume that $|k| = \infty$. Then, by Exercise 1.8.6, $T_p(V)$ is irreducible and we may unambiguously define $\dim T_p(V)$ as in Definition 1.2.15 or as the dimension of a vector space.

1.4.9 Remark   (a) Assume that the vanishing ideal $\mathbf{I}(V)$ is generated by the polynomials $f_1, \ldots, f_m \in k[X_1, \ldots, X_n]$. Then Remark 1.4.4 shows that

$$T_p(V) = \mathbf{V}(\{d_p(f_1), \ldots, d_p(f_m)\}) \subseteq k^n.$$

(b) By Lemma 1.4.3, we have a k-linear isomorphism

$$\Phi \colon T_p(V) \to \mathrm{Der}_k(A[V], k_p), \quad v \mapsto D_v,$$

where D_v is defined by $D_v(\bar{f}) = d_p(f)(v)$. In what follows, we will often identify $T_p(V) = \mathrm{Der}_k(A[V], k_p)$ using that isomorphism.

1.4.10 Example   Assume that k is an infinite field.

(a) Let V be a linear subspace of k^n. Then V is defined by linear polynomials $f_1, \ldots, f_m \in k[X_1, \ldots, X_n]$ with constant term zero. By Exercise 1.8.6, we have $\mathbf{I}(V) = (f_1, \ldots, f_m)$. Now $d_p(f) = f$ for all such polynomials, and so $T_p(V) = V$ for all $p \in V$. In particular, we have $\dim V = \dim T_p(V)$ for all $p \in V$.

This applies, in particular, to the case where $V = k^n$. Then we have $\mathbf{I}(V) = \{0\}$ by Exercise 1.8.2, and so $T_p(k^n) = k^n$ for all $p \in k^n$.

(b) Let $V = \{p\} \subseteq k^n$ be a singleton set. Then, as in Exercise 1.2.19(a), we see that $\mathbf{I}(V) = (X_1 - p_1, \ldots, X_n - p_n)$, and this yields $T_p(\{p\}) = \{0\}$.

(c) Consider the twisted cubic $C = \mathbf{V}(f_1, f_2) \subseteq k^3$, where $f_1 = X_2 - X_1^2$ and $f_2 = X_3 - X_1^3$. By Exercise 1.8.4(a), we have $\mathbf{I}(C) = (f_1, f_2)$. Let $p = (x, x^2, x^3) \in C$. Then we have $d_p(f_1) = -2xX_1 + X_2$ and $d_p(f_2) = -3x^2X_1 + X_3$, and so

$$T_p(C) = \langle (1, 2x, 3x^2) \rangle_k \subseteq k^3.$$

Thus, $T_p(C)$ is a one-dimensional subspace for each $p \in C$.

1.4.11 Theorem *Assume that k is an infinite perfect field, and let $V \subseteq k^n$ be an irreducible (and, hence, non-empty) algebraic set.*

(a) We have $\dim T_p(V) \geqslant \dim V$ for all $p \in V$. Furthermore, the set of all $p \in V$ with $\dim T_p(V) = \dim V$ is non-empty and open.

(b) Let $p \in V$ be such that $\dim T_p(V) = \dim V$. Then there exists some $f \in k[X_1, \ldots, X_n]$ with $f(p) \neq 0$ and regular maps $\psi_j \colon \tilde{V}_f \to k^n$ such that

$$T_{\pi_f(\tilde{q})}(V) = \langle \psi_1(\tilde{q}), \ldots, \psi_d(\tilde{q}) \rangle_k \subseteq k^n \quad \text{for all } \tilde{q} \in \tilde{V}_f,$$

where $\pi_f \colon \tilde{V}_f \to V$ is defined as in §1.1.14.

We will say that a point $p \in V$ is *non-singular* if $\dim T_p(V) = \dim V$; otherwise, p is called *singular*. If all points of V are non-singular, we call V *non-singular*. Later, we will give another characterization of non-singular points; see Proposition 2.3.8.

Proof (a) Assume that $\mathbf{I}(V) = (f_1, \ldots, f_m)$. Since V is irreducible, $A[V]$ is an integral domain; let K be its field of fractions. Then, by Corollary 1.4.6, we have

$$\dim V = n - \operatorname{rank}_K M, \quad \text{where} \quad M := \left(\overline{D_i(f_j)} \right)_{\substack{1 \leqslant i \leqslant n \\ 1 \leqslant j \leqslant m}} \in A[V]^{n \times m}.$$

Now note that $d_p(f_j) = \sum_{i=1}^n \overline{D_i(f_j)}(p) X_i$ for any $p \in V$. Hence, by Remark 1.4.9(a), we have $\dim T_p(V) = n - \operatorname{rank}_k(M(p))$. Since the rank of a matrix is given by the vanishing and non-vanishing of certain minors, it is clear that $\operatorname{rank}_k M(p) \leqslant \operatorname{rank}_K(M)$ for all $p \in V$, with equality for at least some p. Thus, we have $\dim T_p(V) \geqslant \dim V$ for all $p \in V$, where equality holds for some p.

Now set $r = \operatorname{rank}_K M$. Then the set of all $p \in V$ where $\dim T_p(V) = \dim V$ is the complement of the set of all $p \in V$ where all $(r+1) \times (r+1)$ minors of $M(p)$ vanish. Since these minors are given by polynomial expressions in the coefficients of $M(p)$ and, hence, by polynomials in the coordinates of p, we see that the set of all $p \in V$ with $\dim T_p(V) = \dim V$ is open.

(b) Let $d = \dim V$. Since $\dim T_p(V) = d$, there exist $f_1, \ldots, f_{n-d} \in \mathbf{I}(V)$ such that $T_p(V) = \mathbf{V}(\{d_p(f_1), \ldots, d_p(f_{n-d})\})$. We shall assume that the labelling of the variables is chosen such that

$$\det\Big(D_j(f_i)(p)\Big)_{1\leqslant i,j\leqslant n-d} \neq 0. \tag{1}$$

Now consider the matrix (where i is the row index and j is the column index)

$$M_q := \Big(D_j(f_i)(q)\Big)_{\substack{1\leqslant i\leqslant n-d \\ 1\leqslant j\leqslant n}} \quad \text{where } q \in V.$$

Consider the elements of k^n as column vectors and let $v \in k^n$, with components $v_1,\ldots,v_n$. Then the ith component of the product $M_q v \in k^n$ is given by

$$(M_q v)_i = \sum_{j=1}^{n} D_j(f_i)(q)\, v_j = d_q(f_i)(v) \quad \text{where } q \in V. \tag{2}$$

Our assumption (1) means that, if we write $M_q = \begin{pmatrix} A_q & B_q \end{pmatrix}$ with A_q of size $(n-d)\times(n-d)$ and B_q of size $(n-d)\times d$, then $\det A_p \neq 0$. Since the determinant of a matrix is given by a polynomial expression in the coefficients, there exists a non-zero polynomial $f \in k[X_1,\ldots,X_n]$ such that $\det A_q = f(q)$ for all $q \in V$. Since $f(p) = \det A_p \neq 0$, we have $p \in V_f \subseteq V$; furthermore,

$$d = \dim\, \mathbf{V}(\{d_q(f_1),\ldots,d_q(f_{n-d})\}) \quad \text{for all } q \in V_f.$$

For any $q \in V_f$, we have that $T_q(V)$ is contained in the set on the right-hand side of the above identity (by the definition of $T_q(V)$). On the other hand, we have $\dim T_q(V) \geqslant d$ by (a). So we conclude that

$$T_q(V) = \mathbf{V}(\{d_q(f_1),\ldots,d_q(f_{n-d})\}) \quad \text{for all } q \in V_f. \tag{3}$$

Now consider $\tilde{V}_f = \{(q,y) \in V \times k \mid f(q)y = 1\}$. For $1 \leqslant j \leqslant d$, we define

$$\psi_j\colon \tilde{V}_f \to k^n, \quad (q,y) \mapsto j\text{th column of } \begin{bmatrix} -A_q^{-1}B_q \\ I_d \end{bmatrix}.$$

We claim that ψ_j is a regular map. Indeed, using the formula for the inverse of a matrix and recalling the definition of f, we see that there exist polynomials $h_{j1},\ldots,h_{jn} \in k[X_1,\ldots,X_n]$ such that the ith component of $\psi_j(q,y)$ is given by $h_{ji}(q)/f(q) = yh_{ji}(q)$ for all $(q,y) \in \tilde{V}_f$. Thus, ψ_j certainly is regular.

Working out the product $M_q\psi_j(q,y)$, we see that the first $n-d$ components of the result are 0. By (2), this means that

$$d_q(f_i)(\psi_j(q,y)) = 0 \quad \text{for } 1 \leqslant i \leqslant n-d \text{ and } (q,y) \in \tilde{V}_f.$$

Furthermore, $\{\psi_1(q,y),\ldots,\psi_d(q,y)\}$ are certainly linearly independent. Thus, using (3), we see that $T_q(V) = \langle \psi_1(q,y),\ldots,\psi_d(q,y)\rangle_k$ for all $(q,y) \in \tilde{V}_f$. $\qquad\square$

1.4.12 Example Assume that k is algebraically closed and $f \in k[X_1,\ldots,X_n]$ is a non-constant and irreducible polynomial. Then the corresponding hypersurface $H_f \subseteq k^n$ is irreducible, and we have $\mathbf{I}(H_f) = (f)$; see Theorem 1.1.9. Thus, using Remark 1.4.9, we see that $T_p(H_f) = \mathbf{V}(\{d_p(f)\}) \subseteq k^n$ for any $p \in H_f$ that is, $T_p(H_f)$ is defined by one linear equation. We conclude that

$$\dim T_p(II_f) - \begin{cases} n-1 & \text{if } D_i(f)(p) \neq 0 \text{ for some } i, \\ n & \text{if } D_i(f)(p) = 0 \text{ for all } i. \end{cases}$$

Using Exercise 1.8.14, it is easily checked that the set $U := \{p \in H_f \mid D_i(f)(p) = 0 \text{ for all } i\}$ is a proper closed subset of V. Hence we have $\dim T_p(H_f) = n-1$ for all $p \in H_f \setminus U$; note that $\dim H_f = n-1$ by Example 1.2.16(b).

Now consider the special case where $n = 2$ so that $H_f \subseteq k^2$ is an *affine curve*; see Example 1.2.19(b). We claim that all but finitely many points of H_f are non-singular. Indeed, let $U \subseteq H_f$ be the set of singular points. We have seen above that $U \neq H_f$. So we have $\dim U < \dim H_f = 1$ by Proposition 1.2.20. Thus, $\dim U = 0$, and so U is a finite set by Exercise 1.8.10(a) and Proposition 1.1.11.

1.4.13 The differential of a regular map Let $V \subseteq k^n$ and $W \subseteq k^m$ be non-empty algebraic sets and $\varphi\colon V \to W$ be a regular map. Let $p \in V$ and $q := \varphi(p) \in W$. The map

$$d_p\varphi\colon \mathrm{Der}_k(A[V],k_p) \to \mathrm{Der}_k(A[W],k_q), \quad D \mapsto D \circ \varphi^*,$$

where $\varphi^*\colon A[W] \to A[V]$ is the algebra homomorphism in §1.3.3, is called the *differential* of φ. If $V = W$ and $\varphi = \mathrm{id}_V$, then $d_p(\mathrm{id})$ is the identity map on $\mathrm{Der}_k(A[V],k_p)$ for all $p \in V$. Furthermore, if $Z \subseteq k^l$ is another algebraic set and $\psi\colon W \to Z$ is a regular map, then we have $d_p(\psi \circ \varphi) = d_q\psi \circ d_p\varphi$.

In particular, this shows that if $\varphi\colon V \to W$ is an isomorphism, then the differential $d_p\varphi\colon \mathrm{Der}_k(A[V], k_p) \to \mathrm{Der}_k(A[W], k_q)$ is a vector-space isomorphism.

Finally, on the level of $T_p(V) \subseteq k^n$ and $T_q(W) \subseteq k^m$, the differential $d_p\varphi$ is given as follows. Suppose that φ is defined by $f_1, \ldots, f_m \in k[X_1, \ldots, X_n]$. Then, for $v \in T_p(V)$, we have $d_p\varphi(D_v) = D_w$, where $w = (w_1, \ldots, w_n) \in T_q(W)$ is given by $w_i = D_w(\bar{Y}_i) = (D_v \circ \varphi^*)(\bar{Y}_i) = D_v(\bar{f}_i)$ for all j. Thus, the linear map $d_p\varphi\colon T_p(V) \to T_q(W)$ is given by multiplication with the $m \times n$ matrix

$$J_p(\varphi) := \Big(D_j(f_i)(p) \Big)_{\substack{1 \leqslant i \leqslant m \\ 1 \leqslant j \leqslant n}}.$$

The following result shows that the tangent space at a point $p \in V$ only depends on an open neighbourhood of that point.

1.4.14 Lemma *Let $V \subseteq k^n$ be a non-empty algebraic set and $0 \neq \bar{f} \in A[V]$. As in §1.1.14, let $\tilde{V}_f = \{(v, y) \in V \times k \mid f(v)y = 1\} \subseteq k^{n+1}$. Then the map*

$$\pi_f\colon \tilde{V}_f \to V, \quad (v, y) \mapsto v,$$

is regular, and $d_{(p,x)}\pi_f\colon T_{(p,x)}(\tilde{V}_f) \to T_p(V)$ is an isomorphism for $(p, x) \in \tilde{V}_f$.

Proof By §1.4.13, $d_{(p,x)}(\pi_f)$ is given by the matrix $J_{(p,x)}(\varphi)$ which, in our case, is an $n \times (n + 1)$-matrix whose first n columns form the $(n \times n)$-identity matrix and whose last column is zero. Thus, $d_{(p,x)}(\pi_f)\colon T_{(p,x)}(\tilde{V}_f) \to T_p(V)$ is given by projecting onto the first n coordinates. On the other hand, we know that $\mathbf{I}(\tilde{V}_f) = (\mathbf{I}(V), fY - 1) \subseteq k[X_1, \ldots, X_n, Y]$ by Lemma 1.1.15. So we have

$$T_{(p,x)}(\tilde{V}_f) = \{(v, y) \in k^{n+1} \mid v \in T_p(V) \text{ and } d_{(p,x)}(fY - 1)(v, y) = 0\}$$

$$= \{(v, y) \in k^{n+1} \mid v \in T_p(V) \text{ and } y = -x d_p(f)(v)/f(p)\},$$

where the last equality follows from $d_{(p,x)}(fY - 1)(v, y) = x d_p(f)(v) + f(p)y$. Thus, for each $(v, y) \in T_{(p,x)}(\tilde{V}_f)$, we have $v \in T_p(V)$, and y is uniquely determined by v. Consequently, $d(\pi_f)_{(p,x)}$ is an isomorphism. $\qquad\square$

1.4.15 Proposition (differential criterion for dominance)

Let k be an infinite perfect field, $V \subseteq k^n$ and $W \subseteq k^m$ be irreducible algebraic sets, and $\varphi \colon V \to W$ be a regular map. Assume that $d_p\varphi \colon T_p(V) \to T_{\varphi(p)}(W)$ is surjective, where $p \in V$ is non-singular and $\varphi(p) \in W$ is non-singular.

Then φ is dominant; that is, we have $\overline{\varphi(V)} = W$.

Proof We begin by showing that there is an open set $U \subseteq V$ containing p such that the above hypotheses are satisfied for all $x \in U$, that is, x is non-singular, $\varphi(x)$ is non-singular, and $d_x\varphi$ is surjective. We shall obtain U as the intersection of three open subsets.

First, let U_1 be the open set of all non-singular points in V. Next, let U_2 be the preimage under φ of the open set of all non-singular points in W. Finally, we claim that there exists an open set $U_3 \subseteq V$ such that $p \in U_3$ and the image of $T_x(V)$ under $d_x\varphi$ has dimension $\geqslant \dim W$ for all $x \in U_3$. This is seen as follows. By Theorem 1.4.11(b), there exists some $f \in k[X_1, \ldots, X_n]$ with $f(p) \neq 0$ and regular maps $\psi_j \colon \tilde{V}_f \to k^n$ ($1 \leqslant j \leqslant d := \dim V$) such that

$$T_x(V) = \langle \psi_1(\tilde{x}), \ldots, \psi_d(\tilde{x}) \rangle_k \subseteq k^n \quad \text{for all } x \in V_f.$$

Let $\Psi(x)$ be the $n \times d$-matrix whose columns are given by $\psi_1(\tilde{x}), \ldots, \psi_d(\tilde{x})$. Now $d_x\varphi \colon T_x(V) \to T_{\varphi(x)}(W)$ is the linear map given by $v \mapsto J_x(\varphi)v$ ($v \in T_x(V)$) where $J_x(\varphi)$ is the $m \times n$ matrix defined in §1.4.13. Consequently, the image of $T_x(V)$ under $d_x\varphi$ is the k-span of the columns of $J_x(\varphi)\Psi(x)$. As usual, the condition that the rank of $J_x(\varphi)\Psi(x)$ be $\geqslant \dim W$ can be expressed by the non-vanishing of certain determinants in the coordinates of x. But, for $p \in V$, we know that this condition is satisfied. Hence, the set of all $x \in V_f$ for which $J_x(\varphi)\Psi(x)$ has rank $\geqslant \dim W$ is open. This yields the desired set U_3.

Now set $U := U_1 \cap U_2 \cap U_3$. First note that we have $p \in U$ by our hypothesis. Now let $x \in U$. Then x and $\varphi(x)$ are non-singular since $x \in U_1 \cap U_2$. Furthermore, the image of $T_x(V)$ under $d_x\varphi$ is contained in $T_{\varphi(x)}(W)$ and, hence, has dimension at most $\dim W$ (since $\varphi(x) \in U_2$). Since we also have $x \in U_3$, that dimension must be equal to $\dim W$, and so $d_x\varphi$ is surjective. Now we can complete the proof as follows. Let $Z := \overline{\varphi(V)} \subseteq W$. Then Z is closed irreducible (Remark 1.3.2) and we must show that $\dim Z = \dim W$; see Proposition 1.2.20. This is seen as follows. We can factor φ as

$\varphi = \iota \circ \varphi'$, where $\varphi' \colon V \to Z$ is dominant and $\iota \colon Z \hookrightarrow W$ is the inclusion. Now consider the open set S of all non-singular points in Z. Since $\varphi'(V) = \varphi(V)$ is dense in Z, we have $S \cap \varphi(V) \neq \varnothing$. Hence $(\varphi')^{-1}(S)$ is a non-empty open set in V. This open set has non-empty intersection with U, and so there exists some $x \in U$ such that $\varphi(x)$ is non-singular in Z. Now $d_x\varphi$ is the composition of $d_x\varphi' \colon T_x(V) \to T_{\varphi(x)}(Z)$ with the inclusion $T_{\varphi(x)}(Z) \subseteq T_{\varphi(x)}(W)$; see Exercise 1.8.16. The map $d_x\varphi$ is surjective since $x \in U$. Hence we must have $T_{\varphi(x)}(Z) = T_{\varphi(x)}(W)$. Since $\varphi(x) \in Z$ is non-singular, this yields $\dim Z = T_{\varphi(x)}(Z) = \dim T_{\varphi(x)}(W) \geqslant \dim W$, as desired.

$\square$

1.5 The Lie algebra of a linear algebraic group

The purpose of this section is to study the tangent spaces of a linear algebraic group $G \subseteq M_n(k)$. We shall show that the tangent space at the identity element of G carries an additional structure, that of a Lie algebra. (The definition of a Lie algebra is recalled in Example 1.4.2.) As examples, we will determine the Lie algebras of $\mathrm{SL}_n(k)$ and the symplectic and orthogonal groups.

1.5.1 Remark   Let $G \subseteq M_n(k)$ be a linear algebraic group, as in Definition 1.3.9. For $x \in G$, consider the map $\lambda_x \colon G \to G$, $\lambda_x(g) := \mu(x, g)$. By Remark 1.3.12, λ_x is an isomorphism of algebraic sets. So, using §1.4.13, we conclude that $d_g(\lambda_x) \colon T_g(G) \to T_{xg}(G)$ is an isomorphism for all $g \in G$. In particular, taking $x = 1$, we see that $\dim T_g(G) = \dim T_1(G)$ for all $g \in G$.

1.5.2 Proposition   *Let $G \subseteq M_n(k)$ be a linear algebraic group, where k is an infinite perfect field. Then we have $\dim T_g(G) = \dim G$ for all $g \in G$. In particular, if G is irreducible then G is non-singular.*

Proof   Let $G^\circ \subseteq G$ be the irreducible component containing the identity element $1 \in G$. By Exercise 1.8.16, we have $T_g(G^\circ) \subseteq T_g(G)$ for any $g \in G^\circ$. We prove that we have in fact equality. This is seen as follows. There exist $1 = g_1, g_2, \ldots, g_r \in G$ such that $G = \coprod_{i=1}^{r} G^\circ g_i$. By Proposition 1.3.13, the cosets $G^\circ g_i$ are precisely the irreducible components of G. Thus, $G \setminus G^\circ = \bigcup_{i=2}^{r} G^\circ g_i \subseteq G$ is a closed subset of G. Since, by §1.1.7, the operator $\mathbf{I}$ is injective, there exists some

$f \in \mathbf{I}(G \setminus G^\circ)$ such that $f \notin \mathbf{I}(G)$. Let $p \in G^\circ$ be such that $f(p) \neq 0$. Now let $v \in T_p(G)$. This means that $d_p(h)(v) = 0$ for all $h \in \mathbf{I}(G)$. Furthermore, for $h' \in \mathbf{I}(G^\circ)$, we have $h'f \in \mathbf{I}(G)$ and $h'(p) = 0$, and so

$$0 = d_p(h'f)(v) = d_p(h')(v)f(p) + h'(p)\,d_p(f)(v) = d_p(h')(v)f(p)$$

for any $v \in T_p(G)$. Since $f(p) \neq 0$, we conclude that $d_p(h')(v) = 0$. This holds for all $h' \in \mathbf{I}(G^\circ)$, and so $v \in T_p(G^\circ)$. Thus, we have shown that $T_p(G^\circ) = T_p(G)$ for some $p \in G^\circ$. By Remark 1.5.1, the tangent spaces at all elements of G° have the same dimension. Since there exist non-singular points by Theorem 1.4.11, we conclude that $\dim G^\circ = \dim T_p(G^\circ) = \dim T_p(G)$. Finally, we have $\dim G = \dim G^\circ$ by Corollary 1.3.14, and so $\dim G = \dim T_p(G)$. It remains to apply once more Remark 1.5.1 to G and its tangent spaces. $\quad\square$

The following result shows that the tangent space at $1 \in G$ carries naturally the structure of a *Lie algebra*. In the following result, we use the classical fact that $M_n(k)$ is a Lie algebra with product given by $[A, B] = AB - BA$.

1.5.3 Theorem *Let $G \subseteq M_n(k)$ be a linear algebraic group and $T_1(G) \subseteq M_n(k)$ be the corresponding tangent space at $1 \in G$.*

(a) For $A, B \in T_1(G)$, we have $[A, B] := AB - BA \in T_1(G)$. Thus, $T_1(G)$ is a Lie subalgebra of $M_n(k)$.

(b) If $H \subseteq M_m(k)$ is another linear algebraic group and $\varphi \colon G \to H$ is a homomorphism of algebraic groups, then $d_1\varphi \colon T_1(G) \to T_1(H)$ is a homomorphism of Lie algebras.

Proof (a) We begin by introducing a product on $\mathrm{Der}_k(A[G], k_1)$. This will then be transported to $T_1(G)$, using the isomorphism in Remark 1.4.9(b). For $D \in \mathrm{Der}_k(A[G], k_1)$ and $f \in A[G]$, we define $f \star D \in A[G]$ by

$$(f \star D)(x) = D(\lambda_x^*(f)) \quad \text{for } x \in G. \tag{$*$}$$

Here, $\lambda_x^* \colon A[G] \to A[G]$ is the algebra homomorphism induced by the map $\lambda_x \colon G \to G$, $g \mapsto xg$; see Remark 1.3.12. First we claim that

$$fg \star D = f(g \star D) + g(f \star D) \quad \text{for all } f, g \in A[G].$$

Indeed, for any $x \in G$, we have $(fg \star D)(x) = D(\lambda_x^*(f)\lambda_x^*(g)) = \lambda_x^*(f)(1)D(\lambda_x^*(g)) + \lambda_x^*(g)(1)D(\lambda_x^*(f)) = f(x)(g \star D)(x) + g(x)$

$(f \star D)(x)$, as required. Now, for $D, D' \in \mathrm{Der}_k(A[G], k_1)$, we define a map $[D, D'] \colon A[G] \to k$ by

$$[D, D'](f) = D(f \star D') - D'(f \star D) \quad \text{for } f \in A[G]. \qquad (*')$$

A straightforward computation, using $(*)$, yields that

$$
\begin{aligned}
{[D, D'](fg)} &= f(1)(D(g \star D') - D'(g \star D)) + g(1)(D(f \star D') \\
&\quad - D'(f \star D)) \\
&= f(1)[D, D'](g) + g(1)[D, D'](f) \quad \text{for all } f, g \in A[G].
\end{aligned}
$$

Thus, we have $[D, D'] \in \mathrm{Der}_k(A[G], k_1)$. We will now transport this operation to $T_1(G)$, using the isomorphism $\Phi \colon T_1(G) \to \mathrm{Der}_k(A[G], k_1)$, $P = (p_{ij}) \mapsto D_P$; see Remark 1.4.9(b). We have $A[G] = k[X_{ij} \mid 1 \leqslant i, j \leqslant n]/\mathbf{I}(G)$. Let $\bar{X}_{ij}$ denote the image of X_{ij} in $A[G]$. Then D_P is determined by $D_P(\bar{X}_{ij}) = p_{ij}$ for $1 \leqslant i, j \leqslant n$. To complete the proof of (a), we must identify the matrix in $T_1(G)$ corresponding to the derivation $[D_P, D_{P'}]$ where $P, P' \in T_1(G)$. Thus, we must show the following identity for all $D, D' \in \mathrm{Der}_k(A[G], k_1)$:

$$[D, D'](\bar{X}_{ij}) = \sum_{l=1}^{n} \left(D(\bar{X}_{il})D'(\bar{X}_{lj}) - D'(\bar{X}_{il})D(\bar{X}_{lj}) \right).$$

Using the defining formula for $[D, D']$, we see that it is enough to show that

$$\bar{X}_{ij} \star D = \sum_{l=1}^{n} \bar{X}_{il}D(\bar{X}_{lj}) \quad \text{for all } D \in \mathrm{Der}_k(A[G], k_1).$$

To prove this, we argue as follows. Let $x \in G$. Then $(\bar{X}_{ij} \star D)(x) = D(\lambda_x^*(\bar{X}_{ij}))$, and evaluating the right-hand side at x yields $\sum_{l=1}^{n} \bar{X}_{il}(x)D(\bar{X}_{lj})$. Thus, the above identity will hold if the following identity holds:

$$\lambda_x^*(\bar{X}_{ij}) = \sum_{l=1}^{n} \bar{X}_{il}(x)\bar{X}_{lj} \quad \text{for all } x \in G.$$

To see this, let us evaluate both sides on $y \in G$. Then the left-hand side becomes $\lambda_x^*(\bar{X}_{ij})(y) = \bar{X}_{ij}(xy) = \bar{X}_{ij}(\mu(x, y)) = \mu^*(\bar{X}_{ij})(x, y)$. But, using the formula for μ^* in Definition 1.3.9, we have $\mu^*(\bar{X}_{ij}) = \sum_{l=1}^{n} \bar{X}_{il} \otimes \bar{X}_{lj}$, and this yields the desired identity. Note that, since the Jacobi identity is known to hold for the Lie product on $M_n(k)$, it now automatically follows that the product introduced on $\mathrm{Der}_k(A[G], k_1)$ also satisfies the Jacobi identity.

(b) Again, we work with $\mathrm{Der}_k(A[G], k_1)$ and the Lie product defined by $(*')$. First we claim that we have the following identity:

$$\varphi^*(h) \star D = \varphi^*(h \star d_1\varphi(D)) \quad \text{for all } D \in \mathrm{Der}_k(A[G], k_1)$$

$$\text{and } h \in A[H]. \tag{$\dagger$}$$

Indeed, for any $x \in G$, we have $(\varphi^*(h) \star D)(x) = D(\lambda_x^*(\varphi^*(h)))$ and $\varphi^*(h \star d_1\varphi(D))(x) = (h \star d_1\varphi(D))(\varphi(x)) = D(\varphi^*(\lambda_{\varphi(x)}^*(h)))$. Hence it is enough to check that we have $\lambda_x^*(\varphi^*(h)) = \varphi^*(\lambda_{\varphi(x)}^*(h))$. Evaluating both sides at $y \in G$, the left-hand side becomes $\lambda_x^*(\varphi^*(h))(y) = \varphi^*(h)(xy) = h(\varphi(xy))$ and the right-hand side becomes $\varphi^*(\lambda_{\varphi(x)}^*(h))(y) = (\lambda_{\varphi(x)}^*(h))(\varphi(y)) = h(\varphi(x)\varphi(y)) = h(\varphi(xy))$, where the last equality holds since φ is a group homomorphism. Thus, $(\dagger)$ is proved. Now let $D, D' \in \mathrm{Der}_k(A[G], k_1)$ and $h \in A[H]$. Then we have

$$\begin{aligned}
[d_1\varphi(D), d_1\varphi(D')](h) &= D(\varphi^*(h \star d_1\varphi(D'))) - D'(\varphi^*(h \star d_1\varphi(D))) \\
&= D(\varphi^*(h) \star D') - D'(\varphi^*(h) \star D) \quad \text{(by $(\dagger)$)} \\
&= [D, D'](\varphi^*(h)) = d_1\varphi([D, D'])(h).
\end{aligned}$$

Thus, $d_1\varphi$ is a homomorphism of Lie algebras. $\qquad\square$

1.5.4 Remark Identifying $T_1(G) = \mathrm{Der}_k(A[G], k_1)$, we have seen in the above proof that the Lie product on $T_1(G)$ can be characterized entirely in terms of the affine algebra $A[G]$. Indeed, for $D, D' \in \mathrm{Der}_k(A[G], k_1)$, we have

$$[D, D'](f) = D(f \star D') - D'(f \star D) \quad \text{for all } f \in A[G],$$

where $f \star D \in A[G]$ is defined by $(f \star D)(x) = D(\lambda_x^*(f))$ for $x \in G$ and $\lambda_x^* \colon A[G] \to A[G]$ is the algebra homomorphism induced by left translation with x.

1.5.5 Example Let k be an infinite perfect field.

(a) Let $U_n(k)$ be the group of all upper triangular matrices with 1 on the diagonal; see Example 1.3.11. Then we have

$$T_1(U_n(k)) = \{(a_{ij}) \in M_n(k) \mid a_{ij} = 0 \text{ for } 1 \leqslant j \leqslant i \leqslant n\}$$
$$= \{A - I_n \mid A \in U_n(k)\}.$$

Indeed, $U_n(k)$ is defined by the polynomials X_{ij} ($1 \leqslant j < i \leqslant n$) and $X_{ii} - 1$ ($1 \leqslant i \leqslant n$). Computing $d_1(f)$ for these polynomials, we obtain the inclusion '$\subseteq$'. On the other hand, as an algebraic set, we have $U_n(k) \cong k^{n(n-1)/2}$ and so $T_1(U_n(k))$ has dimension $n(n-1)/2$ (see Example 1.4.10). Thus, the inclusion must be an equality. By a similar argument, we see that

$$T_1(U'_n(k)) = \{A - I_n \mid A \in U'_n(k)\},$$

where $U'_n(k)$ is the group of all lower triangular matrices with 1 on the diagonal.

(b) Let $T_n^{(1)}(k)$ be the group of all diagonal matrices in $M_n(k)$ with determinant 1. Then, by a similar argument, we see that

$$T_1(T_n^{(1)}(k)) = \{A \in M_n(k) \mid A \text{ diagonal and } \operatorname{trace}(A) = 0\}.$$

Indeed, $T_n^{(1)}$ is defined by the polynomials X_{ij} ($i \neq j$) and $X_{11} \cdots X_{nn} - 1$. Computing $d_1(f)$ for these polynomials, we obtain the inclusion '$\subseteq$'. To prove the inclusion '$\supseteq$', it is enough to show that $T_n^{(1)}$ is connected of dimension $n - 1$ (see Proposition 1.5.2). Now we note that the following map is an isomorphism:

$$\prod_{i=1}^{n-1} \{(x_i, y_i) \in k^2 \mid x_i y_i = 1\} \longrightarrow T_n^{(1)}(k)$$
$$((x_1, y_1), \ldots, (x_{n-1}, y_{n-1})) \mapsto \operatorname{diag}(x_1, \ldots, x_{n-1}, y_1 \cdots y_{n-1})$$

Since $\{(x, y) \in k^2 \mid xy = 1\}$ is connected of dimension 1 (why?), we conclude that $T_n^{(1)}(k)$ is connected of dimension $n - 1$ (by Proposition 1.3.8).

We now establish some differentiation formulas which will turn out to be useful tools for computing tangent spaces; see Example 1.5.8.

1.5.6 Lemma *Let $G \subseteq M_n(k)$ be a linear algebraic group; then $T_1(G) \subseteq M_n(k)$. Let $\mu\colon G \times G \to G$ and $\iota\colon G \to G$ be the regular maps defining the multiplication and the inversion in G, respectively.*

(a) We have $d_{(1,1)}\mu(A, B) = A + B$ for all $A, B \in T_1(G)$, where we identify $T_{(1,1)}(G \times G) = T_1(G) \oplus T_1(G)$; see Exercise 1.8.16.
(b) We have $d_1\iota(A) = -A$ for all $A \in T_1(G)$.

Proof (a) Let $x = (x_{ij})$ and $y = (y_{ij})$ be two matrices in G. Then $\mu(x, y)$ is the matrix with (i, j)-coordinate $\mu_{ij} := \sum_{l=1}^{n} x_{il}y_{lj}$. So we obtain

$$D_{rs}^{(x)}(\mu_{ij})(I_n, I_n) = D_{rs}^{(y)}(\mu_{ij})(I_n, I_n) = \delta_{ir}\delta_{sj},$$

where $D_{rs}^{(x)}$ and $D_{rs}^{(y)}$ denote partial derivatives with respect to the x-variable and to the y-variable, respectively, which is indexed by (r, s). This yields

$$d_{(I_n, I_n)}\mu_{ij}(A, B) = \sum_{r,s=1}^{n} \delta_{ir}\delta_{sj}(a_{rs} + b_{rs}) = a_{ij} + b_{ij}$$

where $A = (a_{ij})$ and $B = (b_{ij})$ are in $T_1(G)$.
(b) For $x = (x_{ij}) \in G$, let us denote $\iota(x) = (\tilde{x}_{ij})$. Thus, we have $\delta_{ij} = \sum_{l=1}^{n} x_{il}\tilde{x}_{lj}$ for $1 \leqslant i, j \leqslant n$. Denoting by D_{rs} the partial derivative with respect to the variable indexed by (r, s), we obtain

$$0 = \sum_{l=1}^{n}\bigl(D_{rs}(x_{il})\,\tilde{x}_{lj} + x_{il}\,D_{rs}(\tilde{x}_{lj})\bigr)(I_n)$$

$$= \sum_{l=1}^{n}\bigl(\delta_{ir}\delta_{ls}\,\tilde{x}_{lj}(I_n) + \delta_{il}\,D_{rs}(\tilde{x}_{lj})(I_n)\bigr) = \delta_{ir}\delta_{sj} + D_{rs}(\tilde{x}_{ij})(I_n).$$

Thus, we have $D_{rs}(\tilde{x}_{ij})(I_n) = -\delta_{ir}\delta_{sj}$ and so $d_1\iota(A)_{ij} = -\sum_{r,s=1}^{n}\delta_{ir}\delta_{sj}a_{rs} = -a_{ij}$ for all $A = (a_{ij}) \in T_1(G)$. $\qquad\square$

1.5.7 Lemma *Let $G \subseteq M_n(k)$ be a linear algebraic group. Let $\varphi\colon G \to M_n(k)$ be a regular map with $\varphi(1) = I_n$, and let $B \in M_n(k)$ be fixed. Then*

$$\tilde{\varphi}_B\colon G \to M_n(k), \quad g \mapsto gB\varphi(g),$$

is a regular map and $d_1\tilde{\varphi}_B\colon T_1(G) \to T_B(M_n(k)) = M_n(k)$ is given by

$$d_1\tilde{\varphi}_B(A) = AB + B\,d_1\varphi(A) \quad \text{for all } A \in T_1(G).$$

Proof First note that, if $B = b_1 B_1 + b_2 B_2$, where $b_i \in k$ and $B_i \in M_n(k)$, then $\tilde{\varphi}_B = b_1 \tilde{\varphi}_{B_1} + b_2 \tilde{\varphi}_{B_2}$, and so $d_1 \tilde{\varphi}_B = b_1 d_1 \tilde{\varphi}_{B_1} + b_2 d_1 \tilde{\varphi}_{B_2}$. Thus, it will be enough to consider the case where B is an elementary matrix, that is, there exist $1 \leqslant u, v \leqslant n$ such that the (u, v)-entry in B is 1 and all other entries are zero. In this case, the (i, j)-coefficient of $\tilde{\varphi}_B(g) \in M_n(k)$ is given by $g_{iu}\, \varphi(g)_{vj}$. Denoting by D_{rs} the partial derivative with respect to the (r, s)-variable, we obtain

$$D_{rs}(g_{iu}\, \varphi(g)_{vj})(I_n) = \delta_{ri}\delta_{us}\varphi(g)_{vj} + g_{iu}\, D_{rs}(\varphi(g)_{vj})(I_n).$$

Given $A = (a_{rs}) \in T_1(G)$, this yields $d_1 \tilde{\varphi}_B(A)_{ij} = a_{iu}\delta_{vj} + \delta_{iu}d_1\varphi_{vj}(A) = (AB)_{ij} + (B\, d_1\varphi(A))_{ij}$, as required. $\qquad\square$

1.5.8 Example Let k be an infinite perfect field and $G = \mathrm{SL}_n(k)$. With the notation of Example 1.5.5, consider the regular map

$$\varphi \colon U_n(k) \times T_n^{(1)}(k) \times U_n'(k) \to G, \quad (u, h, u') \mapsto uhu'.$$

Since $uhu' = \mu(u, \mu(h, u'))$, Lemma 1.5.6 shows that

$$d_1\varphi \colon T_1(U_n(k)) \oplus T_1(T_n^{(1)}(k)) \oplus T_1(U_n'(k)) \to T_1(G)$$

is given by $(x, h, x') \mapsto x + h + x'$. Now $T_1(U_n(k))$ consists of strictly upper triangular matrices, $T_1(T_n^{(1)}(k))$ consists of diagonal matrices and $T_1(U_n'(k))$ consists of strictly lower triangular matrices. It follows that $d_1\varphi$ is injective and so $\dim T_1(G) \geqslant n^2 - 1$. On the other hand, we certainly have $T_1(G) \subseteq \{A \in M_n(k) \mid d_1(\det)(A) = 0\}$. Now, denoting by D_{rs} the partial derivative with respect to the variable indexed by (r, s), we obtain

$$d_1(\det) = \sum_{\pi \in \mathfrak{S}_n} \mathrm{sgn}(\pi) \sum_{r,s=1}^{n} D_{rs}(X_{1\pi(1)}X_{2\pi(2)} \cdots X_{n\pi(n)})(I_n)\, X_{rs}$$

$$= \sum_{\pi \in \mathfrak{S}_n} \mathrm{sgn}(\pi) \sum_{r,s=1}^{n} \Big(\delta_{s\pi(r)} \prod_{\substack{i=1 \\ i \neq r}}^{n} X_{i\pi(i)}\Big)(I_n)\, X_{rs}$$

$$= \sum_{\pi \in \mathfrak{S}_n} \mathrm{sgn}(\pi) \Big(\sum_{r,s=1}^{n} \delta_{s\pi(r)} \prod_{\substack{i=1 \\ i \neq r}}^{n} \delta_{i\pi(i)}\Big) X_{rs}.$$

Now note that the expression $\prod_{i \neq r} \delta_{i\pi(i)}$ is zero unless $\pi \in \mathfrak{S}_n$ fixes $n - 1$ points. This automatically implies that π must be the identity, and so the above expression simplifies to $d_1(\det) = \sum_{r,s=1}^{n} \delta_{sr} X_{rs} = \sum_{r=1}^{n} X_{rr}$. Thus, we have $d_1(\det)(A) = \mathrm{trace}(A)$ and this gives an upper bound for $\dim T_1(G)$. Combining this with $\dim T_1(G) \geqslant n^2 - 1$, we finally conclude that

$$T_1(\mathrm{SL}_n(k)) = \{A \in M_n(k) \mid \mathrm{trace}(A) = 0\}.$$

By Theorem 1.5.3, the Lie product on $T_1(\mathrm{SL}_n(k))$ is given by $[A, A'] = AA' - A'A$. The Lie algebra $T_1(\mathrm{SL}_n(k))$ is usually denoted by $\mathrm{sl}_n(k)$. The above discussion also implies that $d_1\varphi$ is surjective and so the map φ is dominant, by Theorem 1.4.15.

Our next aim is to determine the Lie algebras of the symplectic and orthogonal groups that we introduced in Section 1.3. The method is roughly the same as that in Example 1.5.8. For this purpose, we need the following preliminary results about some subgroups. Recall the definition of $U_n(k)$ from Example 1.5.5.

1.5.9 Lemma   *The group* $U := U_{2m}(k) \cap \mathrm{Sp}_{2m}(k)$ *(* $m \geqslant 1$ *) consists of all matrices of the form*

$$u(A, S) := \left[\begin{array}{c|c} A & AQ_mS \\ \hline 0 & Q_m(A^{-1})^{\mathrm{tr}}Q_m \end{array}\right],$$

where $A \in U_m(k)$ *and* $S \in M_m(k)$ *is such that* $S = S^{\mathrm{tr}}$. *We have*

$$\dim U = m^2 \ \textit{and} \ U \ \textit{is connected if} \ |k| = \infty.$$

Furthermore, a diagonal matrix belongs to $\mathrm{Sp}_{2m}(k)$ *if and only if the diagonal entries are* $t_1, t_2, \ldots, t_{m-1}, t_m, t_m^{-1}, t_{m-1}^{-1}, \ldots, t_1^{-1}$, *where* $0 \neq t_i \in k$.

Proof   The statements concerning the matrices are proved by a straightforward computation (which we leave as an exercise to the reader). Now assume that k is infinite. Then we note that the map

$$U_m(k) \times \{S \in M_m(k) \mid S = S^{\mathrm{tr}}\} \to U, \quad (A, S) \mapsto u(A, S),$$

is regular and bijective. Furthermore, the inverse is easily seen to be a regular map, too. Thus, the above map is an isomorphism of algebraic sets. Now $U_m(k)$ and $\{S \in M_m(k) \mid S = S^{\mathrm{tr}}\}$ are linear subspaces of $M_n(k)$. Hence these sets are irreducible, and their dimension as an algebraic set equals the dimension as a vector space; see Exercise 1.8.6. This yields the desired formula and the fact U is irreducible, using Proposition 1.3.8(a) and (c). $\qquad\square$

1.5.10 Lemma *Assume that* $\mathrm{char}(k) \neq 2$. *Then the group* $U := U_{2m+1}(k) \cap O_{2m+1}(k)$ *(*$m \geqslant 1$*) consists of all matrices of the form*

$$
u(A, S, w) := \left[\begin{array}{c|c|c}
A & v & AQ_mS \\
\hline
0 & 1 & w \\
\hline
0 & 0 & Q_m(A^{-1})^{\mathrm{tr}}Q_m
\end{array}\right] \qquad (v := -AQ_m w^{\mathrm{tr}}),
$$

where $A \in U_m(k), S \in M_m(k)$, *and* $w \in k^m$ *(row vector) are such that* $S + S^{tr} = -w^{\mathrm{tr}}w$. *We have*

$$
\dim U = m^2 \text{ and } U \text{ is connected if } |k| = \infty.
$$

Furthermore, a diagonal matrix belongs to $O_{2m+1}(k)$ *if and only if the diagonal entries are* $t_1, t_2, \ldots, t_{m-1}, t_m, \pm 1, t_m^{-1}, t_{m-1}^{-1}, \ldots, t_1^{-1}$ *where* $0 \neq t_i \in k$.

Proof This is similar to the proof of the previous lemma. Now the map

$$
U_m(k) \times \{(w, S) \in k^m \times M_m(k) \mid S + S^{\mathrm{tr}} = -w^{\mathrm{tr}}w\} \to U,
$$

given by sending (A, w, S) to $u(A, S, w)$, is an isomorphism of algebraic sets. To compute the dimension of these sets, it remains to note that the map

$$
k^m \times \{S \in M_m(k) \mid S + S^{\mathrm{tr}} = 0\} \to \{(w, S) \in k^m \times M_m(k) \mid
$$
$$
S + S^{\mathrm{tr}} = -w^{\mathrm{tr}}w\},
$$

given by sending (w, S) to $(w, S - w^{\mathrm{tr}}w/2)$, is an isomorphism. $\qquad\square$

Finally, we consider the even-dimensional orthogonal groups, where we use the description using a suitable quadratic form as in §1.3.16.

1.5.11 Lemma   *The group $U := U_{2m}(k) \cap O_{2m}^+(k)$ $(m \geqslant 1)$ consists of all matrices of the form*

$$u(A, S) := \left[\begin{array}{c|c} A & AQ_mS \\ \hline 0 & Q_m(A^{-1})^{tr}Q_m \end{array}\right],$$

where $A \in U_m(k)$ and $S \in M_m(k)$ is such that $S + S^{tr} = 0$ and all diagonal entries of S are zero. We have that

$$\dim U = m^2 - m \text{ and } U \text{ is connected if } |k| = \infty.$$

Furthermore, a diagonal matrix belongs to $O_{2m}^+(k)$ if and only if the diagonal entries are $t_1, t_2, \ldots, t_{m-1}, t_m, t_m^{-1}, t_{m-1}^{-1}, \ldots, t_1^{-1}$, where $0 \neq t_i \in k$.

Proof   Again, this is similar to the proof of Lemma 1.5.9. Just note that, if $\mathrm{char}(k) \neq 2$, then the condition $S + S^{tr} = 0$ automatically implies that all diagonal entries of S are zero. If $\mathrm{char}(k) = 2$, the fact that all diagonal entries of S are zero follows from the additional equations in the description of $O_{2m}^+(k)$ in §1.3.16. Now we have an isomorphism of algebraic sets

$$U_m(k) \times \left\{ S \in M_m(k) \,\middle|\, \begin{array}{l} S + S^{tr} = 0 \text{ and} \\ s_{ii} = 0 \text{ for } 1 \leqslant i \leqslant m \end{array} \right\} \to U,$$

given by sending (A, S) to $u(A, S)$. The dimension is computed as before. $\qquad\square$

1.5.12 Remark   Let $G \subseteq M_n(k)$ be one of the classical groups:

$$G = \begin{cases} \mathrm{SO}_{2m+1}(k), & n = 2m + 1, \mathrm{char}(k) \neq 2, \\ \mathrm{Sp}_{2m}(k), & n = 2m, \text{ any characteristic}, \\ \mathrm{SO}_{2m}^+(k), & n = 2m, \text{ any characteristic}; \end{cases}$$

see Example 1.3.15 and §1.3.16. We set

$$n_0 := \begin{cases} Q_{2m}^- & \text{if } n = 2m \text{ and } G = \mathrm{Sp}_{2m}(k), \\ Q_n & \text{otherwise.} \end{cases}$$

This matrix has the property that conjugation by n_0 transforms an upper triangular matrix into a lower triangular matrix. We have

$$A \in G \;\Rightarrow\; n_0 A n_0^{-1} \in G \quad (\text{note that } n_0^2 = \pm I_n).$$

This is easily shown by an explicit computation which we leave as an exercise. Thus, the results in Lemmas 1.5.9–11 also hold for the groups of lower unitriangular matrices in G. Now set

$$U := U_n(k) \cap G \qquad \text{and} \qquad H := \{A \in G \mid A \text{ diagonal}\}.$$

Assume that k is infinite. Then we have already remarked in Lemmas 1.5.9–11 that U is connected; similarly, $U' := n_0 U n_0^{-1}$ is connected. Finally,

$$H \text{ is connected and } \dim H = m.$$

This is seen by an argument similar to that which was used to deal with the group of diagonal matrices in $\mathrm{SL}_n(k)$; see Example 1.5.5.

1.5.13 Theorem   *Let k be an infinite perfect field. Then we have*

$$T_1(\mathrm{SO}_n(k)) \;=\; \{A \in M_n(k) \mid A^{tr}Q_n + Q_n A = 0\} \quad (\mathrm{char}(k) \neq 2),$$
$$T_1(\mathrm{Sp}_{2m}(k)) = \{A \in M_{2m}(k) \mid A^{tr}Q_{2m}^- + Q_{2m}^- A = 0\}$$
$$(\textit{any characteristic}),$$

where Q_n and Q_{2m}^- are defined as in §1.3.15. Furthermore, if $\mathrm{char}(k) = 2$, *then*

$$T_1(\mathrm{SO}_{2m}^+(k)) = \left\{ A \in M_{2m}(k) \,\middle|\, \begin{array}{c} A^{tr}Q_{2m} + Q_{2m}A = 0 \text{ and} \\ (A^{tr}P_{2m} + P_{2m}A)_{rr} = 0 \\ \text{for } 1 \leqslant r \leqslant 2m \end{array} \right\},$$

where P_{2m} and $\mathrm{SO}_{2m}^+(k)$ are defined as in §1.3.16.

Proof   Let $Q \in M_n(k)$ be invertible. Writing out the equation $A^{tr}QA = Q$ for all matrix entries, we get a system of n^2 quadratic polynomials in n^2 variables. By a computation similar to that in Example 1.5.8, we obtain $d_{I_n}(f)$ for each of these polynomials. This yields that

$$T_1(\Gamma_n(Q, k)) \subseteq \{A \in M_n(k) \mid A^{tr}Q + QA = 0\}.$$

Now assume that $\mathrm{char}(k) = 2$ and that $Q \in M_{2m}(k)$ is alternating. As in §1.3.16, write $Q = P + P^{tr}$, where $f = \sum_{ij} p_{ij} X_i X_j$. Let $\{e_1, \ldots, e_{2m}\}$ be the standard basis of k^{2m}. For each i, the identity $f(Ae_i) = f(e_i)$ yields a polynomial equation for the coefficients of A.

Computing d_{I_n} for these polynomials, we obtain

$$T_1(\Gamma_{2m}^+(f,k)) \subseteq \left\{ A \in M_{2m}(k) \,\middle|\, \begin{array}{l} A^{\mathrm{tr}}Q + QA = 0 \text{ and} \\ (A^{\mathrm{tr}}P + PA)_{rr} = 0 \text{ for } 1 \leqslant r \leqslant 2m \end{array} \right\}.$$

Thus, in each case, we have already established one inclusion. To show that these inclusions are equalities, we use a dimensional argument. Let us give the details for $G = \mathrm{Sp}_{2m}(k)$. (The proof for the other cases is completely analogous.) Let

$$U := U_{2m}(k) \cap \mathrm{Sp}_{2m}(k) \quad \text{and} \quad H := \{A \in \mathrm{Sp}_{2m}(k) \mid A \text{ diagonal}\}.$$

Now note that the map $\gamma \colon A \mapsto n_0 A n_0^{-1}$ defines an isomorphism of $\mathrm{Sp}_{2m}(k)$; see Remark 1.5.12. Let $U' := \gamma(U)$ and consider the regular map

$$\varphi \colon U \times H \times U' \to \mathrm{Sp}_{2m}(k), \quad (u, h, u') \mapsto uhu'.$$

Since $uhu' = \mu(u, \mu(h, u'))$, Lemma 1.5.6 shows that

$$d_1\varphi \colon T_1(U) \oplus T_1(H) \oplus T_1(U') \to T_1(\mathrm{Sp}_{2m}(k)), \quad (x, u, x') \mapsto x + u + x'.$$

Now, we certainly have $T_1(U) \subseteq T_1(U_{2m}(k))$, and so Example 1.5.5(a) shows that $T_1(U)$ consists of strictly upper triangular matrices. Similarly, $T_1(H)$ consists of diagonal matrices. Finally, note that γ transforms an upper triangular matrix into a lower triangular matrix. Consequently, $T_1(U')$ consists of strictly lower triangular matrices. It follows that $d_1\varphi$ is injective and, hence,

$$\dim T_1(\mathrm{Sp}_{2m}(k)) \geqslant \dim(T_1(U) \oplus T_1(H) \oplus T_1(U'))$$
$$= 2\dim T_1(U) + \dim T_1(H),$$

where the last equality holds since U and U' are isomorphic. Using Lemma 1.5.9, Remark 1.5.12, and Proposition 1.5.2, we conclude that $\dim T_1(\mathrm{Sp}_{2m}(k)) \geqslant 2m^2 + m$. On the other hand, we have already seen that $T_1(\mathrm{Sp}_{2m}(k)) \subseteq \{A \in M_{2m}(k) \mid A^{\mathrm{tr}}Q_{2m}^- + Q_{2m}^- A = 0\}$ and the space on the right hand side has dimension $2m^2 + m$. Thus, we must have $\dim T_1(\mathrm{Sp}_{2m}(k)) = 2m^2 + m$, as required. $\qquad\square$

1.5.14 Corollary *Let k be an infinite perfect field. Then we have*

$$\dim \mathrm{SO}_{2m+1}(k) = 2m^2 + m \quad (any\ m \geqslant 1,\ \mathrm{char}(k) \neq 2),$$
$$\dim \mathrm{Sp}_{2m}(k) = 2m^2 + m \quad (any\ m \geqslant 1,\ any\ characteristic),$$
$$\dim \mathrm{SO}_{2m}^+(k) = 2m^2 - m \quad (any\ m \geqslant 1,\ any\ characteristic).$$

Proof This follows from Proposition 1.5.13 and Proposition 1.5.2.

$\square$

1.5.15 Groups of Lie type For readers familiar with the basic theory of semisimple complex Lie algebras (see, for example, Humphreys (1972)), we just give a very rough sketch of Chevalley's construction of *groups of Lie type*.

Let $\mathfrak{g}$ be a semisimple Lie algebra over $\mathbb{C}$ and let $\mathfrak{h} \subseteq \mathfrak{g}$ be a Cartan subalgebra. Then we have a corresponding Cartan decomposition of $\mathfrak{g}$; see §8 of Humphreys (1972). Thus, there exists a finite subset $\Phi \subseteq \mathrm{Hom}(\mathfrak{h}, \mathbb{C}) \setminus \{0\}$ and a basis $B = \{h_1, \ldots, h_n; e_\alpha, \alpha \in \Phi\}$ of $\mathfrak{g}$ where $h_1, \ldots, h_n$ form a basis of $\mathfrak{h}$ and $[h_i, e_\alpha] = \alpha(h_i)e_\alpha$ for all i and all $\alpha \in \Phi$. Now Chevalley has shown that B can be chosen such that the structure constants of $\mathfrak{g}$ with respect to B lie in $\mathbb{Z}$; see §25.2 of Humphreys (1972). Hence, for any field k, we obtain a Lie algebra over k by setting $\mathfrak{g}_k := k \otimes_{\mathbb{Z}} \langle B \rangle_{\mathbb{Z}}$.

Now, given $\alpha \in \Phi$ and $t \in \mathbb{C}$, then $t\,\mathrm{ad}(e_\alpha)$ is a nilpotent endomorphism of $\mathfrak{g}$, and so we have $\exp(t\,\mathrm{ad}(e_\alpha)) \in \mathrm{SL}(\mathfrak{g})$; see §2.3 of Humphreys (1972). Moreover, it can be shown that the entries of the matrix of $\exp(t\,\mathrm{ad}(e_\alpha))$ with respect to B can be written as polynomials in t with coefficients in $\mathbb{Z}$. This makes it possible to define an automorphism $x_\alpha(t) \in \mathrm{SL}(\mathfrak{g}_k)$, by transferring the above exponential construction to k; see §25.5 of Humphreys (1972). Then we define

$$G(k) := \langle x_\alpha(t) \mid \alpha \in \Phi, t \in k \rangle \subseteq \mathrm{SL}(\mathfrak{g}_k),$$

and call this the *Chevalley group* or the *group of Lie type* over k associated with $\mathfrak{g}$. Now Chevalley's construction shows that, for a fixed $\alpha \in \Phi$, the map $x_\alpha \colon k \to \mathrm{SL}(\mathfrak{g}_k), t \mapsto x_\alpha(t)$, is a regular map; see §3 of Steinberg (1967). This implies, using a general criterion which we shall establish later in Theorem 2.4.6, that $G(k)$ is a closed connected subgroup of $\mathrm{SL}(\mathfrak{g}_k)$ when k is algebraically closed. Then one can also show that $\mathfrak{g}_k$ indeed is the Lie algebra of $G(k)$, in the sense of Theorem 1.5.3. For more details see Steinberg (1967) and Carter (1972).

1.6 Groups with a split *BN*-pair

We now study an important class of groups, the groups with a so-called *BN*-pair or Tits system. As mentioned in the introduction of this chapter, this concept has turned out to be extremely useful. Our aim is to determine such BN-pairs for all the groups of classical type introduced in §1.3.15 and §1.3.16. As an immediate application, we can determine exactly which of these groups are connected. In Chapter 3, we will exhibit a deep property of the subgroup B involved in a BN-pair: it is a Borel subgroup in the sense of Definition 3.4.1.

1.6.1 Definition (Tits)   Let G be an abstract group. We say that G is a *group with a BN-pair* or that G admits a *Tits system* if there are subgroups $B, N \subseteq G$ such that the following conditions are satisfied.

(BN1) G is generated by B and N.

(BN2) $H := B \cap N$ is normal in N, and the quotient $W := N/H$ is a group generated by a set S of elements of order 2.

(BN3) $n_s B n_s \neq B$ if $s \in S$ and n_s is a representative of s in N.

(BN4) $n_s B n \subseteq B n_s n B \cup B n B$ for any $s \in S$ and $n \in N$.

(BN5) $\displaystyle\bigcap_{n \in N} n B n^{-1} = H$.

The group W is called the *Weyl group* of G. (BN5) is called the *saturation axiom*. Sometimes, it is omitted from the definition of a BN-pair, but we shall find it convenient to include this condition here.

Before we give some examples, we shall establish some elementary properties of a group with a BN-pair. For any $w \in W$, we set $C(w) := B n_w B$, where $n_w \in N$ is a representative of w in N. Since any two representatives of w lie in the same coset of $H \subseteq B$, we see that $C(w)$ does not depend on the choice of the representative. The sets $C(w)$ are called *Bruhat cells*.

We define a length function on W as follows. We set $l(1) = 0$. If $w \neq 1$, we define $l(w)$ to be the length of a shortest possible expression of w as a product of generators in S. (Note that we don't have to take into account inverses, since $s^2 = 1$ for all $s \in S$.) Thus, any $w \in W$ can be written in the form $w = s_1 \cdots s_p$, where $p = l(w)$ and $s_i \in S$ for all i. Such an expression (which is by no means unique) will be called a *reduced expression* for w. Note that, given a non-unit $w \in W$, we can always find some $s \in S$ such that $l(sw) < l(w)$. (Take, for example, the first factor in a reduced expression for w.) This will be used frequently in proofs by induction on $l(w)$.

1.6.2 Lemma *Let $J \subseteq S$ and define $W_J \subseteq W$ to be the subgroup generated by J. Let $N_J \subseteq N$ be the preimage of W_J under the natural map $N \to W$. Then $P_J := BN_J B$ is a subgroup of G.*

Proof The set P_J is closed under inversion, so it only remains to show that it is closed under multiplication. Now, we certainly have $BP_J \subseteq P_J$. Hence it will be enough to show that $N_J P_J \subseteq P_J$. Since $N_J = \langle H; n_s, s \in J \rangle$, it will be enough to show that $n_s P_J \subseteq P_J$ for all $s \in J$. But this is clear by (BN4). $\qquad\square$

1.6.3 Proposition (Bruhat decomposition) *We have the double-coset decomposition*

$$G = \coprod_{w \in W} Bn_w B.$$

Furthermore, given $s \in S$ and $w \in W$, we have $l(sw) = l(w) \pm 1$ and

$$Bn_s B.Bn_w B = \begin{cases} Bn_s n_w B & \text{if } l(sw) = l(w) + 1, \\ Bn_s n_w B \cup Bn_w B & \text{if } l(sw) = l(w) - 1. \end{cases}$$

Proof Taking $J = S$ in Lemma 1.6.2, we see that $G = BNB$ and so $G = \bigcup_{w \in W} Bn_w B$. Now let $y, w \in W$. Then we must show that $Bn_y B = Bn_w B$ implies $y = w$. We will prove this by induction on $\min\{l(y), l(w)\}$. We may assume without loss of generality that $l(y) \leqslant l(w)$. Now, if $l(y) = 0$, then $y = 1$ and so $B = Bn_y B = Bn_w B$. This yields $n_w \in B \cap N = H$ and so $w = 1$, as required. Now suppose that $l(y) > 0$. Then we can write $y = sx$, where $s \in S$ and $x \in W$ are such that $l(y) = l(x) + 1$. Then we have $n_s n_x B \subseteq Bn_y B = Bn_w B$ and so, using (BN4):

$$n_x B \subseteq n_s Bn_w B \subseteq Bn_s n_w B \cup Bn_w B.$$

Hence $Bn_x B = Bn_s n_w B$ or $Bn_x B = Bn_w B$. By induction, we have $x = sw$ or $x = w$. The latter case is impossible since $l(x) < l(y) \leqslant l(w)$. Thus, we must have $x = sw$, and so $y = sx = w$, as required.

Now consider the multiplication rule. Let $s \in S$ and $w \in W$. First note that we certainly have $l(w) - 1 \leqslant l(sw) \leqslant l(w) + 1$. Now assume first that $l(sw) \geqslant l(w)$. We show by induction on $l(w)$ that $Bn_s B.Bn_w B = Bn_s n_w B$. If $l(w) = 0$, then $w = 1$ and so there is nothing to prove. Now assume that $l(w) > 0$ and write $w = yt$, where $t \in S$ and $y \in W$ are such that $l(w) = l(y) + 1$. By (BN4), we already know that $Bn_s B.Bn_w B = Bn_s n_w B$ or $n_s Bn_w \cap Bn_w B \neq \varnothing$. Assume, if possible, that we were in the second case. Then $n_s Bn_w \cap Bn_w B \neq \varnothing$ and so $n_s Bn_y \cap Bn_w Bn_t \neq \varnothing$. Now $l(sy) \geqslant l(y)$ and so, by induction, we have $n_s Bn_y \subseteq Bn_s n_y B$. We conclude that $Bn_s n_y B \cap Bn_w Bn_t \neq \varnothing$. Now, by inverting the relation in (BN4), we see that there is also a 'right-handed' version of (BN4). This yields that $Bn_s n_y B \cap Bn_w n_t B \neq \varnothing$ or $Bn_s n_y B \cap Bn_w B \neq \varnothing$. Thus, by the first part of the proof, we have $sy = wt$ or $sy = w$. The former equation gives $sy = yt^2 = y$ and so $s = 1$, contradicting (BN3). The latter gives $sw = y$ and so $l(sw) = l(y) < l(w)$, a contradiction. Thus, our assumption was wrong, and so we must have $Bn_s B.Bn_w B = Bn_s n_w B$, as required.

Now assume that $l(sw) \leqslant l(w)$. By (BN4), we have $n_s Bn_s \subseteq B \cup Bn_s B$. Using (BN3), this implies that $n_s Bn_s \cap Bn_s B \neq \varnothing$. Consequently, we also have $n_s B \cap Bn_s Bn_s \neq \varnothing$ and $n_s Bn_w \cap Bn_s Bn_s n_w \neq \varnothing$. Now, we have $l(s(sw)) = l(w) \geqslant l(sw)$ and so $Bn_s Bn_s n_w \subseteq Bn_w B$ by the previous case. This means $n_s Bn_w \cap Bn_w B \neq \varnothing$. On the other hand, we certainly have $n_s n_w \in Bn_s B.n_w B$. So (BN4) shows that $Bn_s B.Bn_w B = Bn_s n_w B \cup Bn_w B$. $\qquad\square$

The sharpened multiplication rules for Bruhat cells in Proposition 1.6.3 are extremely strong statements. We shall give a number of applications.

1.6.4 Corollary *We have* $\mathrm{N}_G(B) = B$.

Proof Let $g \in \mathrm{N}_G(B)$. Then, by the Bruhat decomposition, we can write $g = bn_w b'$, where $b, b' \in B$ and $w \in W$. This yields $n_w \in \mathrm{N}_G(B)$. Now let $s \in S$ be such that $l(sw) < l(w)$. By Proposition 1.6.3, we have $n_s Bn_w \cap Bn_w B \neq \varnothing$ and so $n_s \in Bn_w Bn_w^{-1} B \subseteq \mathrm{N}_G(B)$, contradicting (BN3). Thus, we have $w = 1$, as required. $\qquad\square$

1.6.5 Corollary *We have* $S = \{1 \neq w \in W \mid B \cup Bn_w B \subseteq G$ *is a subgroup*$\}$.

Proof Let $s \in S$. Then $P_{\{s\}} = B \cup Bn_s B$ is a subgroup by Lemma 1.6.2. Conversely, let $1 \neq w \in W$ be such that $B \cup Bn_w B$ is a subgroup. Let $s \in S$ be such that $l(sw) < l(w)$. Then, as in the proof of Corollary 1.6.4, we have $n_s \in Bn_w Bn_w^{-1} B \subseteq B \cup Bn_w B$ and so $s = w$ by Proposition 1.6.3. $\qquad\square$

1.6.6 Corollary *Let G be a group with a BN-pair, where W is finite. Let $n_0 \in N$ be such that the image of n_0 has maximal length in W. Then we have*

$$G = \langle B, n_0 B n_0^{-1} \rangle.$$

Proof Let $w_0 \in W$ denote the image of n_0. Then, for any $s \in S$, we have $l(sw_0) \leqslant l(w_0)$ and so, using Proposition 1.6.3, $n_0 \in Bn_s B \cdot Bn_0 B$. This yields $n_s \in Bn_0 Bn_0^{-1} B \subseteq \langle B, n_0 B n_0^{-1} \rangle$ and so $G = \langle B, N \rangle \subseteq \langle B, n_0 B n_0^{-1} \rangle$. $\qquad\square$

1.6.7 Remark Assume that G is an affine algebraic group and that G has a BN-pair, where W is finite and B is a closed connected subgroup of G. Then $B \subseteq G^\circ$ and $n_0 B n_0^{-1} \subseteq G^\circ$, and so $G = G^\circ$ by Corollary 1.6.6. This will be one of our tools for determining which of the classical groups are connected.

1.6.8 Corollary (Exchange condition) *Let $w \in W$ and write $w = s_1 \cdots s_p$, where $s_i \in S$ and $p = l(w)$. Let $s \in S$ and assume that $l(sw) < l(w)$. Then*

$$sw = s_1 \cdots s_{i-1} s_{i+1} \cdots s_p \quad \text{for some } 1 \leqslant i \leqslant p.$$

Proof As in the proof of Corollary 1.6.4, $n_s \in Bn_w B n_w^{-1} B = C(w).C(w^{-1})$. Hence, by Exercise 1.8.22(a), there exists some $0 \leqslant q \leqslant p$ such that

$$s = xw^{-1}, \quad \text{where } x = s_{i_1} \cdots s_{i_q} \text{ and } 1 \leqslant i_1 < \cdots < i_q \leqslant p.$$

Now consider a reduced expression for x, and evaluate the product xw^{-1} by multiplying one by one the factors in that expression for x; at each step, the length either increases or decreases by 1. As a result, we see that $l(xw^{-1}) \geqslant l(w) - l(x)$. Since $l(x) \leqslant q$ and $s = xw^{-1}$, we deduce that $1 \geqslant p - q$ and so $q \geqslant p - 1$. If we had $q = p$, then $x = w$ and so $s = 1$, which is absurd. Hence we have $q = p - 1$, and so there exists some $1 \leqslant i \leqslant p$ such that $x = s_1 \cdots s_{i-1} s_{i+1} \cdots s_p$. This yields the desired assertion. $\qquad\square$

1.6.9 Example Let k be a field and $G = \mathrm{GL}_n(k)$. We set

$B_n(k) :=$ subgroup of all upper triangular matrices in $\mathrm{GL}_n(k)$,

$N_n(k) :=$ subgroup of all monomial matrices in $\mathrm{GL}_n(k)$.

(A matrix is called *monomial* if it has exactly one non-zero entry in each row and and each column.) Our aim will be to show that the groups $B_n(k)$ and $N_n(k)$ form a BN-pair in G. For this purpose, we introduce some special matrices and subgroups. We have

$$T_n(k) := B_n(k) \cap N_n(k) = \text{subgroup of all diagonal matrices}$$

$$\text{in } \mathrm{GL}_n(k)$$

and $B_n(k) = U_n(k).T_n(k)$, where $U_n(k)$ is the group of all upper unitriangular matrices, as in Example 1.3.11. Note that $U_n(k) \cap T_n(k) = \{1\}$ and $U_n(k) \subseteq B_n(k)$ is a normal subgroup.

For $1 \leqslant i \leqslant n-1$, let n_i be the permutation matrix which is obtained by interchanging the ith and the $(i+1)$th row in I_n. More generally, if $w \in \mathfrak{S}_n$ is any permutation, let n_w be the matrix which is obtained by permuting the rows of I_n as specified by w. (Thus, the ith row of I_n is the $w(i)$th row of n_w.) Then we have $N_n(k) = \{hn_w \mid h \in T_n(k), w \in \mathfrak{S}_n\}$ and so

$$W := N_n(k)/T_n(k) \cong \mathfrak{S}_n \quad \text{and} \quad n_i \in N_n(k) \leftrightarrow \sigma_i \in \mathfrak{S}_n,$$

where σ_i denotes the basic transposition $\sigma_i = (i, i+1)$.

Next, for $1 \leqslant i, j \leqslant n$, let E_{ij} be the 'elementary' matrix with coefficient 1 at the position (i, j) and 0 otherwise. We define

$$X_{ij} := \{I_n + \lambda E_{ij} \mid \lambda \in k\}, \quad \text{where } 1 \leqslant i, j \leqslant n \text{ and } i \neq j.$$

We write $X_i := X_{i,i+1}$ and $X_{-i} := X_{i+1,i}$ for $1 \leqslant i \leqslant n-1$. These are all subgroups of G. Furthermore, let

$$V_i := \text{group of all matrices in } U_n(k) \text{ with } 0 \text{ at position } (i, i+1).$$

By straightforward matrix computations, it is readily checked that the following relations hold.

(a) $U_n(k) = X_i.V_i = V_i.X_i$ and $V_i \cap X_i = \{1\}$ for $1 \leqslant i \leqslant n-1$.
(b) $n_w X_{ij} n_w^{-1} = X_{w(i),w(j)}$ for all $w \in \mathfrak{S}_n$ and $i \neq j$.
(c) $n_i X_i n_i^{-1} = X_{-i}$ and $n_i V_i n_i^{-1} = V_i$ for any $1 \leqslant i \leqslant n-1$.

1.6.10 Proposition *In the above setting, the groups $B_n(k)$ and $N_n(k)$ form a BN-pair in $G = \mathrm{GL}_n(k)$, with Weyl group $W \cong \mathfrak{S}_n$.*

Proof We must show that the five axioms in Definition 1.6.1 are satisfied. Let $B := B_n(k), N := N_n(k), H := T_n(k) = B \cap N$, and $U := U_n(k)$.

(BN1): Let $g \in G$ and choose $b \in B$ to maximize the total number of zeros at the beginnings of all the rows of bg. These beginnings must all be of different lengths since otherwise we could subtract a multiple of some row from an earlier row, that is, modify b, and increase the total number of zeros. It follows that, for some $w \in \mathfrak{S}_n$, we have $n_w bg \in B$, as required.

(BN2): The group H is normal in N, and $W = N/H \cong \mathfrak{S}_n$ is generated by the involutions n_i $(1 \leqslant i \leqslant n - 1)$.

(BN3): Let $1 \leqslant i \leqslant n - 1$. By Example 1.6.9(a), we have $B = X_i.V_i.T_n(k)$. Conjugating by n_i and using Example 1.6.9(c) yields that $n_i B n_i = n_i B n_i^{-1} = n_i X_i n_i^{-1}.n_i V_i n_i^{-1}.H = X_{-i}.V_i.H \not\subseteq B$.

(BN4): First we show that $n_i B n_i \subseteq B \cup B n_i B$. For this purpose, we begin with the following identity in $\mathrm{GL}_2(k)$, where $0 \neq t \in k$:

$$\begin{bmatrix} 1 & 0 \\ t & 1 \end{bmatrix} = \begin{bmatrix} 1 & t^{-1} \\ 0 & 1 \end{bmatrix} \cdot \begin{bmatrix} 0 & 1 \\ 1 & 0 \end{bmatrix} \cdot \begin{bmatrix} 1 & -t \\ 0 & 1 \end{bmatrix} \begin{bmatrix} t & 0 \\ 0 & -t^{-1} \end{bmatrix}.$$

Now let $1 \leqslant i \leqslant n - 1$. Then we have an embedding

$$\varphi_i \colon \mathrm{GL}_2(k) \hookrightarrow \mathrm{GL}_n(k), \quad A \mapsto \left[\begin{array}{c|c|c} I_{i-1} & 0 & 0 \\ \hline 0 & A & 0 \\ \hline 0 & 0 & I_{n-i-1} \end{array} \right].$$

Applying φ_i to the above identity, we obtain

$$X_{-i} \subseteq \{1\} \cup X_i n_i X_i H. \tag{$*$}$$

Now we write $B = V_i.X_i.H$. Then we have $n_i B n_i = n_i B n_i^{-1} = V_i.n_i X_i n_i^{-1}.H = V_i.X_{-i}.H$ by Example 1.6.9. Using $(*)$, this yields

$$n_i B n_i \subseteq V_i.(\{1\} \cup X_i n_i X_i H).H \subseteq V_i.H \cup U n_i X_i H \subseteq B \cup B n_i B.$$

Next, let $w \in W$ and $1 \leqslant i \leqslant n - 1$. Assume first that $w^{-1}(i) < w^{-1}(i + 1)$. Then, using Example 1.6.9(b), we have $n_w^{-1} X_i n_w = X_{w^{-1}(i), w^{-1}(i+1)} \subseteq U$. Furthermore, by Example 1.6.9(c), we also have $n_i V_i \subseteq V_i n_i$. Thus, we obtain

$$n_i B n_w \subseteq n_i V_i X_i H n_w \subseteq V_i n_i n_w.n_w^{-1} X_i n_w H \subseteq V_i n_i n_w U H$$
$$\subseteq B n_i n_w B.$$

Finally, assume that $w^{-1}(i) > w^{-1}(i + 1)$ and set $y := \sigma_i w$. Then we have $y^{-1}(i) = w^{-1}(i + 1) < w^{-1}(i) = y^{-1}(i + 1)$. Hence, by the previous argument, we have $n_i B n_y \subseteq B n_i n_y B = B n_w B$. This yields that

$$n_i B n_w = n_i B n_i n_y \subseteq (B \cup B n_i B) n_y \subseteq B n_y \cup B n_i B n_y$$
$$\subseteq B n_i n_w B \cup B n_w B.$$

Combining the above statements, we obtain that $n_i B n_w \subseteq B n_w B \cup B n_i n_w B$ for all $w \in \mathfrak{S}_n$ and all $i \in \{1, \dots, n-1\}$, as required.

(BN5): Let $B^- \subseteq G$ be the subgroup of all lower triangular matrices. Then we have $B^- = n_0 B n_0^{-1}$, where $n_0 \in N$ is the permutation matrix corresponding to $(1, n)(2, n-1)(3, n-2) \cdots \in \mathfrak{S}_n$. Thus, we have $\bigcap_{n \in N} n B n^{-1} \subseteq B \cap B^- = H$. $\qquad\square$

1.6.11 Remark Let $G' = \mathrm{SL}_n(k)$ and set $B' := B_n(k) \cap G'$ and $N' := N_n(k) \cap G'$. Since G' is a normal subgroup of $\mathrm{GL}_n(k)$ and $U_n(k) \subseteq G'$, we can use Exercise 1.8.23 to conclude that B' and N' form a BN-pair in G', with Weyl group $W \cong \mathfrak{S}_n$. Furthermore, we have $B_n(k) = U_n(k).T_n^{(1)}(k)$ with $T_n^{(1)}(k)$ as in Example 1.5.5(b).

Now assume that k is an infinite field. As we have seen in Example 1.5.5, the groups $U_n(k)$ and $T_n^{(1)}(k)$ are connected. Hence so is $B_n(k) = U_n(k).T_n^{(1)}(k)$. (Indeed, the natural map $U_n(k) \times T_n^{(1)}(k) \to B_n(k)$ given by multiplication is surjective; so it remains to use Proposition 1.3.8(a) and Remark 1.3.2.) Thus, Remark 1.6.7 gives yet another proof of the fact that $\mathrm{SL}_n(k)$ is connected.

1.6.12 Remark Consider once more the BN-pair for $G = \mathrm{GL}_n(k)$. We have already remarked in Example 1.6.9 that $B_n(k) = U_n(k).T_n(k)$, where $U_n(k)$ is normal and $U_n(k) \cap T_n(k) = \{1\}$. Thus $B_n(k)$ is a semidirect product of $T_n(k)$ with $U_n(k)$. Actually, something stronger holds: Let $g \in G$ and $u \in U_n(k)$ be such that $gug^{-1} \in B_n(k)$. Then gug^{-1} is a triangular matrix all of whose eigenvalues are 1. Thus, we must have $gug^{-1} \in U_n(k)$. A similar argument also applies to the BN-pair in $G' = \mathrm{SL}_n(k)$. Thus, we see that these BN-pairs are split in the sense of the following definition.

1.6.13 Definition Let G be a group with a BN-pair. We say that this is a *split BN*-pair if there exists a normal subgroup $U \subseteq B$ such that the following conditions are satisfied.

(a) Let $H = B \cap N$ as in Definition 1.6.1. Then we have $B = UH$ and $U \cap H = \{1\}$. Thus, B is the semidirect product of H with U.

(b) For any $n \in N$, we have $n^{-1} U n \cap B \subseteq U$.

We mention here that all BN-pairs that we will encounter in this book are split.

We shall now establish another important property of these groups, namely, a sharp form of the Bruhat decomposition. This will require a number of preparations. For the remainder of this section, we shall assume that

$$G \text{ is a group with a } BN\text{-pair where } W \text{ is finite.}$$

1.6.14 Proposition (the longest element) *There exists a unique element w_0 in W of maximal length. We have $w_0^2 = 1$, and $w_0 \in W$ is characterized by the property that $l(sw_0) < l(w_0)$ for all $s \in S$.*

Proof Since W is finite, there exists some element $w_0 \in W$ such that $l(w) \leqslant l(w_0)$ for all $w \in W$. We claim that

$$l(ww_0) = l(w_0) - l(w) \qquad \text{for all } w \in W. \tag{$*$}$$

We prove this by induction on $l(w)$. If $l(w) = 0$, then $w = 1$ and there is nothing to prove. Now assume that $l(w) = p+1$ where $p \geqslant 0$ and write $w = s_1 \cdots s_p s$, where $s, s_i \in S$. Then $q := l(s_1 \cdots s_p w_0) = l(w_0) - p$ by induction. Now take a reduced expression $s_1 \cdots s_p w_0 = t_1 \cdots t_q$, where $t_j \in S$. Then

$$w_0 = s_p \cdots s_2 s_1 t_1 t_2 \cdots t_q$$

is a reduced expression for w_0. Now, since $l(sw_0) < l(w_0)$, the exchange condition in Corollary 1.6.8 shows that there are two possibilities:

(1) $sw_0 = s_p \cdots s_{i+1} s_{i-1} \cdots s_2 s_1 t_1 t_2 \cdots t_q$ for some $i \in \{1, \ldots, p\}$

or

(2) $sw_0 = s_p \cdots s_2 s_1 t_1 t_2 \cdots t_{j-1} t_{j+1} \cdots t_q$ for some $j \in \{1, \ldots, q\}$.

If (1) holds, then $ss_p \cdots s_2 s_1 = s_p \cdots s_{i+1} s_{i-1} \cdots s_1$ and so $l(w) = l(w^{-1}) \leqslant p - 1$, a contradiction. Thus, (2) must hold and this yields $ww_0 = t_1 t_2 \cdots t_{j-1} t_{j+1} \cdots t_q$. Hence, we have $l(ww_0) = q - 1 = l(w_0) - p - 1 = l(w_0) - l(w)$, as desired.

Thus, $(*)$ is proved. Setting $w = w_0$, we see that $l(w_0^2) = 0$ and so $w_0^2 = 1$. Now let $w \in W$ be another element such that $l(w) = l(w_0)$. Then $l(ww_0) = 0$ by $(*)$ and so $w = w_0^{-1} = w_0$, as required. Finally, assume that $w \in W$ is such that $l(sw) < l(w)$ for all $s \in S$. Assume, if possible, that $l(w) < l(w_0)$. Then we have $l(ww_0) = l(w_0) - l(w) > 0$ by $(*)$ and so $l(sww_0) < l(ww_0)$ for some $s \in S$. Using $(*)$ once more, we obtain $l(sw) > l(w)$, a contradiction. $\qquad\square$

1.6.15 Lemma *Let $s \in S$ and $w \in W$ be such that $l(sw) = l(w) + 1$. Then*

$$n_s B n_s \cap B n_w B n_w^{-1} \subseteq B.$$

Proof Assume this to be false. Then, since $n_s B n_s \subseteq B \cup B n_s B$, we have $B n_s B \cap B n_w B n_w^{-1} \neq \varnothing$ and so $B n_s B n_w \cap B n_w B \neq \varnothing$. But we also have $B n_s B n_w \subseteq B n_s n_w B \cup B n_w B$, and so we must have $B n_s B n_w \not\subseteq B n_s n_w B$, contradicting Lemma 1.6.3. $\qquad\square$

It will now be convenient to introduce the following notation. We set

$$B^w := n_w^{-1} B n_w \quad \text{and} \quad B_w := B^{w_0 w} = B \cap n_{w_0 w}^{-1} B n_{w_0 w}$$

for any $w \in W$. (Note that B^w, B_w do not depend on the choice of the representative $n_w \in N$.)

1.6.16 Lemma *Let $y, w \in W$ be such that $l(yw) = l(y) + l(w)$. Then*

$$B \cap B^{yw} \subseteq B \cap B^w.$$

Furthermore, we have $B \cap B^{w_0} = H$.

Proof We proceed by induction on $l(y)$. If $l(y) = 0$, then $y = 1$ and there is nothing to prove. Next assume that $l(y) = 1$; then $y = s \in S$. By Lemma 1.6.15, we have

$$n_w (B \cap B^{sw}) n_w^{-1} = n_w (B \cap n_w^{-1} n_s^{-1} B n_s n_w) n_w^{-1}$$

$$= n_w B n_w^{-1} \cap n_s B n_s \subseteq B$$

and so $B \cap B^{sw} \subseteq B \cap B^w$, as required. Now assume that $l(y) > 1$ and choose $s \in S$ such that $l(ys) < l(y)$. We set $y' := ys$ and $w' := sw$. Then $yw = y'sw = y'w'$ and $l(y'w') = l(y') + l(w')$. So, by induction, we have $B \cap B^{yw} \subseteq B \cap B^{w'} = B \cap B^{sw}$. By the case $l(y) = 1$, we have $B \cap B^{sw} \subseteq B \cap B^w$, as required. Now consider the statement concerning $B \cap B^{w_0}$. Let $y \in W$. By relation $(*)$ in the proof of Proposition 1.6.14, we can write $w_0 = xy$ with $x \in W$ such that

$l(w_0) = l(x) + l(y)$. Then $B \cap B^{w_0} = B \cap B^{xy} \subseteq B^y$. This holds for all $y \in W$ and so $B \cap B^{w_0} \subseteq \bigcap_{y \in W} B^y \subseteq \bigcap_{n \in N} n^{-1}Bn = H$, using (BN5). $\qquad\square$

1.6.17 Lemma *Let $s \in S$ and $w \in W$ be such that $l(ws) = l(w) + 1$. Then the following hold.*

    (a) $B = B_s.B_{w_0 s} = B_{w_0 s}.B_s$ *and* $H \subsetneqq B_s$.
    (b) $B_s \subseteq B_{w_0 w}$.
    (c) $B_{ws} = B_s.B_w^s = B_w^s.B_s$.

Proof (a) Let $y := w_0 s$; then we have $l(ys) = l(y) + 1$ and so $l(sy^{-1}) = l(y^{-1}) + 1$. By Proposition 1.6.3, this yields $n_s B n_y^{-1} \subseteq B n_s n_y^{-1} B$ and so

$$B \subseteq n_s^{-1} B n_s . n_y^{-1} B n_y = B^s B^y.$$

Let $b \in B$. Then $b = b'b''$ with $b' \in B^s$ and $b'' \in B^y$. Now $b' \in n_s B n_s \cap B n_y^{-1} B n_y \subseteq B$ by Lemma 1.6.15, and so b', b'' both lie in B. Hence,

$$B = (B \cap B^s).(B \cap B^y) = B_{w_0 s}.B_s.$$

Taking inverses also yields $B = B_s.B_{w_0 s}$, as required. Furthermore, it is clear that $H \subseteq B_s$. If we had $B_s = H$, then $B = B_{w_0 s} = B \cap B^s$ and so $n_s B n_s = B$, contradicting (BN3).

(b) By relation $(*)$ in the proof of Proposition 1.6.14, we can write $w_0 = yws$, where $y \in W$ is such that $l(w_0) = l(y) + l(ws) = l(y) + l(w) + 1$. Then $w_0 s = yw$, where $l(w_0 s) = l(w_0) - 1 = l(y) + l(w)$. So Lemma 1.6.16 yields

$$B_s = B \cap B^{w_0 s} = B \cap B^{yw} \subseteq B \cap B^w = B_{w_0 w}.$$

(c) We have $B = B_s.B_{w_0 s}$ by (a). Now note that $l(w_0 wss) = l(w_0) - l(w) = l(w_0) - l(ws) + 1 = l(w_0 ws) + 1$ and so $B_s \subseteq B_{ws}$, using (b). This yields

$$B_{ws} = B_{ws} \cap B = B_{ws} \cap B_s B_{w_0 s} = B_s B_{ws} \cap B_s B_{w_0 s} = B_s(B_{ws} \cap B_{w_0 s}).$$

It remains to show that $B_{ws} \cap B_{w_0 s} = B_w^s$. First note that

$$B_{ws} \cap B_{w_0 s} = B \cap B^{w_0 ws} \cap B^s = n_s^{-1}(B^s \cap B^{w_0 w} \cap B)n_s$$

$$= n_s^{-1}(B^s \cap B_w)n_s.$$

Now $l(w_0 ws) = l(w_0) - l(w) - 1 = l(w_0 w) - 1$, and so we can write $w_0 w = xs$ where $x \in W$ is such that $l(xs) = l(x) + 1$. By

Lemma 1.6.16, this implies that $B_w = B \cap B^{w_0 w} \subseteq B^s$. Hence we have $B_{ws} \cap B_{w_0 s} = n_s^{-1} B_w n_s$ as required. $\square$

1.6.18 Proposition   *We have* $B = B_w.B_{w_0 w} = B_{w_0 w}.B_w$ *for all* $w \in W$.

Proof   We begin with the following statement.

$$B_{sw} = B_w.B_s^w \quad \text{for } s \in S \text{ and } w \in W \text{ such that } l(sw) = l(w) + 1.$$
$$(*)$$

We will prove this by induction on $l(w)$. If $l(w) = 0$, then $w = 1$ and so $B_1 = B \cap B^{w_0} = H$ by Lemma 1.6.16. Thus, we have $B_s = B_1.B_s$ as required. Now let $l(w) > 0$ and choose $t \in S$ such that $l(wt) < l(w)$. We set $y := wt$. Then $sw = syt$ and $l(sy) = l(y) + 1$. So we obtain

$$
\begin{aligned}
B_{sw} = B_{syt} &= B_{sy}^t.B_t \quad \text{(by Lemma 1.6.17(c))} \\
&= n_t^{-1}(B_y.B_s^y)n_t B_t = (B_s^y.B_y)^t B_t \quad \text{(by induction)} \\
&= B_s^{yt}.B_y^t.B_t = B_s^{yt}.B_{yt} = B_s^w.B_w \quad \text{(by Lemma 1.6.17(c))},
\end{aligned}
$$

as required. Now we prove the factorization $B = B_w.B_{w_0 w}$. Again, we will do this by induction on $l(w)$. If $l(w) = 0$, then $w = 1$ and $B_{w_0} = B$. Thus, we certainly have $B = B_1.B_{w_0}$. Now let $l(w) > 0$ and write $w = sy$, where $s \in S$ and $y \in W$ are such that $l(sy) = l(y) + 1$. Then, using $(*)$, we obtain

$$B_w.B_{w_0 w} = B_{sy}.B_{w_0 sy} = B_y.B_s^y.B_{w_0 sy}.$$

Now note that $B_{w_0 sy} = B \cap B^{sy} = B^{sy}_{w_0(sy)^{-1}}$. This yields

$$
\begin{aligned}
B_w.B_{w_0 w} &= B_y.B_s^y.B^{sy}_{w_0(sy)^{-1}} = B_y.n_y^{-1}\big(B_s.B^s_{w_0(sy)^{-1}}\big)n_y \\
&= B_y.n_y^{-1} B_{w_0(sy)^{-1}s} n_y = B_y.B^y_{w_0 y^{-1}}
\end{aligned}
$$

by Lemma 1.6.17(c). Now, we note again that $B^y_{w_0 y^{-1}} = B_{w_0 y}$. Thus, by induction, the right-hand side of the above identity equals $B_y.B_{w_0 y} = B$. $\square$

The following result will turn out to be most useful. For example, it leads to an order formula for finite groups with a split BN-pair; see Exercise 1.8.21(c) for an example. Another application will be given in the following section, where we construct BN-pairs for the symplectic and orthogonal groups.

1.6.19 Theorem (sharp form of the Bruhat decomposition)
*Let G be a group with a split BN-pair, and assume that W is finite.
We set*

$$U_w := U \cap n_{w_0w}^{-1} U n_{w_0w} \quad \text{for } w \in W$$

*(Note again that U_w does not depend on the choice of the representatives.) Then every element $g \in Bn_wB$ is uniquely expressible in the
form $g = bn_w u$, where $b \in B, w \in W$, and $u \in U_w$. Thus, we have*

$$G = \coprod_{w \in W} B\, n_w\, U_w \quad \text{with uniqueness of expressions.}$$

Furthermore, we have $U_w \neq \{1\}$ for all non-unit $w \in W$.

Proof We shall write $U^w := n_w^{-1} U n_w$ for all $w \in W$. First we
show that $B_w = U_w H$ for all $w \in W$. This is seen as follows. Since
H is normal in N, and since $B = UH$, we have $B \cap B^w = UH \cap
U^w H \supseteq (U \cap U^w)H$. We must show that the reverse inclusion also
holds. So let $g \in B \cap B^w$. Then we can write $g = n_w^{-1} u n_w h = u'h'$,
where $u, u' \in U$ and $h, h' \in H$. Since $g \in B$, we conclude that
$n_w^{-1} u n_w \in n_w^{-1} U n_w \cap B \subseteq U$ by condition (b) in Definition 1.6.13.
Thus, using condition (a), we conclude that $n_w^{-1} u n_w = u'$ and $h = h'$.
This yields $g \in (U \cap U^w)H$ as required. Replacing w by w_0w, we see
that $B_w = U_w H$.

Now Proposition 1.6.18 yields the factorization $U = U_w . U_{w_0w} =
U_{w_0w} . U_w$ for all $w \in W$. Further note that

$$n_w U_{w_0w} n_w^{-1} = n_w (U \cap n_w^{-1} U n_w) n_w^{-1} = n_w U n_w^{-1} \cap U \subseteq U,$$

and so $Bn_w B = Bn_w HU = Bn_w U = Bn_w U_{w_0w} U_w = Bn_w U_{w_0w}
n_w^{-1} n_w U_w = Bn_w U_w$. Thus, we have

$$G = \coprod_{w \in W} Bn_w U_w,$$

and it remains to prove the uniqueness of expressions. Thus, let
$g \in Bn_w U_w$ and write $g = bn_w u = b'n_w u'$, where $b, b' \in B$ and
$u, u' \in U_w$. Then we have

$$n_w u(u')^{-1} n_w^{-1} = b^{-1}b' \in n_w(U \cap U^{w_0w})n_w^{-1} \cap B \subseteq B \cap U^{w_0}.$$

Thus, it will be enough to show that $B \cap U^{w_0} = \{1\}$. Now, by
Lemma 1.6.16, we have $B \cap U^{w_0} \subseteq B \cap B^{w_0} = H$. But, by Definition 1.6.13(b), we also have $B \cap U^{w_0} \subseteq U$, and so that intersection
must be trivial, as desired.

Finally, assume that $U_w = \{1\}$. Then we have $B_w = H$. If $w \neq 1$, then there exists some $s \in S$ such that $l(ws) = l(w) - 1$. By Lemma 1.6.17, we have $B_w = B_s.B_{ws}^s \supseteq B_s \supsetneq H$, a contradiction. $\qquad\square$

1.6.20 Remark In the above proof, we have seen that there is a factorization

$$U = U_w.U_{w_0w} = U_{w_0w}.U_w \qquad \text{for all } w \in W.$$

Thus, the map $U_w \times U_{w_0w} \to U$ given by multiplication is surjective. We claim that it is also injective. Indeed, we have

$$U_w \cap U_{w_0w} \subseteq n_w^{-1}(n_{w_0}^{-1}Un_{w_0} \cap U)n_w \subseteq n_w^{-1}Hn_w = H,$$

where we used Lemma 1.6.16. Thus, since the left hand side lies in U, we conclude that $U_w \cap U_{w_0w} = \{1\}$ and this yields the injectivity of the above map.

1.6.21 Coxeter groups Finally, we mention the fact that $W = N/H$ is always a *Coxeter group*; that is, we have the following presentation of W:

$$W = \langle s \in S \mid s^2 = 1; (st)^{m_{st}} = 1 \ (s \neq t \text{ in } S \text{ with } m_{st} < \infty) \rangle,$$

where $m_{st} \geqslant 2$ denotes the order of $st \in W$. This is deduced from the fact that the exchange condition in Lemma 1.6.8 holds in W. Since this result will not be used explicitly in this book, we refer to Chapter IV, §1, Theorem 1 of Bourbaki (1968) for the proof. Furthermore, we remark that there is a complete classification of the finite Coxeter groups. To describe the result, we encode the above presentation in a graph, called the *Coxeter* graph of W. It has vertices labelled by the elements of S, and two vertices labelled by $s \neq t$ are joined by an edge if $m_{st} \geqslant 3$. Moreover, if $m_{st} \geqslant 4$, we label the edge by m_{st}. If the graph is connected, we say that W is an *irreducible Coxeter group*. The graphs corresponding to the irreducible finite Coxeter groups are given in Table 1.2; the orders of the corresponding groups are given as follows.

Type	A_n	B_n	D_n	$I_2(m)$	H_3
Order	$(n+1)!$	$2^n n!$	$2^{n-1} n!$	$2m$	120

Type	H_4	F_4	E_6	E_7	E_8
Order	14400	1152	51840	2903040	696729600

Table 1.2   Coxeter graphs of irreducible finite Coxeter groups

The numbers on the vertices correspond to a chosen labelling of the elements of S.

In general, if $\Gamma_1, \ldots, \Gamma_r$ are the connected components of the Coxeter graph of W, we have a corresponding direct-product decomposition $W = W_1 \times \cdots \times W_r$, where $W_i = \langle S_i \rangle$ and $S_i \subseteq S$ is the subset labelling the vertices in Γ_i; furthermore, each W_i is a Coxeter group with generating set S_i. For more details, see Chapter VI, no. 4.1, of Bourbaki (1968).

For example, the Weyl group of $\mathrm{GL}_n(k)$ is $\mathfrak{S}_n$, whose Coxeter graph is that of type A_{n-1} in Table 1.2, where the vertex labelled by i corresponds to the generator $\sigma_i = (i, i+1)$. We shall say that $\mathrm{GL}_n(k)$ has a *BN*-pair of type A_{n-1}.

1.7   *BN*-pairs in symplectic and orthogonal groups

In this section, we will determine *BN*-pairs for the symplectic and orthogonal groups introduced in §1.3.15 and §1.3.16. For the groups $\mathrm{SO}_{2m+1}(k)$ ($\mathrm{char}(k) \neq 2$) and $\mathrm{Sp}_{2m}(k)$, we obtain a *BN*-pair by a general construction from the *BN*-pair in $\mathrm{GL}_n(k)$. For the even-dimensional orthogonal groups $\mathrm{SO}_{2m}^+(k)$, we verify the *BN*-pair axioms along the lines of the proof of Proposition 1.6.10. As an application, we obtain the result that all these groups are connected.

1.7.1 Groups with a split *BN*-pair and automorphisms   Let G be a group with a split *BN*-pair having a finite Weyl group W, and let $\varphi\colon G \to G$ be a bijective homomorphism of (abstract) groups satisfying the following conditions.

(BN$^\varphi$1) We have $\varphi(U) = U, \varphi(H) = H$ and $\varphi(N) = N$, where $B = U.H$ and $H = B \cap N$ as in Definition 1.6.13.

(BN$^\varphi$2) Every coset Hn (where $n \in N$) such that $\varphi(Hn) \subseteq Hn$ contains an element which is fixed by φ.

Since φ is a group homomorphism, the fixed point set $G^\varphi := \{g \in G \mid \varphi(g) = g\}$ is a subgroup of G. Similarly, the fixed point sets B^φ, N^φ are subgroups of G^φ. We would like to show that B^φ, N^φ form a BN-pair in G^φ.

First, since $B = UH$ is a semidirect product, we conclude that we have a similar decomposition of B^φ:

$$B^\varphi = U^\varphi.H^\varphi \quad \text{and} \quad H^\varphi = B^\varphi \cap N^\varphi. \tag{1}$$

Then it is also clear that $n^{-1}U^\varphi n \cap B^\varphi \subseteq U^\varphi$ for all $n \in N^\varphi$. Thus, the splitness conditions in Definition 1.6.13 are satisfied. Next, consider $W = N/H$. Since N and H are invariant under φ, we obtain an induced homomorphism $\bar{\varphi}\colon W \to W, Hn \mapsto H\varphi(n)$. Since $\varphi(N) = N$, we note that $\bar{\varphi}$ is surjective and, hence, an isomorphism (since W is finite). Then φ induces a permutation of the Bruhat cells; we have

$$\varphi(C(w)) = C(\bar{\varphi}(w)) \quad \text{for all } w \in W. \tag{2}$$

Now let $s \in S$. Then $B \cup C(\bar{\varphi}(s)) = \varphi(B \cup C(s))$ is a subgroup of G and so $\bar{\varphi}(s) \in S$; see Corollary 1.6.5. Thus, $\bar{\varphi}$ leaves S invariant. Furthermore, we have a natural injective map $N^\varphi/H^\varphi \to W = N/H, H^\varphi n \mapsto Hn$. By (BN$^\varphi$2), the image of this map is precisely the group of fixed points $W^{\bar{\varphi}} = \{w \in W \mid \bar{\varphi}(w) = w\}$. We shall identify $W^{\bar{\varphi}}$ with N^φ/H^φ using the above map. Thus, we have

$$W^{\bar{\varphi}} = N^\varphi/H^\varphi \quad \text{and} \quad l(\bar{\varphi}(w)) = l(w) \text{ for all } w \in W. \tag{3}$$

Let $\bar{S}$ be the set of orbits of $\bar{\varphi}$ on S. For each $J \in \bar{S}$, we consider the subgroup $W_J \subseteq W$ generated by J. By Exercise 1.8.24(b), there is a unique longest element $w_J \in W_J$. The invariance of S under $\bar{\varphi}$ and the uniqueness of w_J imply that $w_J \in W^{\bar{\varphi}}$ for all $J \in \bar{S}$.

Furthermore, by Exercise 1.8.25, we have

$$W^{\bar{\varphi}} = \langle w_J \mid J \in \bar{S} \rangle.$$

Thus, we have all the ingredients for a BN-pair: the subgroups B^{φ}, N^{φ}, and the set of generators $\{w_J \mid J \in \bar{S}\}$ for the group $W^{\bar{\varphi}} = N^{\varphi}/H^{\varphi}$.

1.7.2 Proposition *In the set-up of §1.7.1, we have $G^{\varphi} = \langle B^{\varphi}, N^{\varphi} \rangle$ and*

$$G^{\varphi} = \coprod_{w \in W^{\bar{\varphi}}} U^{\varphi} H^{\varphi} n_w U_w^{\varphi} \quad \text{with uniqueness of expressions,}$$

where $n_w \in N^{\varphi}$ for any $w \in W^{\bar{\varphi}}$. For $J \in \bar{S}$ and $w \in W^{\bar{\varphi}}$, we have

$$n_J B^{\varphi} n_w \subseteq B^{\varphi} n_J n_w B^{\varphi} \cup B^{\varphi} n_w B^{\varphi}, \quad \text{where } n_J := n_{w_J} \in W^{\bar{\varphi}}.$$

Moreover, if $U_{w_J}^{\varphi} \neq \{1\}$ for all $J \in \bar{S}$, then B^{φ} and N^{φ} form a split BN-pair in G^{φ}, with Weyl group $W^{\bar{\varphi}}$.

Proof We have already remarked in §1.7.1 that the splitness conditions in Definition 1.6.13 hold. We will now show that all axioms for a BN-pair are satisfied, except possibly (BN3) in Definition 1.6.1. The assumption that $U_{w_J}^{\varphi} \neq \{1\}$ for all $J \in \bar{S}$ will be needed to show that (BN3) holds.

Now, by eqn 1.7.1(3), every $w \in W^{\bar{\varphi}}$ has a representative $n_w \in N^{\varphi}$. In combination with eqn 1.7.1(2) and the Bruhat decomposition for G, this yields that

$$G^{\varphi} = \coprod_{w \in W^{\bar{\varphi}}} \left(B n_w B\right)^{\varphi}.$$

Now let us take into account the sharp form of the Bruhat decomposition. Since the length function is invariant under $\bar{\varphi}$, the longest element $w_0 \in W$ certainly is fixed by $\bar{\varphi}$. Consequently, we have

$$\varphi(U_w) \subseteq U_w \quad \text{for any } w \in W^{\bar{\varphi}}.$$

Thus, given $g \in G^{\varphi}$ and $w \in W^{\bar{\varphi}}$ such that $g \in B n_w B$, we can write $g = u h n_w u'$, where $u \in U, h \in H$, and $u' \in U_w$. Since U, H, and

U_w are all invariant under φ, the uniqueness of the above expression implies that u, h, and u' are all fixed by φ. This yields the decomposition

$$G^\varphi = \coprod_{w \in W^{\bar\varphi}} U^\varphi H^\varphi n_w U_w^\varphi \quad \text{and} \quad (Bn_wB)^\varphi = U^\varphi H^\varphi n_w U_w^\varphi$$

$$= B^\varphi n_w B^\varphi. \qquad (\dagger)$$

We have uniqueness of expressions since this already holds in G. Furthermore, we now see that $G^\varphi = \langle B^\varphi, N^\varphi \rangle$ as required. Next, we have already seen in §1.7.1 that $B^\varphi = U^\varphi H^\varphi$ where $H^\varphi = B^\varphi \cap N^\varphi$. Furthermore, we have

$$H^\varphi \subseteq \bigcap_{n \in N^\varphi} nB^\varphi n^{-1} \subseteq (B \cap n_0^{-1} Bn_0)^\varphi = H^\varphi.$$

Thus, (BN5) is satisfied. Next, we have $N^\varphi/H^\varphi = W^{\bar\varphi}$, and this group is generated by elements of order 2; thus (BN2) holds. Let us now consider (BN4). Let $J \in \bar{S}$ and $w \in W^{\bar\varphi}$. Writing $w_J = s_1 \cdots s_n$ where $s_i \in J$, and applying the rule in Exercise 1.8.22(a), we obtain

$$C(w_J).C(w) \subseteq \bigcup_{w' \in W_J} C(w'w).$$

Now let us take fixed points under φ. We have $C(w'w)^\varphi = \varnothing$ if $\bar\varphi(w'w) \neq w'w$. Since $w \in W^{\bar\varphi}$, we have $\bar\varphi(w'w) = w'w$ if and only if $\bar\varphi(w') = w'$. But, by Exercise 1.8.25, we have $W_J^{\bar\varphi} = \{1, w_J\}$ and so

$$n_J B^\varphi n_w \subseteq \big(C(w_J).C(w)\big)^\varphi \subseteq \bigcup_{w' \in W_J} C(w'w)^\varphi$$

$$= C(w)^\varphi \cup C(w_J w)^\varphi = B^\varphi n_J n_w B^\varphi \cup B^\varphi n_w B^\varphi$$

as required, where the last equality holds again by ($\dagger$). Thus, all the required assertions are proved. It remains to check that (BN3) holds under the given hypothesis. So let $J \in \bar{S}$ and assume, if possible, that $n_J B^\varphi n_J = B^\varphi$. Since $w_J^2 = 1$, we can write this as $n_J B^\varphi n_J^{-1} = B^\varphi$. Now, since $U_{w_J}^\varphi \subseteq B^\varphi$, we also have $n_J U_{w_J}^\varphi n_J^{-1} \subseteq B^\varphi$. On the other hand, recalling the definition of U_{w_J} from Theorem 1.6.19, we see that $n_J U_{w_J} n_J^{-1} \subseteq n_J U n_J^{-1} \cap U^{w_0} \subseteq U^{w_0}$. Thus, we have $n_J U_{w_J}^\varphi n_J^{-1} \subseteq B^\varphi \cap U^{w_0} \subseteq B \cap B^{w_0} = H$, where the last equality holds by Lemma 1.6.16. Since n_J normalizes H, we conclude that $U_{w_J}^\varphi \subseteq H$ and so $U_{w_J}^\varphi = \{1\}$, contrary to our assumption. $\qquad\square$

1.7.3 Example We have a split BN-pair in $\mathrm{GL}_n(k)$ (where k may be any field) by taking the groups $B_n(k) = U_n(k).T_n(k) \subseteq \mathrm{GL}_n(k)$ and $N_n(k) \subseteq \mathrm{GL}_n(k)$; see Proposition 1.6.10. Let n_0 be the matrix defined in Remark 1.5.12; this matrix is used to define the groups $\mathrm{O}_n(k)$ (if $\mathrm{char}(k) \neq 2$) or $\mathrm{Sp}_{2m}(k)$ (if $n = 2m$ is even). To place ourselves in the setting of §1.7.1, we consider the map

$$\varphi\colon \mathrm{GL}_n(k) \to \mathrm{GL}_n(k), \quad A \mapsto n_0^{-1}(A^{\mathrm{tr}})^{-1} n_0.$$

Then φ is a bijective homomorphism of algebraic groups, and we have

$$\mathrm{GL}_n(k)^\varphi = \begin{cases} \mathrm{Sp}_{2m}(k) & \text{if } n = 2m \text{ is even (any characteristic)}, \\ \mathrm{O}_{2m}(k) & \text{if } \mathrm{char}(k) \neq 2 \text{ and } n = 2m, \\ \mathrm{O}_{2m+1}(k) & \text{if } \mathrm{char}(k) \neq 2 \text{ and } n = 2m + 1. \end{cases}$$

We wish to show that the two conditions in §1.7.1 are satisfied. It is readily checked that $\varphi(U_n(k)) = U_n(k), \varphi(T_n(k)) = T_n(k)$, and $\varphi(N_n(k)) = N_n(k)$. Now let us identify the Weyl group of $\mathrm{GL}_n(k)$ with $\mathfrak{S}_n$; we have

$$\mathfrak{S}_n = \langle \sigma_1, \dots, \sigma_{n-1} \rangle, \quad \text{where } \sigma_i = (i, i+1) \text{ for } 1 \leqslant i \leqslant n-1. \quad \text{(a)}$$

Then the induced automorphism $\bar\varphi\colon \mathfrak{S}_n \to \mathfrak{S}_n$ is given by conjugation with the permutation $w_0 := (1,n)(2,n-1)(3,n-2)\cdots \in \mathfrak{S}_n$. Thus, we are in the setting of Exercise 1.8.26 which shows that $\mathfrak{S}_n^{\bar\varphi} = \langle t, s_1, \dots, s_{m-1} \rangle$, where

$$\begin{cases} s_i = \sigma_{m-i}\sigma_{m+i} & \text{and } t = \sigma_m & \text{if } n = 2m, \\ s_i = \sigma_{m-i}\sigma_{m+i+1} & \text{and } t = \sigma_m\sigma_{m+1}\sigma_m & \text{if } n = 2m+1. \end{cases} \quad \text{(b)}$$

Now, for each element in $\mathfrak{S}_n^{\bar\varphi}$, we have to find a representative in $N_n(k)^\varphi$. It will actually be enough to do this for the above generators of $\mathfrak{S}_n^{\bar\varphi}$. One checks that such a representative for the generator t is given by

$$\left[\begin{array}{c|c|c} I_{m-1} & 0 & 0 \\ \hline 0 & J & 0 \\ \hline 0 & 0 & I_{m-1} \end{array} \right] \in \mathrm{GL}_n(k)^\varphi,$$

where J is one of the matrices

$$\begin{bmatrix} 0 & 1 \\ -1 & 0 \end{bmatrix} \quad \text{or} \quad \begin{bmatrix} 0 & 1 \\ 1 & 0 \end{bmatrix} \quad \text{or} \quad \begin{bmatrix} 0 & 0 & 1 \\ 0 & -1 & 0 \\ 1 & 0 & 0 \end{bmatrix}$$

according to whether $\mathrm{GL}_n(k)^\varphi$ is $\mathrm{Sp}_{2m}(k)$ or $\mathrm{O}_{2m}(k)$ or $\mathrm{O}_{2m+1}(k)$, respectively. Furthermore, representatives for the generators s_i can be found in the image of the embedding $\mathrm{GL}_m(k) \hookrightarrow \mathrm{GL}_n^\varphi$ defined by sending $A \in \mathrm{GL}_m(k)$ to

$$
\left[\begin{array}{c|c} A & 0 \\ \hline 0 & A' \end{array}\right]
\quad \text{or} \quad
\left[\begin{array}{c|c|c} A & 0 & 0 \\ \hline 0 & 1 & 0 \\ \hline 0 & 0 & A' \end{array}\right],
\quad \text{where } A' = Q_m(A^{-1})^{\mathrm{tr}} Q_m,
$$

according to whether n is even or odd. (We leave the details of these calculations as an exercise to the reader.) Thus, the two conditions in §1.7.1 are satisfied.

1.7.4 Theorem   *In the above setting, let $G := \mathrm{GL}_n(k)^\varphi$ and assume that $G = \mathrm{Sp}_{2m}(k)$ (any characteristic) or $G = \mathrm{O}_{2m+1}(k)$ $(\mathrm{char}(k) \neq 2)$. Then*

$$
B := B_n(k) \cap G \quad \text{and} \quad N := N_n(k) \cap G
$$

form a split BN-pair in G, where W^φ is a Coxeter group of type B_m; see Table 1.2. We have $B = UH$, where $H = B \cap N$ and $U = U_n(k) \cap G$.

   Now assume that k is an infinite field. Then the following hold:

   (a) If $G = \mathrm{Sp}_{2m}(k)$, then B is a closed connected subgroup.
   (b) If $G = \mathrm{O}_{2m+1}(k)$, let $B' := B \cap \mathrm{SO}_{2m+1}(k)$ and $N' := N \cap \mathrm{SO}_{2m+1}(k)$. Then B' and N' form a split BN-pair in $\mathrm{SO}_{2m+1}(k)$. The group B' is a closed connected subgroup of $\mathrm{SO}_{2m+1}(k)$.
   (c) The groups $\mathrm{Sp}_{2m}(k)$ and $\mathrm{SO}_{2m+1}(k)$ are connected.

Proof   We have already checked in Example 1.7.3 that the assumptions $(\mathrm{BN}^\varphi 1)$ and $(\mathrm{BN}^\varphi 2)$ in §1.7.1 are satisfied. Furthermore, using the description of the groups U_w in Exercise 1.8.21, one checks that $U_{w_J}^\varphi \cong k$ for every $\bar\varphi$-orbit J. Thus, all the hypotheses of Proposition 1.7.2 are satisfied. This yields that B and N form a split BN-pair in G. The identification of W^φ is given by Exercise 1.8.26(a).

   Now assume that k is an infinite field. If $G = \mathrm{Sp}_{2m}(k)$, then the groups U and H are closed and connected by Remark 1.5.12. Hence so is B (by an argument similar to that in Remark 1.6.11). Thus, (a) is proved. Next assume that $G = \mathrm{O}_{2m+1}(k)$ and let $G' := \mathrm{SO}_{2m+1}(k)$. Then G' is a closed normal subgroup of G containing U. Thus, by Exercise 1.8.23, the groups B' and N' form a split BN-pair in G'. The descriptions of U and H in Lemma 1.5.10 show that U and

$H' := H \cap G'$ are closed and connected. Hence so is $B' = U.H'$ (by the same argument as above). This proves (b). Finally, (c) is a consequence of Remark 1.6.7. $\qquad\qquad\square$

We now turn to the even-dimensional orthogonal groups, defined in §1.3.16:

$$O_{2m}^+(k) = \{A \in M_{2m}(k) \mid f_{2m}(Av) = f_{2m}(v) \quad \text{for all } v \in k^{2m}\}.$$

If the characteristic of k is not 2, then we could in fact apply a similar argument as in Theorem 1.7.4 to exhibit a BN-pair. However, this methods fails in characteristic 2 since then $O_{2m}^+(k)$ is not defined as the fixed-point set under an automorphism. Instead, we shall proceed as in the proof of Proposition 1.6.10, by verifying directly the BN-pair axioms. Incidentally, this works simultaneously for the case of odd characteristic as well. One of the key problems will be to identify the group $O_{2m}^+(k)^\circ$. Recall from §1.3.15 and §1.3.16 that we define

$$\mathrm{SO}_{2m}^+(k) := \begin{cases} O_{2m}^+(k) \cap \mathrm{SL}_{2m}(k) & \text{if } \mathrm{char}(k) \neq 2, \\ O_{2m}^+(k)^\circ & \text{if } \mathrm{char}(k) = 2. \end{cases}$$

In the subsequent discussion, the following matrix will play a special role:

$$t_m := \left[\begin{array}{c|cc|c} I_{m-1} & \multicolumn{2}{c|}{0} & 0 \\ \hline 0 & 0 & 1 & 0 \\ & 1 & 0 & \\ \hline 0 & \multicolumn{2}{c|}{0} & I_{m-1} \end{array}\right] \in O_{2m}^+(k).$$

1.7.5 An intermediate group   We place ourselves in the setting of Example 1.7.3 where $n = 2m$. Then $\mathrm{GL}_{2m}(k)^\varphi$ is given by $O_{2m}(k)$ (if $\mathrm{char}(k) \neq 2$) or $\mathrm{Sp}_{2m}(k)$ (if $\mathrm{char}(k) = 2$). Thus, if we set $G_m := O_{2m}^+(k)$, then we have $G_m \subseteq \mathrm{GL}_{2m}(k)^\varphi$ in both cases. Now let

$$B_m := B_{2m}(k) \cap G_m, \quad N_m := N_{2m}(k) \cap G_m, \quad U_m := U_{2m}(k) \cap G_m.$$

Using Lemma 1.5.11, we see that $T_m := T_{2m}(k)^\varphi \subseteq G$ in both cases. Furthermore, we have $B_m = U_m T_m$ and $T_m = B_m \cap N_m$. Considering the generators of $\mathfrak{S}_{2m}^{\bar\varphi}$ in Example 1.7.3, we also see that

$N_{2m}(k)^\varphi \subseteq G_m$ in both cases, and so $N_m = N_{2m}(k)^\varphi$. Thus, we have

$$W_m := N_m/T_m \cong \mathfrak{S}_{2m}^{\varphi} = \langle t, s_1, \ldots, s_{m-1} \rangle.$$

Now let $W'_m \subseteq W_m$ be the subgroup of index 2 defined in Exercise 1.8.26(c), and let $\tilde{N}_m \subseteq N_m$ be the preimage of W'_m (under the map $N_m \to W_m$). We set

$$\tilde{G}_m := \langle B_m, \tilde{N}_m \rangle \subseteq G_m.$$

We have the following two properties:

(a) The matrix t_m (as defined above) lies in N_m, and it normalizes B_m.

(b) We have $n_0 B_m n_0^{-1} \subseteq \tilde{G}$, where n_0 is the matrix defined in Example 1.7.3.

Indeed, (a) is checked by a straightforward computation, using the descriptions of U_m, T_m in Lemma 1.5.11. Now consider (b). It is readily checked that $n_0 \in N_{2m}(k)^\varphi \subseteq N_m$. Now, if m is even, then $n_0 \in \tilde{N}_m \subseteq \tilde{G}_m$ and so we trivially have $n_0 B_m n_0^{-1} \subseteq G_m$. On the other hand, if m is odd, then $n_0 \notin \tilde{N}_m$ but $n_0 t_m \in \tilde{N}_m$. Then, using (a), we have $n_0 B_m n_0^{-1} = n_0 t_m B t_m^{-1} n_0^{-1} \subseteq \tilde{G}_m$ as required.

1.7.6 Lemma *In the above setting, the groups B_m and $\tilde{N}_m$ form a split BN-pair in $\tilde{G}_m$, with Weyl group W'_m. Furthermore, we have $t_m \notin \tilde{G}_m$.*

Proof First note that the splitness conditions in Definition 1.6.13 are satisfied (by the same argument as in Remark 1.6.12). So we must verify the axioms in Definition 1.6.1. The condition (BN1) is clear by definition, and (BN2) holds by Exercise 1.8.26(c):

$$W'_m = \langle u, s_1, \ldots, s_{m-1} \rangle, \quad \text{where } u := t s_1 t.$$

Furthermore, we have seen in the proof of §1.7.5(b) that there exists some $\tilde{n} \in \tilde{N}_m$ such that $n_0 B_m n_0^{-1} = \tilde{n} B_m \tilde{n}^{-1}$. Then we have $\bigcap_{n \in \tilde{N}_m} n B_m n^{-1} \subseteq B_m \cap n_0 B_m n_0^{-1} = T_m$; thus (BN5) holds. It remains to verify (BN3) and (BN4). For this purpose, we work with the embedding $\mathrm{GL}_m(k) \hookrightarrow \tilde{G}_m$ given by

$$A \mapsto \hat{A} := \left[\begin{array}{c|c} A & 0 \\ \hline 0 & A' \end{array} \right], \quad \text{where } A' = Q_m (A^{-1})^{\mathrm{tr}} Q_m.$$

We set $n_i = \hat{\sigma}_{m-i} \in \tilde{N}_m$ for $1 \leqslant i \leqslant m-1$, where we regard $\sigma_1, \ldots, \sigma_{m-1}$ as generators for $\mathfrak{S}_m \subseteq \mathrm{GL}_m(k)$. Then $n_i \in \tilde{N}_m$ is a representative of $s_i \in W_m'$. Note also that $n_u := t_m n_1 t_m \in \tilde{N}_m$ is a representative of u. Furthermore, using Lemma 1.5.11, it is easily checked that

$$U_m = V_m.\hat{U}_m(k) = \hat{U}_m(k).V_m \quad \text{and}$$

$$n_i V_m n_i = V_m \text{ for } 1 \leqslant i \leqslant m-1, \tag{1}$$

where

$$V_m := \left\{ \left[\begin{array}{c|c} I_m & Q_m S \\ \hline 0 & I_m \end{array} \right] \ \middle| \ \begin{array}{l} S = (s_{ij}) \in M_m(k) \text{ such that} \\ S = -S^{\mathrm{tr}} \text{ and } s_{ii} = 0 \text{ for all } i \end{array} \right\}.$$

Now consider (BN3). Assume, if possible, that $n_i B_m n_i = B$ for some $i \in \{1, \ldots, m-1\}$. Since n_i normalizes T_m and V_m, this implies that $n_i \hat{U}_m(k) n_i = \hat{U}_m(k)$ and so $\sigma_{m-i} U_m(k) \sigma_{m-i} = U_m(k)$, contradicting (BN3) for $\mathrm{GL}_m(k)$; see Proposition 1.6.10. Thus, we have $n_i B_m n_i \neq B_m$ for $1 \leqslant i \leqslant m-1$. Since t_m normalizes B_m, we also have $n_u B_m n_u = t_m n_1 t_m B_m t_m n_1 t_m \neq B_m$. Thus, (BN3) holds.

Finally, in order to verify (BN4), we proceed as in the proof of Proposition 1.6.10. For $1 \leqslant i \leqslant m-1$, we define

$$X_{s_i} := \{ I_{2m} + \lambda(E_{m-i,m-i+1} - E_{m+i,m+i+1}) \mid \lambda \in k \} \subseteq U_m,$$

$$V_{s_i} := \{ (a_{ij}) \in U_m \mid a_{m-i,m-i+1} = a_{m+i,m+i+1} = 0 \}.$$

It is readily checked that these are subgroups such that $U_m = X_{s_i}.V_{s_i} = V_{s_i}.X_{s_i}$ and $X_{s_i} \cap V_{s_i} = \{1\}$. In fact, we have $X_{s_i} = \hat{X}_{m-i}$ and $V_{s_i} = \hat{V}_{m-i} V_m$ with $X_{m-i}, V_{m-i} \subseteq \mathrm{GL}_m(k)$ as in Example 1.6.9. First we claim that

$$n_i V_{s_i} n_i = V_{s_i} \quad \text{and} \quad n_i X_{s_i} n_i \subseteq \{1\} \cup X_{s_i} n_i X_{s_i} T_m$$

$$\text{for } 1 \leqslant i \leqslant m-1. \tag{2}$$

Indeed, using the above descriptions of X_{s_i} and V_{s_i}, the proof of (2) can be reduced to the analogous statements for $X_{m-i}, V_{m-i} \subseteq \mathrm{GL}_m(k)$, where they are verified in Example 1.6.9; note also that $n_i V_m n_i = V_m$. Using (2), we obtain

$$n_i B_m n_i = n_i X_{s_i} V_{s_i} T_m n_i = n_i X_{s_i} n_i V_{s_i} T_m \subseteq n_i X_{s_i} n_i B_m$$

$$\subseteq (\{1\} \cup X_{s_i} n_i X_{s_i} T_m) B_m \subseteq B_m \cup B_m n_i B_m.$$

Next we wish to show that $n_i B_m n_w \subseteq B_m n_i n_w B_m$, where $w \in W_m'$ is such that $l(\sigma_{m-i} w) > l(w)$; here, l is the usual length function on $\mathfrak{S}_{2m}$. For this purpose, we first prove:

$$n_w^{-1} X_{s_i} n_w \subseteq U_m \quad \text{for all } w \in W_m' \text{ such that } l(\sigma_{m-i}w) > l(w).$$

(3)

This is seen as follows. By the invariance of the length function under $\bar{\varphi}$, we also have $l(\sigma_{m+i}w) > l(w)$. Using Example 1.6.9(b), we obtain

$$n_w^{-1} X_{s_i} n_w \subseteq \{I_{2m} + \lambda(E_{w^{-1}(m-i),\,w^{-1}(m-i+1)}$$
$$- E_{w^{-1}(m+i),\,w^{-1}(m+i+1)}) \mid \lambda \in k\},$$

and this is contained in $U_{2m}(k)$ by the characterization of the length condition in Exercise 1.8.21(a). Thus, we have $n_w^{-1} X_{s_i} n_w \subseteq U_{2m}(k) \cap G_m = U_m$ as claimed.

Now we can proceed as follows. Let $1 \leqslant i \leqslant m$ and $w \in W_m'$ be such that $l(\sigma_{m-i}w) > l(w)$. Then, using (2) and (3), we obtain

$$n_i B_m n_w = n_i V_{s_i} X_{s_i} T_m n_w \subseteq V_{s_i} n_i n_w n_w^{-1} X_{s_i} n_w$$
$$\subseteq V_{s_i} n_i n_w U_m T_m \subseteq B_m n_i n_w B_m.$$

Next, let $w \in W_m'$ be such that $l(\sigma_{m-i}w) < l(w)$. Then set $y := s_i w \in W_m'$. If we had $l(\sigma_{m-i}y) < l(y)$, then we could apply the argument in Exercise 1.8.25(*) and conclude that $y = s_i y'$, where $y' \in W_m'$ is such that $l(y) = l(s_i) + l(y')$. But then we automatically have $y' = w$ and so $l(s_i w) > l(w)$. This in turn would imply $l(\sigma_{m-i}w) > l(w)$, contrary to our assumption. Thus, we must have $l(\sigma_{m-i}y) > l(y)$. So we can apply the previous discussion and conclude that

$$n_i B_m n_w = n_i B_m n_i n_y \subseteq B_m n_y \cup B_m n_i B_m n_y$$
$$\subseteq B_m n_y \cup B_m n_i n_y B_m = B_m n_i n_w B_m \cup B_m n_w B_m.$$

Thus, we have shown (BN4) for all the generators s_i. Finally, using the fact that t_m normalizes B_m, we also have for any $w \in W_m'$:

$$n_u B_m n_w = t_m n_1 t_m B_m n_w \subseteq t_m(n_1 B_m n_{twt})t_m$$
$$\subseteq t_m(B_m n_1 n_{twt} B_m \cup B_m n_{twt} B_m)t_m$$
$$= t_m B_m n_1 n_{twt} B_m t_m \cup t_m B_m n_{twt} B_m t_m$$
$$= B_m t_m n_1 t_m n_w B_m \cup B_m n_w B_m$$
$$= B_m n_u n_w B_m \cup B_m n_w B_m.$$

Thus, we have verified all the axioms for a BN-pair. To show that $t_m \notin \tilde{G}_m$, we can now argue as follows. If we had $t_m \in \tilde{G}_m$, then

$t_m \in \mathrm{N}_{\tilde{G}_m}(B_m) = B_m$ by Corollary 1.6.4 and §1.7.5(a). This is absurd, since t_m evidently is not a triangular matrix. $\qquad\square$

The next step consists of clarifying the relation between $\tilde{G}_m$ and G_m°.

1.7.7 Lemma   *Assume that k is an infinite field. Then we have $G_m^\circ = \tilde{G}_m$, and this group has index 2 in G_m.*

Proof   First we show that $G_m^\circ \subseteq \tilde{G}_m$. To see this, set $X := U_m' \cdot T_m \cdot U_m$. Since U_m, T_m, U_m' are connected (see Remark 1.5.12), we have $X \subseteq \langle U_m, T_m, U_m' \rangle \subseteq G_m^\circ$. Conversely, let $g \in G_m^\circ$ and consider the subset $gX \subseteq G_m^\circ$. By Exercise 1.8.27, $X \subseteq G_m^\circ$ is open. Hence so is gX (since left multiplication by g is an isomorphism of affine varieties). Since G_m° is irreducible, this implies $X \cap gX \neq \varnothing$ and so $g \in X^{-1} \cdot X$. Thus, we have shown that

$$G_m^\circ = \langle U_m, T_m, U_m' \rangle.$$

On the other hand, by §1.7.5(b), we have $U_m, T_m, U_m' \subseteq \tilde{G}_m$. We conclude that $G_m^\circ \subseteq \tilde{G}_m$, as claimed. Using Lemma 1.7.6, we obtain

$$G_m^\circ \subseteq \tilde{G}_m \subsetneq G_m.$$

Hence, in order to complete the proof, it will be enough to show that G_m° has index at most 2 in G_m. Since G_m° is a normal subgroup, it will be enough to show that

$$G_m = \langle G_m^\circ, t_m \rangle = G_m^\circ \cup t_m G_m^\circ.$$

We will do this by induction on m. For $m = 1$, this holds by the explicit description of $\mathrm{O}_2^+(k)$ in §1.3.15 and §1.3.16. Now let $m > 1$. The group G_m acts naturally on k^{2m} (column vectors). Let $g \in G_m$ and set $x := g.e_1$ where $e_1 \in k^{2m}$ is the first vector of the standard basis of k^{2m}. Then x is a non-zero vector of k^{2m} such that $f_{2m}(x) = f_{2m}(g.e_1) = f_{2m}(e_1) = 0$. We claim that there exists some $g_1 \in \langle G_m^\circ, t_m \rangle$ such that $g_1.e_1 = x$. To see this, assume first that x has the property that $x_1 \neq 0$. Then we can find a suitable diagonal matrix $h \in T_m \subseteq G_m^\circ$ such that the first component of $h.x_1$ is 1. So we may assume without loss of generality that $x_1 = 1$. Then consider the matrix

$$
u(x) := \left[
\begin{array}{c|ccc|c}
1 & 0 & \cdots & 0 & 0 \\
\hline
x_2 & & & & 0 \\
\vdots & & I_{2(m-1)} & & \vdots \\
x_{2m-1} & & & & 0 \\
\hline
x_{2m} & x_{2m-1} & \cdots & x_2 & 1
\end{array}
\right] .
$$

It is readily checked that $u(x) \in G_m$. (To see this, one needs the fact that $f_{2m}(x) = 0$.) We have $u(x) \in U'_m \subseteq G^{\circ}_m$ and $u(x).e_1 = x$, as desired. Now, if $x_1 = 0$, then there exists some $g' \in \langle G^{\circ}_m, t_m \rangle$ such that the first component of $g'.x$ is non-zero. To see this, first note that we have a natural embedding

$$
\mathrm{SL}_m(k) \to \mathrm{O}^{+}_{2m}(k), \quad A \mapsto \hat{A} := \left[
\begin{array}{c|c}
A & 0 \\
\hline
0 & A'
\end{array}
\right] \in G_m,
$$

where $A' = Q_m (A^{-1})^{\mathrm{tr}} Q_m$. Since $\mathrm{SL}_m(k)$ is connected (see Example 1.6.11), this implies that $\hat{A} \in G^{\circ}_m$ for all $A \in \mathrm{SL}_m(k)$. Thus, if $x_i \neq 0$ for some $i \leqslant m$, we can find some $g' \in G^{\circ}_m$ in the image of the above embedding such that the first component of $g'.x_i$ is non-zero. Otherwise, we have $x_i \neq 0$ for some $i > m$ and so there exists some $g' \in G^{\circ}_m$ in the image of the above embedding such that the $(m+1)$-component of $g'.x$ is non-zero. Then the mth component of $t_m g'.x$ will be non-zero, and we can apply the previous discussion. In any case, as desired, we have shown that there exists some $g_1 \in \langle G^{\circ}_m, t_m \rangle$ such that $g_1.e_1 = x = g.e_1$. Then $g_1^{-1} g = (a_{ij}) \in G_m$ is a matrix such that $a_{i1} = 0$ for all $i > 1$. It is easily checked that then we automatically have $a_{2m,j} = 0$ for all $j < 2m$, and so

$$
g_1^{-1} g = \left[
\begin{array}{c|ccc|c}
1 & & * & & * \\
0 & & & & \\
\vdots & & g_2 & & * \\
0 & & & & \\
\hline
0 & 0 & \cdots & 0 & 1
\end{array}
\right] , \qquad \text{where } g_2 \in G_{m-1}.
$$

Now we have a natural embedding

$$
G_{m-1} \hookrightarrow G_m, \quad A \mapsto \tilde{A} :=
\left[
\begin{array}{c|ccc|c}
1 & 0 & \cdots & 0 & 0 \\
\hline
0 & & & & 0 \\
\vdots & & A & & \vdots \\
0 & & & & 0 \\
\hline
0 & 0 & \cdots & 0 & 1
\end{array}
\right].
$$

Under this embedding, we certainly have $\tilde{t}_{m-1} = t_m$ and $\tilde{G}^\circ_{m-1} \subseteq G^\circ_m$. This yields $\tilde{g}_2^{-1} g_1^{-1} g \in U_m \subseteq G^\circ_m$. Now, we have $g_2 \in \langle G^\circ_{m-1}, t_{m-1} \rangle$ by induction, and so $\tilde{g}_2 \in \langle G^\circ_m, t_m \rangle$. This yields $g \in \langle G^\circ_m, t_m \rangle$ as required. $\qquad\square$

1.7.8 Theorem   *Assume that k is infinite, and let $G'_m := \mathrm{SO}_{2m}^+(k)$ $(m \geqslant 1)$ be the even-dimensional orthogonal group, as defined in §1.3.16. Then*

$$
B'_m := B_{2m}(k) \cap G'_m \quad \text{and} \quad N'_m := N_{2m}(k) \cap G'_m
$$

form a split BN-pair in G'_m, where $W \cong W'_m$ is of type D_m; see Table 1.2. We have $B'_m = U'_m T'_m$, where $T'_m = B'_m \cap N'_m$ and $U'_m := U_{2m}(k) \cap G'$. The group B'_m is closed and connected. Furthermore, $G'_m = \mathrm{O}_{2m}^+(k)^\circ$, and this group has index 2 in $\mathrm{O}_{2m}^+(k)$.

Proof   By Lemma 1.7.7, G°_m has index 2 in G_m. This implies that $G'_m = G^\circ_m$. Indeed, this is clear by definition if $\mathrm{char}(k) = 2$; on the other hand, if $\mathrm{char}(k) \neq 2$, then G'_m is a closed subgroup of index 2 in G_m and so $G^\circ_m \subseteq G'_m$ by Proposition 1.3.13. Since G°_m has index 2, we conclude that $G^\circ_m = G'_m$.

Using Lemma 1.7.7, we also obtain $G'_m = G^\circ_m = \tilde{G}_m$. Furthermore, it is easily checked that $N'_m = \tilde{N}_m$ and $B_m = B'_m$. Thus, Lemma 1.7.6 yields that B'_m and N'_m form a split BN-pair in G'_m. The fact that B'_m is closed and connected is shown as in the proof of Theorem 1.7.4. $\qquad\square$

1.7.9 Summary: BN-pairs in classical groups Let k be an infinite field and $G \subseteq \mathrm{GL}_n(k)$ be one of the classical groups:

$$
G = \begin{cases}
\mathrm{SL}_n(k), & \text{any } n, \text{ any characteristic;} \\
\mathrm{SO}_{2m+1}(k), & n = 2m+1, \mathrm{char}(k) \neq 2; \\
\mathrm{Sp}_{2m}(k), & n = 2m, \text{ any characteristic;} \\
\mathrm{SO}_{2m}^{+}(k), & n = 2m, \text{ any characteristic;}
\end{cases}
$$

see §1.3.15 and §1.3.16. Let $B := B_n(k) \cap G$ and $N = N_n(k) \cap G$, where $B_n(k)$ and $N_n(k)$ are defined in Example 1.6.9. Furthermore, let $U := U_n(k) \cap G$ and $H := T_n(k) \cap G$. Then the following hold.

(a) The groups B and N form a split BN-pair in G; we have $B = UH$.

(b) The group U is closed, connected, and normal in B; we have $\dim U = l(w_0)$. The group H is closed and connected, and consists of diagonal matrices.

(c) The group B is closed and connected. Furthermore, G is connected.

For the proofs, see once more Remark 1.6.11 and Theorems 1.7.4 and 1.7.8. The fact that $\dim U = l(w_0)$ follows from Exercises 1.8.21 and 1.8.26. It is also easily checked that U is nilpotent and B is solvable; we will come back to this point in Example 2.4.9 and Definition 3.4.5. See also Example 4.2.6 where we will describe the finite classical groups.

1.8 Bibliographic remarks and exercises

The material in the first five sections on algebraic sets, tangent spaces, and so on is more or less standard. The main differences to standard text books on algebraic geometry (e.g. Mumford (1988) or Shafarevich (1994)), are (1) that we tried to keep the prerequisites to a minimum (in particular, we only prove a version of Hilbert's nullstellensatz for hypersurfaces), (2) that we address algorithmic questions (Groebner bases), and (3) that we illustrate the general theory from an early point on by examples from the theory of algebraic groups.

For further reading on Groebner bases and algorithmic aspects, see Cox *et al.* (1992) and Greuel and Pfister (2002). Example 1.3.6

and Exercise 1.8.9 are taken from Chapter 3, §3 of Cox *et al.* (1992). The discussion of the Hilbert polynomial and the dimension follows Chapter 9 of Cox *et al.* (1992). This polynomial plays an important role in the general dimensional theory of noetherian rings; see Chapter 11 of Atiyah and Macdonald (1969). The results on algebraic groups are all given by Borel (1991), Humphreys (1991), or Springer (1998). The proof of Theorem 1.4.11(b) is taken from §A.3 of Goodman and Wallach (1998).

For more on classical groups, see Dieudonné (1971), Taylor (1992) (especially for finite ground fields), Goodman and Wallach (1998) (over ground fields of characteristic 0), Chapter 7 of Aschbacher (2000), and Grove (2002). Here, we just develop this material to the point where we can show the existence of BN-pairs in these groups; see Section 1.7.

The general construction of groups of Lie type in §1.5.15 was discovered by C. Chevalley in the 1950s; see Humphreys (1972) for an introduction and Steinberg (1967) and Carter (1972) for detailed expositions of this work.

BN-pairs were introduced by Tits (1962). A survey on groups with a BN-pair, associated geometric structures ('buildings'), and Tits' classification of such groups (where W is finite and $|S| > 2$) can be found in Chapter 15 of Carter (1972). The discussion of the sharp form of the Bruhat decomposition follows §2.5 of Carter (1985); here, we have tried to reduce the required prerequisites to a minimum. Our definition of split BN-pairs is slightly weaker than the usual one; see, for example, §30 of Gorenstein *et al.* (1994). Proposition 1.7.2 is adapted from §13.5.4 of Carter (1972).

The argument for the proof of (BN1) in Proposition 1.6.10 is taken from p. 36 of Steinberg (1967). For the determination of the BN-pairs in the symplectic and orthogonal groups (except in characteristic 2), we followed §2.14 of Steinberg (1974). For a more direct approach, see Chapter IV, §2, Exercise 20 of Bourbaki (1968). The problem with orthogonal groups in characteristic 2 is that, in some way, one has to show that $O_{2m}^+(k)^\circ$ has index 2 in $O_{2m}^+(k)$. Usually, this is based on the so-called *Dickson invariant*, which is the group homomorphism

$$D\colon O_{2m}^+(k) \to \mathbb{Z}/2\mathbb{Z} \subseteq k, \quad (a_{ij}) \mapsto \sum_{i,j=1}^{m} a_{i,2m+1-j}\, a_{2m+1-i,j};$$

see §II.10 of Dieudonné (1971). The proof uses the Clifford algebra associated with $O_{2m}^+(k)$; see also Grove (1982) and Exercise 14.5 of

Aschbacher (2000). A different approach is offered by Dye (1977), who shows that $D(A) \equiv \mathrm{rank}_k(A - I_{2m}) \pmod 2$ for all $A \in \mathrm{O}_{2m}^+(k)$. Our argument in Theorem 1.7.8 provides yet an alternative approach, based on the theory of groups with a BN-pair. For the existence of BN-pairs in groups of Lie type in general, see §3 of Steinberg (1967) and §8.2 of Carter (1972).

1.8.1 Exercise Let A be a commutative noetherian ring (with 1). Show that every submodule of a finitely generated A-module is again finitely generated.

[*Hint.* Every finitely generated A-module is a quotient of the free A-module A^n for some $n \geqslant 1$. So it is enough to consider A^n. If you need further help, see Proposition 6.5 of Atiyah and Macdonald (1969).]

1.8.2 Exercise Let k be any field.

(a) Assume that $|k| = \infty$, and let $f \in k[X_1, \ldots, X_n]$. Show that $f = 0$ if and only if $f(x) = 0$ for all $x \in k^n$.

(b) Assume that k is algebraically closed. Show that $|k| = \infty$. Let $n \geqslant 2$ and let $f \in k[X_1, \ldots, X_n]$ be non-constant. Show that $|V(\{f\})| = \infty$.

1.8.3 Exercise Assume that k is an infinite field.

(a) Let $f, g \in k[X, Y]$ be non-constant polynomials which have no common irreducible factor. Then show that $\mathbf{V}(\{f, g\})$ is a finite set.

(b) Assume that $V \subsetneqq k^2$ is an irreducible algebraic set. Show that either V is a singleton set or there exists an irreducible polynomial $f \in k[X, Y]$ such that $V = \mathbf{V}(\{f\})$.

The statement in (a) is a first step toward *Bézout's* theorem on the intersection of curves in projective space; see Fischer (1994) or Chapter 5 of Fulton (1969).

1.8.4 Exercise We consider the polynomial ring $R = k[X_1, \ldots, X_n, Y_1, \ldots, Y_m]$, where k is an infinite field. Let $g_1, \ldots, g_n \in k[Y_1, \ldots, Y_m] \subseteq R$ and consider the ideal $I = (X_1 - g_1, \ldots, X_n - g_n) \subseteq R$.

(a) Show that every $f \in R$ is of the form $f = g + h$, where $g \in I$ and $h \in k[Y_1, \ldots, Y_m]$. Deduce from this that $R/I \cong k[Y_1, \ldots, Y_m]$ and $\mathbf{I}(\mathbf{V}(I)) = I$.

(b) Given the **LEX** order with $Y_m \preceq \cdots \preceq Y_1 \preceq X_n \preceq \cdots \preceq X_1$, show that $(\mathrm{LT}(f) \mid 0 \neq f \in I)$ is generated by $G = \{X_1, \ldots, X_n\}$, and that G is a Groebner basis of I.

1.8.5 Exercise Let $I \subseteq k[X_1, \ldots, X_n]$ and choose a lexicographic order such that $X_n \preceq X_{n-1} \preceq \cdots \preceq X_1$. Let G be a Groebner basis of I. Then show that, for any $i \in \{1, \ldots, n\}$, we have $I \cap k[X_i, X_{i+1}, \ldots, X_n] = (G \cap k[X_i, X_{i+1}, \ldots, X_n])$.

1.8.6 Exercise Assume that $|k| = \infty$. Let us regard k^n as a k-vector space in the usual way, and let $V \subseteq k^n$ be a linear subspace of k^n.

(a) Show that $V = \mathbf{V}(\{f_1, \ldots, f_m\})$, where each f_i is a polynomial of the form $f_i = \sum_{j=1}^{n} a_{ij} X_j$ with $a_{ij} \in k$. Show that V is irreducible.

(b) Using Gaussian elimination, show that we can assume the polynomials f_i to be of the following special form. We have $m \leqslant n$, and there exist $1 \leqslant j_1 < j_2 < \cdots < j_m \leqslant n$ such that $f_i = X_{j_i} + h_i$, where h_i is a linear polynomial (with constant term zero) in the variables X_j ($j > j_i$ and $j \notin \{j_1, \ldots, j_m\}$).

(c) Use Exercise 1.8.4 to show that $\mathbf{I}(V) = (f_1, \ldots, f_m)$ and that $A[V]$ is a polynomial ring in $n - m$ variables. Deduce that the dimension of V as an algebraic set equals the dimension of V as a vector space.

1.8.7 Exercise Let k be algebraically closed, and consider the algebraic set $V = \mathbf{V}(\{X_1^2 - X_2 X_3, X_1 X_3 - X_1\}) \subseteq k^3$. Show that V has three irreducible components. Determine their affine algebras and show that $\dim V = 1$.

1.8.8 Exercise Let $V \subseteq k^n$ and $W \subseteq k^m$ be non-empty algebraic sets. Consider the direct product $V \times W \subseteq k^{n+m}$; see §1.3.7. Show that the projection maps

$$\mathrm{pr}_1 \colon V \times W \to V \quad \text{and} \quad \mathrm{pr}_2 \colon V \times W \to W$$

are open maps, that is, they send open sets to open sets.

[*Hint.* It is enough to prove this for a principal open set. Now a principal open subset of $V \times W$ is a subset of the form $\{(v, w) \in V \times W \mid \sum_{i=1}^{r} f_i(v) g_i(w) \neq 0\}$, where $f_1, \ldots, f_r \in k[X_1, \ldots, X_n]$ and $g_1, \ldots, g_r \in k[Y_1, \ldots, Y_m]$. Then check explicitly that images of such a set under pr_1 and pr_2 are open.]

1.8.9 Exercise We place ourselves in the setting of §1.3.6, where $k \subseteq \mathbb{C}$.

(a) Let $\varphi \colon k^2 \to k^3$ be defined by $\varphi(u, v) = (uv, v, u^2)$ for $u, v \in k$, and set $W_\varphi = \overline{\varphi(k^2)} \subseteq k^3$, the so-called *Whitney umbrella*. Consider the ideal

$$\hat{I} := (X - UV, Y - V, Z - U^2) \subseteq k[U, V, X, Y, Z].$$

Compute a Groebner basis of $\hat{I}$ with respect to the order **LEX** with $U \succeq V \succeq X \succeq Y \succeq Z$. Then use Exercise 1.8.5 to show that $I := \hat{I} \cap k[X, Y, Z] = (X^2 - Y^2 Z)$. By §1.3.6, we have $\mathbf{V}(I) = W_\varphi$. Show that $\varphi(k^2) = W_\varphi$ if $k = \mathbb{C}$, and that $\varphi(k^2) \subsetneqq W_\varphi$ if $k = \mathbb{R}$; describe the 'missing' points for $k = \mathbb{R}$.

(b) Now let $\varphi \colon k^2 \to k^3$ be defined by $\varphi(u, v) = (uv, uv^2, u^2)$ for $u, v \in k$, and set $V_\varphi = \overline{\varphi(k^2)} \subseteq k^3$. Use the same technique as in (a) to show that $V_\varphi = \mathbf{V}(X^4 - Y^2 Z) \subseteq k^3$. Show that $\varphi(k^2) \subsetneqq V_\varphi$, and determine the 'missing' points in the cases where $k = \mathbb{R}$ or $k = \mathbb{C}$.

1.8.10 Exercise Assume that k is algebraically closed, and let $V \subseteq k^n$ be an irreducible algebraic set. Show that the following hold.

(a) We have $\dim V = 0$ if and only if V is a singleton set.
(b) We have $\dim V = n - 1$ if and only if V is a hypersurface.
(c) We have $\dim V = n$ if and only if $V = k^n$.

[*Hint.* (a) By Example 1.2.19(a), we have $\dim V = 0$ if $|V| = 1$. To prove the converse, use Proposition 1.2.20. (b) By Example 1.2.16(b), a hypersurface in k^n has dimension in $n - 1$. Conversely, assume that $\dim V = n - 1$. This means that $\partial_k(A[V]) < \partial_k(k[X_1, \ldots, X_n]) = n$ and so $\mathbf{I}(V) \neq \{0\}$. Let $0 \neq f \in \mathbf{I}(V)$. Since $\mathbf{I}(V)$ is a prime ideal, we may assume that f is irreducible. Thus, we have $V = \mathbf{V}(\mathbf{I}(V)) \subseteq H_f$, where $H_f \subseteq k^n$ is irreducible. Now use Proposition 1.2.20. (c) By Example 1.2.16(a), we have $\dim k^n = n$. The reverse implication follows from Proposition 1.2.20.]

1.8.11 Exercise This exercise is concerned with a special case of *localization*. (In fact, this is all we shall need to know about localization.) Let A be a commutative ring (with 1) and suppose $0 \neq f \in A$. Let X be an indeterminate over A, and set $A_f := A[X]/(fX - 1)$. Composing the natural inclusion $A \subseteq A[X]$ with the canonical projection onto A_f, we obtain a ring homomorphism $\iota \colon A \to A_f$.

(a) Show that $\ker(\iota) = \{a \in A \mid af^n = 0 \text{ for some } n \geqslant 0\}$ (where we set $f^0 = 1$). Thus, ι is injective if and only if f is not a zero divisor in A. Furthermore, we have that f is nilpotent if and only if $1 = 0$ in A_f. If f is not nilpotent, then $\iota(f) \in A_f$ is invertible, with inverse being given by the residue class of X in A_f.

(b) Assume that f is not nilpotent. Then show that we have the following universal property: if $\alpha \colon A \to B$ is any homomorphism into a commutative ring B (with 1) such that $\alpha(f) \in B$ is invertible, then there exists a unique ring homomorphism $\tilde{\alpha} \colon A_f \to B$ such that $\alpha = \tilde{\alpha} \circ \iota$.

(c) Let $I \subseteq A$ be an ideal such that $f^n \notin I$ for all $n \geqslant 0$. Let $\bar{f}$ be the image of f in A/I and I_f be the ideal generated by $\iota(I)$ in A_f. Show that $(A/I)_{\bar{f}} \cong A_f/I_f$. Furthermore, if I is a prime ideal and $\bar{f}$ is not invertible in A_f, then I_f also is a prime ideal.

(d) Assume that A is an integral domain with field of fractions K. Let $\alpha \colon A \hookrightarrow K$ be the inclusion. Show that $\tilde{\alpha} \colon A_f \to K$ is an injective ring homomorphism with image given by $\{a/f^n \mid a \in A, n \geqslant 0\} \subseteq K$.

[*Hint.* (a) Let $a \in A$ and write out explicitly the condition that $a \in (fX - 1)$. Apply this also to $a = 1$. (b) Given α, we have a unique ring homomorphism $\alpha' \colon A[X] \to B$ such that $\alpha'(X) = \alpha(f)^{-1}$ and the restriction of α' to A is α. Since $fX - 1 \in \ker(\alpha')$, we get an induced map $\tilde{\alpha} \colon A_f \to B$ as desired. (c) Consider the canonical map $\pi \colon A \to A/I$; check that we have an induced map $\pi' \colon A_f \to (A/I)_{\bar{f}}$ and consider its kernel. (d) To show that $\tilde{\alpha}$ is injective, check that the ring $A' := \{a/f^n \mid a \in A, n \geqslant 0\} \subseteq K$ (together with the inclusion $A \subseteq A'$) also satisfies the universal property in (b).]

1.8.12 Exercise This exercise contains some standard results about algebraic field extensions and the transcendence degree. Let k be a field and A be a finitely generated k-algebra. Assume that A is an integral domain, and let K be its field of fractions. Then $K \supseteq k$ is a finitely generated field extension, and $\partial_k(K)$ is also called the *transcendence degree* of K over k.

(a) Show that, if $A = k[a_1, \ldots, a_m]$, then $\partial_k(A) \leqslant m$ and one can select an algebraically independent subset of size $\partial_k(A)$ from $\{a_1, \ldots, a_m\}$.

(b) Assume that $z_1, \ldots, z_d \in K$ are algebraically independent and that every element of K is algebraic over $k(z_1, \ldots, z_d)$. Then $d = \partial_k(K)$.

(c) Let $F \subseteq K$ be a subfield containing k. Then $\partial_k(K) = \partial_k(F) + \partial_F(K)$.

[*Hint.* (a) Assume that $A = k[a_1, \ldots, a_m]$. Then we have a canonical surjection $k[X_1, \ldots, X_m] \to A$ such that $X_i \mapsto a_i$ for all i; let I be the kernel of that map. Then use Proposition 1.2.18. The results (b) and (c) are standard; see for example Chapter X of Lang (1984).]

1.8.13 Exercise Let k be an infinite field and $V \subseteq k^n$ be an irreducible algebraic set. Let $f \in k[X_1, \ldots, X_n]$ be such that $f \notin \mathbf{I}(V)$. Show that $\tilde{V}_f$ is irreducible and $\dim \tilde{V}_f = \dim V$.

[*Hint.* Use Exercise 1.8.11 and Proposition 1.2.18.]

1.8.14 Exercise Denote by $\partial f / \partial X_i$ the usual partial derivative of a polynomial $f \in k[X_1, \ldots, X_n]$ with respect to X_i. Show that the following holds.

(a) If the characteristic of k is 0 and $\partial f / \partial X_i = 0$ for all i, then f is constant.

(b) Assume that k is perfect of characteristic $p > 0$. If $\partial f / \partial X_i = 0$ for all i, then there exists some $g \in k[X_1, \ldots, X_n]$ such that $f = g^p$.

(c) Assume that k is perfect and $0 \neq f \in k[X_1, \ldots, X_n]$ is non-constant irreducible. Conclude from (a) and (b) that $\partial f / \partial X_i \neq 0$ for some i.

1.8.15 Exercise This exercise is used in the proof of Proposition 1.4.5. Let k be a perfect field and K be a finitely generated extension field. Let $d = \partial_k(K)$. Then show that there exist $z_1, \ldots, z_{d+1} \in K$ such that:

(a) $K = k(z_1, \ldots, z_{d+1})$;
(b) $\{z_1, \ldots, z_d\}$ are algebraically independent over k;
(c) z_{d+1} is separable algebraic over $k(z_1, \ldots, z_d)$.

Proceed as follows. Let us write $K = k(a_1, \ldots, a_n)$ with $a_i \in K$. By Exercise 1.8.12, we have $d \leqslant n$ and we may assume that $a_1, \ldots, a_d$ are algebraically independent. We use induction on $n - d$. If $n = d$, then $K = k(a_1, \ldots, a_d)$ and we may take $a_{d+1} = 1$. Then all the conditions are satisfied. Now assume that $d < n$ and proceed by the following steps.

Step 1. The set $\{a_1, \ldots, a_{d+1}\}$ cannot be algebraically independent, so there exists some non-zero $F \in k[X_1, \ldots, X_{d+1}]$ such that

$F(a_1, \ldots, a_{d+1}) = 0$. Show that we can assume that F is irreducible and that $\partial f / \partial X_i \neq 0$ for some i.

Step 2. Show that $a_1, \ldots, a_{i-1}, a_{i+1}, \ldots, a_{d+1}$ are algebraically independent. Then let $L \subseteq K$ be the subfield generated by these elements. Show that a_i is separable algebraic over L. Thus, we are done if $n = d + 1$.

Step 3. Now assume that $n > d + 1$. Then a_i and a_{d+2} are algebraic over L and a_i is separable. So the 'primitive-element theorem' shows that there exists some $b \in L(a_i, a_{d+2})$ with $L(b) = L(a_i, a_{d+2})$. Thus,

$$K = k[a_1, \ldots, a_{i-1}, a_{i+1}, \ldots, a_{d+1}, b, a_{d+3}, \ldots, a_n],$$

that is, we have decreased the number of generators by 1.

For more details see Appendix 5, Proposition 1 of Shafarevich (1994).

1.8.16 Exercise   Let $V \subseteq k^n$ and $W \subseteq k^m$ be algebraic sets.

(a) Let $\varphi \colon V \to W$ be a regular map. Let $Z := \overline{\varphi(V)} \subseteq W, p \in V$, and $q := \varphi(p) \in Z$. Then show that $d_p\varphi(T_p(V)) \subseteq T_q(Z) \subseteq T_q(W)$.

(b) Let $p \in V$ and $q \in W$. Identifying $k^{n+m} = k^n \times k^m$ as usual, show that $T_{(p,q)}(V \times W) = T_p(V) \oplus T_q(W)$.

[*Hint.* (a) We have a canonical factorization $\varphi = \iota \circ \varphi'$, where $\iota \colon Z \to W$ is the inclusion and $\varphi' \colon V \to W, v \mapsto \varphi(v)$. Check that $d_q\iota \colon T_q(Z) \to T_q(W)$ is just the inclusion. (b) Use Proposition 1.3.8(b).]

1.8.17 Exercise   Assume that k is an algebraically closed field.

(a) Compute all non-singular points of $V = \mathbf{V}(\{f\}) \subseteq k^2$ for the following polynomials in $k[X, Y]$: $Y^2 - X^2(X + 1)$, $Y^2 - X(X^2 + 1), Y^2 - X^3$. The same for $f = X^2 - Y^2 Z \in k[X, Y, Z]$ and the corresponding algebraic set $V = \mathbf{V}(\{f\}) \subseteq k^3$, where $k = \mathbb{C}$; see also Exercise 1.8.9(a).

(b) Let $V = \mathbf{V}(\{ZX, YZ\}) \subseteq k^3$. Show that there exist points $p \in V$ with $\dim T_p(V) < \dim V$. How do you interpret this result?

1.8.18 Exercise   Assume that k is perfect and infinite. Let $V \subseteq k^n$ be an irreducible algebraic set. Assume that $\dim V = d$. Then show that there exist polynomials $f_1, \ldots, f_{n-d} \in k[X_1, \ldots, X_n]$ such that V is an irreducible component of $\mathbf{V}(\{f_1, \ldots, f_{n-d}\})$.

[*Hint.* By Theorem 1.4.11, there exists some non-singular $p \in V$. Since $T_p(V)$ is a linear subspace of k^n, this means that there exist $f_1, \ldots, f_{n-d} \in \mathbf{I}(V)$ such that $T_p(V) = \mathbf{V}(\{d_p(f_1), \ldots, d_p(f_{n-d})\}) \subseteq k^n$. Now we have $V \subseteq \mathbf{V}(\{f_1, \ldots, f_{n-d}\})$ and so V is contained in some irreducible component V' of $\mathbf{V}(\{f_1, \ldots, f_{n-d}\})$. Then use Proposition 1.2.20 and Theorem 1.4.11 to conclude that $V = V'$.]

1.8.19 Exercise Let k be an algebraically closed field of characteristic $\neq 2$ and consider the orthogonal group $\mathrm{SO}_3(k)$ as defined in §1.3.15. Let $\omega \in k$ be such that $\omega^2 = -2$. Then show that we have a unique homomorphism of algebraic groups $\varphi \colon \mathrm{SL}_2(k) \to \mathrm{SO}_3(k)$ such that

$$\varphi \colon \begin{bmatrix} 1 & t \\ 0 & 1 \end{bmatrix} \mapsto \begin{bmatrix} 1 & \omega t & t^2 \\ 0 & 1 & -\omega t \\ 0 & 0 & 1 \end{bmatrix} \quad \text{and} \quad \begin{bmatrix} 1 & 0 \\ t & 1 \end{bmatrix} \mapsto \begin{bmatrix} 1 & 0 & 0 \\ \omega t & 1 & 0 \\ t^2 & -\omega t & 1 \end{bmatrix}$$

for all $t \in k$. Show that φ is surjective and $\ker(\varphi) = \{\pm I_2\}$.

[*Hint.* First check that $\mathrm{SL}_2(k)$ is generated by the matrices specified above. To show the existence of φ, consider the natural action of $\mathrm{SL}_2(k)$ on the polynomial ring $k[X, Y]$; we have

$$\begin{aligned} g.X &= aX + cY \\ g.Y &= bX + dY \end{aligned}, \quad \text{where} \quad g = \begin{bmatrix} a & b \\ c & d \end{bmatrix} \in \mathrm{SL}_2(k).$$

(See, for example, Example V.5.13 of Huppert (1983).) The subspace $V_2 \subseteq k[X, Y]$ of homogeneous polynomials of degree 2 is invariant under this action. Now express the action of $g \in \mathrm{SL}_2(k)$ on V_2 with respect to the basis $\{\omega^{-1}X^2, XY, \omega^{-1}Y^2\}$.]

1.8.20 Exercise Let k be an algebraically closed field and $G = \mathrm{O}_4^+(k)$, as defined in §1.3.16. Show that the following maps are well-defined injective homomorphisms of algebraic groups:

$$\varphi_1 \colon \mathrm{SL}_2(k) \to G, \qquad \begin{bmatrix} a & b \\ c & d \end{bmatrix} \mapsto \left[\begin{array}{cc|cc} a & 0 & b & 0 \\ 0 & a & 0 & -b \\ \hline c & 0 & d & 0 \\ 0 & -c & 0 & d \end{array}\right],$$

$$\varphi_2 \colon \mathrm{SL}_2(k) \to G, \qquad A \mapsto \left[\begin{array}{c|c} A & 0 \\ \hline 0 & A' \end{array}\right], \qquad \text{where } A' = Q_2(A^{-1})^{\mathrm{tr}}Q_2.$$

Let $G_1 := \varphi_1(\mathrm{SL}_2(k))$ and $G_2 := \varphi_2(\mathrm{SL}_2(k))$. Show that G_1 and G_2 are closed connected normal subgroups of G° and that we have

$$G^\circ = G_1.G_2 \quad \text{where} \quad G_1 \cap G_2 = \{\pm I_4\}.$$

Furthermore, we have $g_1 g_2 = g_2 g_1$ for all $g_1 \in G_1$ and $g_2 \in G_2$.

1.8.21 Exercise Let k be a field and let $G = \mathrm{GL}_n(k)$. By Proposition 1.6.10 and Remark 1.6.12, we know that the groups $B_n(k)$ and $N_n(k)$ form a split BN-pair in G, with Weyl group $W \cong \mathfrak{S}_n$.

(a) Let $1 \leqslant i \leqslant n-1$ and $w \in \mathfrak{S}_n$. Then show that

$$l(\sigma_i w) > l(w) \quad \Leftrightarrow \quad w^{-1}(i) < w^{-1}(i+1).$$

Deduce that $l(w) = |\{(i,j)\,|\,1 \leqslant i < j \leqslant n \text{ and } w(i) > w(j)\}|$ and that $w_0 = (1,n)(2,n-1)(3,n-2)\cdots$ is the *longest element* in $\mathfrak{S}_n$.

(b) Let $w \in \mathfrak{S}_n$ and show that U_w (see Theorem 1.6.19) is given by

$$U_w = \langle X_{ij} \mid 1 \leqslant i < j \leqslant n \text{ and } w(i) > w(j)\rangle.$$

(c) Assume that $k = \mathbb{F}_q$ is a finite field with q elements. Show that $|U_w| = q^{l(w)}$. Then use the sharp form of the Bruhat decomposition to show that

$$|\mathrm{GL}_n(\mathbb{F}_q)| = q^{n(n-1)/2}(q-1)^n \sum_{w \in \mathfrak{S}_n} q^{l(w)}.$$

[*Hint.* (a) In the proof of Proposition 1.6.10, we have seen that, if $w^{-1}(i) < w^{-1}(i+1)$, then $n_i B n_w \subseteq B n_i n_w B$. Now use Proposition 1.6.3 to conclude that this implies $l(\sigma_i w) > l(w)$. Conversely, if $w^{-1}(i) > w^{-1}(i+1)$, then set $y := \sigma_i w$ and apply the previous argument.

(b) Let X_i be defined as in Example 1.6.9. Show that $X_i = U_{\sigma_i}$. Then use induction on $l(w)$ and relation $(*)$ in the proof of Proposition 1.6.18. For a more direct approach, see also Theorem 5.10 of Taylor (1992).]

1.8.22 Exercise Let G be a group with a BN-pair. The purpose of this exercise is to establish some multiplication rules for Bruhat cells. Let us write $C(w) := B n_w B$ for $w \in W$. (Recall that $B n_w B$

does not depend on the choice of the representative $n_w \in N$.) We certainly have

$$C(1) = B, \quad C(ww') \subseteq C(w).C(w') \quad \text{and} \quad C(w^{-1}) = C(w)^{-1}.$$

(a) Let $s_1, \ldots, s_p \in S$. Then show that

$$C(s_1 \cdots s_p).C(w) \subseteq \bigcup_{\substack{1 \leqslant i_1 < \cdots < i_q \leqslant p \\ \text{for some } q \in \{0, \ldots, p\}}} C(s_{i_1} \cdots s_{i_q} w).$$

(b) Let $x, y \in W$ and assume that $l(xy) = l(x) + l(y)$. Show that

$$C(x).C(y) = C(xy).$$

[*Hint.* (a) This is proved by an easy induction on p; see also Chapter IV, §2, of Bourbaki (1968). (b) It is enough to prove the following property. Let $w \in W$ and write $w = s_1 \cdots s_p$, where $s_i \in S$ and $l(w) = p$. Then $C(w) = C(s_1) \ldots C(s_p)$. Use a repeated application of the first rule in Proposition 1.6.3.]

1.8.23 Exercise Let G be a group with a BN-pair and assume that $B = U.H$, where $U \subseteq B$ is a normal subgroup such that $U \cap H = \{1\}$. Let $G' \subseteq G$ be any normal subgroup containing U, and set

$$B' := B \cap G', \quad N' := N \cap G', \quad H' := H \cap G', \quad W' := N'/H'.$$

Show that $G = G'H$ and $N = N'H$. Then show that B' and N' form a BN-pair in G' and that the inclusion $N' \hookrightarrow N$ induces an isomorphism $W' \cong W$. Also note that, if the BN-pair in G is split, then so is the BN-pair in G'.

[*Hint.* To show that $G = G'H$, argue as follows. The subgroup $G'H \subseteq G$ contains B and is normalized by N. Thus, we also have $n_s B n_s^{-1} \subseteq G'.H$ for all $s \in S$. Use Proposition 1.6.3 to conclude that $N \subseteq G'.H$. So we have $G = G'H$. This also implies that $N = N'H$. For further details, see the proof of Chapter IV, §2, Theorem 5 of Bourbaki (1968).]

1.8.24 Exercise Let G be a group with a BN-pair, where the Weyl group W is finite. Let S be the set of generators of W and $J \subseteq S$. Let $W_J := \langle J \rangle \subseteq W$.

(a) Let $w \in W_J$. Show that there exists a reduced expression $w = s_1 \cdots s_p$, where $p = l(w)$ and $s_i \in J$ for all i.

(b) Show that there exists a unique element of maximal length $w_J \in W_J$. Show that $w_J^2 = 1$, and that $w_J \in W_J$ is characterized by the condition that $l(sw_J) < l(w_J)$ for all $s \in J$.

[*Hint.* (a) Since $w \in W_J$, we can write $w = s_1 \cdots s_m$, where $s_i \in J$. If $m > l(w)$, then there exists some $i < m$ such that $l(s_i s_{i+1} \cdots s_m) < l(s_{i+1} \cdots s_m) = m - i$. By Corollary 1.6.8, there exists some $i \leqslant j \leqslant m$ such that $s_i s_{i+1} \cdots s_m = s_i \cdots s_{j-1} s_{j+1} \cdots s_m$. This yields a new expression for w with strictly less than m factors from J. Now continue until the expression is reduced. (b) Argue similarly as in the proof of Proposition 1.6.14.]

1.8.25 Exercise Let G be a group with a BN-pair, where the Weyl group W is finite. Let $\theta \colon W \to W$ be a group automorphism such that $\theta(S) = S$, where S is the set of generators for W. Let $\bar{S}$ be the set of θ-orbits on S.

(a) Show that $W^\theta = \langle w_J \mid J \in \bar{S} \rangle$.
(b) Show that $W_J^\theta = \{1, w_J\}$ for any $J \in \bar{S}$.

[*Hint.* (a) First establish the following fact. Let $J \in \bar{S}$ and $w \in W^{\bar{\varphi}}$. Assume that $l(sw) < l(w)$ for some $s \in J$. Then show that

$$w = w_J w', \quad \text{where } w' \in W^\theta \text{ and } l(w) = l(w_J) + l(w'). \qquad (*)$$

Argue as follows. First note that $l(\theta(s)w) = l(sw) < l(w)$. Thus, we actually have $l(sw) < l(w)$ for all $s \in J$. Now consider all expressions of the form $w = yw'$, where $y \in W_J, w' \in W$ and $l(w) = l(y) + l(w')$. (Such expressions certainly exist; take, for example, $y := 1$ and $w' := w$.) Among all such expressions, choose one where y has maximal possible length. Then argue as in the proof of Proposition 1.6.14 (using the exchange condition) to conclude that $l(sy) < l(y)$ for all $s \in J$. By Exercise 1.8.24(b), this implies that $y = w_J$ is the longest element in W_J. Thus, $(*)$ is proved. An easy induction on $l(w)$ then shows that every $w \in W^\theta$ can be written in the form $w = w_{J_1} \cdots w_{J_r}$, where $J_i \in \bar{S}$ and $l(w) = l(w_{J_1}) + \cdots + l(w_{J_r})$.

(b) Let $w' \in W_J^\theta$ and assume that $w' \neq 1$. Then there exists some $s \in J$ such that $l(sw') < l(w')$. By $(*)$, we can write $w' = w_J x$, where $l(w') = l(w_J) + l(x)$. But, since $w' \in W_J$, we have $l(w') \leqslant l(w_J)$ and so $w' = w_J$. Thus, we have $W_J^\theta = \{1, w_J\}$ as desired.]

1.8.26 Exercise　The purpose of this exercise is to study the fixed-point set of the group $\mathfrak{S}_n$ under the automorphism $\varphi\colon \mathfrak{S}_n \to \mathfrak{S}_n$ given by conjugation with $w_0 = (1,n)(2,n-1)(3,n-2)\cdots \in \mathfrak{S}_n$. (These results are needed in Theorems 1.7.4 and 1.7.8.) We have $\mathfrak{S}_n = \langle \sigma_1, \ldots, \sigma_{n-1} \rangle$, where $\sigma_i = (i, i+1)$. Furthermore, $\varphi(\sigma_i) = \sigma_{n-i}$ for $1 \leqslant i \leqslant n-1$ and so φ leaves the set of generators invariant. The orbits of φ on that set of generators are given by

$$\{\sigma_1, \sigma_{2m-1}\}, \{\sigma_2, \sigma_{2m-2}\}, \ldots, \{\sigma_{m-1}, \sigma_{m+1}\}, \{\sigma_m\} \quad \text{if } n = 2m,$$

$$\{\sigma_1, \sigma_{2m}\}, \{\sigma_2, \sigma_{2m-1}\}, \ldots, \{\sigma_{m-1}, \sigma_{m+2}\}, \{\sigma_m, \sigma_{m+1}\} \quad \text{if } n = 2m+1.$$

We set $W_m := \langle t, s_1, \ldots, s_{m-1} \rangle$, where

$$\begin{cases} s_i = \sigma_{m-i}\sigma_{m+i} & \text{and } t = \sigma_m & \text{if } n = 2m, \\ s_i = \sigma_{m-i}\sigma_{m+i+1} & \text{and } t = \sigma_m\sigma_{m+1}\sigma_m & \text{if } n = 2m+1. \end{cases}$$

It is readily checked that the following relations hold:

$$t^2 = s_1^2 = \cdots = s_{m-1}^2 = 1, \quad ts_1ts_1 = s_1ts_1t, \quad ts_i = s_it \text{ for } i > 1,$$

$$s_is_j = s_js_i \text{ if } |i-j| \geqslant 2, \quad s_is_{i+1}s_i = s_{i+1}s_is_{i+1} \text{ for } i \geqslant 1.$$

Now show the following statements.

(a) We have $\mathfrak{S}_n^{\varphi} = W_m$ and $|W_m| = 2^m m!$. Conclude that W_m is a Coxeter group of type B_m (see Table 1.2). Show that w_0 also is the longest element in W_m, and that its length (with respect to the generators of W_m) is m^2.

(b) Show that there exists a unique group homomorphism $\varepsilon_t\colon W_m \to \{\pm 1\}$ such that $\varepsilon_t(t) = -1$ and $\varepsilon_t(s_i) = 1$ for $1 \leqslant i \leqslant m-1$.

(c) Let $W_m' := \ker(\varepsilon_t) \subseteq W_m$; then W_m' has index 2 in W_m. Show that $W_m' = \langle ts_1t, s_1, \ldots, s_{m-1} \rangle$ if $m \geqslant 2$. Conclude that W_m' is a Coxeter group of type D_m. Show that the length of the longest element in W_m' is $m(m-1)$.

[*Hint.* The relations are checked by a straightforward computation. (a) To show $\mathfrak{S}_n^{\varphi} = W_m$, apply Exercise 1.8.25. On the other hand, we have $W_m = C_{\mathfrak{S}_n}(w_0)$ and it is easy to compute the order of that centralizer. To obtain an expression for w_0 in terms of the generators of W_m, define $t_0 := t$ and $t_i := s_it_{i-1}s_i$ for $1 \leqslant i \leqslant m-1$. Then check that $w_0 = t_0t_1\cdots t_{m-1}$. (b) Restrict the sign homomorphism $\varepsilon\colon \mathfrak{S}_n \to \{\pm 1\}$ to W_m. (c) W_m' is the set of all elements of W_m which can be written as a product of the generators $t, s_1, \ldots, s_{m-1}$ with an even number of occurrences of t. Use the above relations to rearrange such a product in order to see that W_m' is generated as desired. To obtain the length of the longest element, note that either $w_0 \in W_m'$ or $tw_0 \in W_m'$.]

1.8.27 Exercise The principal minors of a matrix $A = (a_{ij}) \in M_n(k)$ are the determinants

$$d_m := \det \begin{bmatrix} a_{11} & \cdots & a_{1m} \\ \vdots & & \vdots \\ a_{m1} & \cdots & a_{mm} \end{bmatrix} \quad \text{for } 1 \leqslant m \leqslant n.$$

Let $\delta \in k[X_{ij} \mid 1 \leqslant i, j \leqslant n]$ be the polynomial which gives the product of the principal minors of $A \in M_n(k)$ for $1 \leqslant m \leqslant n$. Now let $G \subseteq \mathrm{GL}_n(k)$ be one of the classical groups defined in §§1.3.15–16; thus, we have

$$G = \begin{cases} \mathrm{SL}_n(k), & \text{any } n, \\ \mathrm{SO}_{2m+1}(k), & n = 2m + 1, \mathrm{char}(k) \neq 2, \\ \mathrm{Sp}_{2m}(k), & n = 2m, \text{ any characteristic}, \\ \mathrm{O}_{2m}^+(k), & n = 2m, \text{ any characteristic}. \end{cases}$$

Let U and U' be the subgroups of all upper unitriangular matrices and all lower unitriangular matrices in G, respectively. Let T be the subgroup of all invertible diagonal matrices in G. Show that

$$U' \cdot T \cdot U = \{g \in G \mid \delta(g) \neq 0\} \quad \text{is an open subset of } G.$$

[*Hint.* First note that the principal minors of a matrix $A \in M_n(k)$ do not change if we multiply A on the left by a lower unitriangular matrix or on the right by an upper unitriangular matrix. Thus, $U \cdot T \cdot U' \subseteq \{g \in G \mid \delta(g) \neq 0\}$. To prove the reverse inclusion, first consider $G = \mathrm{SL}_n(k)$. Let $A = (a_{ij})$ be a matrix in G such that $\delta(A) \neq 0$. We have to solve the following system of equations:

$$\begin{bmatrix} 1 & 0 & \cdots & 0 \\ u_{21}' & 1 & \ddots & \vdots \\ \vdots & \ddots & \ddots & 0 \\ u_{n1}' & \cdots & u_{n,n-1}' & 1 \end{bmatrix} \begin{bmatrix} h_1 & 0 & \cdots & 0 \\ 0 & h_2 & \ddots & \vdots \\ \vdots & \ddots & \ddots & 0 \\ 0 & \cdots & 0 & h_n \end{bmatrix} \begin{bmatrix} 1 & u_{21} & \cdots & u_{n1} \\ 0 & 1 & \ddots & \vdots \\ \vdots & \ddots & \ddots & u_{n,n-1} \\ 0 & \cdots & 0 & 1 \end{bmatrix} = \begin{bmatrix} a_{11} & \cdots & a_{1n} \\ \vdots & & \vdots \\ a_{n1} & \cdots & a_{nn} \end{bmatrix}$$

with unknowns u_{ij}, u_{ij}' and h_i. We can solve this system recursively. The fact that all principal minors of A are non-zero is used in the following way. First, we have $h_1 = a_{11} \neq 0$. Then, for $i > 1$, we have $u'_{i1}h_1 = a_{i1}$ and so we determine u'_{i1}. Similarly, for $j > 1$, we have $h_1 u_{j1} = a_{1j}$ and so we can determine u_{j1}. Now assume that the first $j - 1$ rows and columns of the above matrices have already been determined, where $h_1, \ldots, h_{j-1}$ are all non-zero. We have an equation $h_j + u'_{j,j-1}h_{j-1}u_{j,j-1} + \cdots + u'_{j1}h_1 u_{j1} = a_{jj}$, which can be solved uniquely for h_j. In particular, we have now determined all coefficients which belong to the first j rows and columns. To show that $h_j \neq 0$, we consider the subsystem of equations made up of these rows and columns; this subsystem looks like the original system written in matrix form above, with n replaced by j. The right-hand side has a non-zero determinant by assumption. Hence all the coefficients $h_1, \ldots, h_j$ must be non-zero. Thus, the assertion holds for $G = \mathrm{SL}_n(k)$. Next note that, given $g \in G$ such that $g = u'hu$, where $u \in U$, $u' \in U'$, and $h \in T$, then the factors u, u', and h are unique. Now use Example 1.7.3 to deduce the desired assertion for $\mathrm{SO}_{2m+1}(k)$ ($\mathrm{char}(k) \neq 2$), $\mathrm{Sp}_{2m}(k)$ (any characteristic), and $\mathrm{O}_{2m}(k) = \mathrm{O}_{2m}^+(k)$ ($\mathrm{char}(k) \neq 2$). It remains to consider the case where $G = \mathrm{O}_{2m}^+(k)$ and $\mathrm{char}(k) = 2$. We have $G \subseteq \tilde{G} = \mathrm{Sp}_{2m}(k)$. Let $g \in G$ be such that $\delta(g) \neq 0$. Then we can write $g = \tilde{u}'h\tilde{u}$, where $\tilde{u}'$ and $\tilde{u}$ are lower and upper unitriangular in $\tilde{G}$, respectively, and h is diagonal in $\tilde{G}$. Next note that $h \in G$ by Lemmas 1.5.9 and 1.5.11. Furthermore, check that $\tilde{u} = vu$, where $u \in G$ and

$$v = \left[\begin{array}{c|c} I_m & Q_m S \\ \hline 0 & I_m \end{array}\right], \quad \text{with } S \in M_m(k) \text{ diagonal.}$$

Similarly, we have $\tilde{u}' = u'v'$, where $u' \in G$. This yields $v'hv \in G$. Now check explicitly, using the defining equations for G, that this implies $v = v' = 1$.]

2

Affine varieties and finite morphisms

In this chapter, we come to those results on affine varieties and algebraic groups which rely in an essential way on the hypothesis that the ground field k is algebraically closed. We will see that, under this hypothesis, the image of a morphism between affine varieties is a relatively 'thick' subset, more precisely, it contains an open subset of its closure. This fact will turn out to be a key tool.

The basic algebraic notion which underlies these results is that of an integral extension of rings; see Section 2.1. As a first application, we obtain the general form of Hilbert's Nullstellensatz which will reveal the geometric meaning of many of the purely algebraic constructions in Chapter 1. Before we go on to study morphisms, we address another point. One disadvantage of the set-up in Chapter 1 is that it always refers to an embedding of an algebraic set in some affine space k^n. Instead, one would just like to work with an intrinsic description of an affine variety and the algebra of regular functions on it. In Section 2.1, following Steinberg, we will give such an abstract definition of affine varieties. Then, in Section 2.2, we prove Chevalley's theorem on the image of a morphism.

Section 2.3 is concerned with the problem of finding criteria which guarantee that a bijective morphism between affine varieties actually is an isomorphism. In general, this is a very difficult and subtle problem, especially over fields of positive characteristic. We shall prove some weak versions of 'Zariski's main theorem' which are sufficient for many applications to algebraic groups.

In Section 2.4, we come to first applications to algebraic groups. We show that a group generated by a family of closed irreducible subsets is itself irreducible. This is an important tool for proving the connectedness of a group. In Section 2.5, we establish some general results concerning algebraic group actions on affine varieties. One of the main results shows that there are always closed orbits for the action of an algebraic group. Finally, in Section 2.6, we illustrate these ideas by studying in detail the conjugation action of $\mathrm{SL}_n(k)$ on unipotent elements.

2.1 Hilbert's Nullstellensatz and abstract affine varieties

In the previous chapter, we have introduced various notions related to an algebraic set $V \subseteq k^n$: the dimension, regular maps, and the tangent space. However, we have also seen characterizations entirely in terms of the algebra $A[V]$: see Proposition 1.2.18 for the dimension, Proposition 1.3.4 for regular maps, and Remark 1.4.9(b) for the tangent space. All this suggests that there is a purely axiomatic description of algebraic sets in terms of algebras of functions. Our aim here is to give a precise sense to this idea. First we establish Hilbert's nullstellensatz, whose proof relies on some properties of integral ring extensions.

Let A and B be commutative rings (with 1) such that $A \subseteq B$. We say that $b \in B$ is *integral* over A, if there is an equation of 'integral dependence'

$$b^n + a_{n-1}b^{n-1} + \cdots + a_1 b + a_0 = 0, \quad \text{where } n \geqslant 1 \text{ and } a_i \in A \text{ for all } i.$$

We say that B is integral over A if every element of B is integral over A.

2.1.1 Lemma *Let $b \in B$. Then the following conditions are equivalent.*

(a) b is integral over A.

(b) The subring $A[b] \subseteq B$ generated by b is a finitely generated A-module.

(c) There exists a subring $C \subseteq B$ such that $A[b] \subseteq C$ and C is finitely generated as an A-module.

Proof '$(a) \Rightarrow (b)$': We have $A[b] = \{f(b) \mid f \in A[X]\}$, where X is an indeterminate. By assumption, there exists a monic polynomial $g := a_0 + a_1 X + \cdots + a_{n-1}X^{n-1} + X^n \in A[X]$ with $n \geqslant 1$ such that $g(b) = 0$. Then we can divide any $f \in A[X]$ by g, leaving a remainder r such that $f = qg + r$, where $q, r \in A[X]$ and $r = 0$ or $\deg(r) < n$. Consequently, $f(b) = q(b)g(b) + r(b) = r(b)$, and every element in $A[b]$ is an A-linear combination of $1, b, b^2, \ldots, b^{n-1}$.

'$(b) \Rightarrow (c)$': This is trivial (set $C := A[b]$).

'$(c) \Rightarrow (a)$': Let $c_1, \ldots, c_n \in C$ be such that $C = \sum_{i=1}^{n} Ac_i$. Since $b \in C$ and C is a ring, we have $bc_i \in C$ for all i. So there exist $a_{ij} \in A$ such that $bc_i = \sum_{j=1}^{n} a_{ij}c_j$ for all i. Setting $M = (a_{ij})_{1 \leqslant i,j \leqslant n} \in M_n(A)$, we can write these equations in matrix form: $Mv = bv$, where $v \in A^n$ is the column vector whose ith component is c_i. Thus, we have $(M - bI_n)v = 0$, where I_n is the $n \times n$ identity matrix. By Cramer's rule, this implies that $c_i \det(M - bI_n) = 0$ for all i. Now, we have $1 \in C$ and so there exist $a_i \in A$ such that $1 = \sum_{i=1}^{n} a_i c_i$. Multiplying this equation with $\det(M - bI_n)$, we obtain $\det(M - bI_n) = 0$. Expanding the determinant yields the required equation of integral dependence for b. $\qquad\square$

2.1.2 Corollary *Let $A \subseteq B$ be commutative rings.*

(a) If $B = A[b_1, \ldots, b_n]$, where each $b_i \in B$ is integral over $A[b_1, \ldots, b_{i-1}]$, then B is a finitely generated A-module and B is integral over A.

(b) The set $\overline{A}_B := \{b \in B \mid b \text{ is integral over } A\}$ is a subring of B and is called the integral closure of A in B.

(c) Let $C \subseteq B$ be a subring such that $A \subseteq C$. If C is integral over A, and B is integral over C, then B is integral over A.

(d) If B is a field, and B is integral over A, then A also is a field.

Proof (a) This immediately follows from Lemma 2.1.1, using induction on n.

(b) Let $b, b' \in B$ be integral over A, and consider the subring $A[b, b'] \subseteq B$. By (a), $A[b, b']$ is a finitely generated A-module. So every element in $A[b, b']$ is integral over A. In particular, $b \pm b'$ and bb' are integral over A.

(c) Let $b \in B$. Since b is integral over C, there exists an equation $b^m + c_{m-1}b^{m-1} + \cdots + c_1 b + c_0 = 0$, where $m \geqslant 1$ and $c_i \in C$. Using (a), we see that the subring $C' := A[b, c_0, c_1, \ldots, c_{m-1}] \subseteq B$ is a finitely generated A-module. Hence, by Lemma 2.1.1, $b \in C'$ is integral over A, as desired.

(d) Let $0 \neq a \in A$. Since B is a field, there exists some $b \in B$ such that $1 = ab$. Now B is integral over A, and so we have an equation $b^n + a_{n-1}b^{n-1} + \cdots + a_1 b + a_0 = 0$, where $n \geqslant 1$ and $a_i \in A$. Multiplying with a^{n-1} and using $ab = 1$, we obtain $b = -(a_{n-1} + a_{n-2}a + \cdots + a_0 a^{n-1}) \in A$, as required. $\qquad\square$

2.1.3 Example   We say that A is *integrally closed* in B if $\overline{A}_B = A$. For example, if A is a factorial domain with field of fractions K, then A is integrally closed in K. (We leave the proof as an exercise to the reader.) In particular, the polynomial ring $k[X_1, \ldots, X_n]$ is integrally closed in its field of fractions.

2.1.4 Theorem (Noether normalization)   *Let A be a finitely generated k-algebra.[1] Then there exist algebraically independent elements $a_1, \ldots, a_d \in A$ such that A is integral over the subring $k[a_1, \ldots, a_d]$.*

Proof   Let us write $A = k[a_1, \ldots, a_n]$ with $a_i \in A$. We now proceed by an induction on n. If $n = 0$, then $A = k$ and there is nothing to prove. Now assume that $n > 0$. If $a_1, \ldots, a_n$ are algebraically independent, we are done. So let us assume that $a_1, \ldots, a_n$ are not algebraically independent. Then there exists some non-constant $F \in k[X_1, \ldots, X_n]$ such that $F(a_1, \ldots, a_n) = 0$. Renumbering the variables if necessary, we may assume that some term of F involves the variable X_n. Writing $F = \sum_\alpha a_\alpha X^\alpha$, let r be an integer which is bigger than any component of any α such that $a_\alpha \neq 0$. Then, setting $N(\alpha) := \sum_{i=0}^{n-1} \alpha_{n-i} r^i$ for any $\alpha = (\alpha_1, \ldots, \alpha_n)$, we have $N(\alpha_1) \neq N(\alpha_2)$ for any $\alpha_1 \neq \alpha_2$ such that $a_{\alpha_1} \neq 0$ and $a_{\alpha_2} \neq 0$. (Check this!) We now proceed as follows. We set $r_i := r^{n-i}$ for $1 \leqslant i \leqslant n-1$, and define $Y_i := X_i - X_n^{r_i}$ for $1 \leqslant i \leqslant n-1$. We claim that, if we set $N = \max\{N(\alpha) \mid a_\alpha \neq 0\}$, then F has the form

$$F = \lambda X_n^N + \sum_{i=0}^{N-1} g_i X_n^i, \quad \text{where } 0 \neq \lambda \in k \text{ and } g_i \in k[Y_1, \ldots, Y_{n-1}].$$

This follows from the fact that all $N(\alpha)$ for $a_\alpha \neq 0$ are different and that

$$X^\alpha = X_1^{\alpha_1} \cdots X_n^{\alpha_n} = (Y_1 + X_n^{r_1})^{\alpha_1} \cdots (Y_{n-1} + X_n^{r_{n-1}})^{\alpha_{n-1}} X_n^{\alpha_n}$$

$$= X_n^{N(\alpha)} + \sum_{i=0}^{N(\alpha)-1} h_i X_n^i \quad \text{with } h_i \in k[Y_1, \ldots, Y_{n-1}].$$

Furthermore, note that N is the degree of F in X_n. Now set $y_i := a_i - a_n^{r_i}$ for $1 \leqslant i \leqslant n-1$, and consider the subring

---

[1] Recall that the term k-algebra means a commutative associative k-algebra with 1.

$R := k[y_1, \ldots, y_{n-1}] \subseteq A$. We define $g := F(y_1, \ldots, y_{n-1}, X_n) \in R[X_n]$. Then $g \neq 0$ and $g(a_n) = 0$. Dividing g by λ, we see that a_n is integral over R. Furthermore, each $a_i = y_i + a_n^{r_i}$ $(1 \leqslant i \leqslant n)$ is also integral over R and so A is integral over R by Corollary 2.1.2(a). By induction, we can find $r_1, \ldots, r_e \in R$ which are algebraically independent and such that R is integral over the subring $k[r_1, \ldots, r_e]$. Then A is integral over $k[r_1, \ldots, r_e]$ by Corollary 2.1.2(c). $\qquad\square$

2.1.5 Theorem (Hilbert's nullstellensatz, weak form)

Assume that k is algebraically closed. Then the maximal ideals in $k[X_1, \ldots, X_n]$ are precisely the ideals of the form $(X_1 - v_1, \ldots, X_n - v_n)$, where $v_i \in k$. More generally, if A is any finitely generated k-algebra, then $A/\mathfrak{m} \cong k$ for every maximal ideal $\mathfrak{m}$ in A.

Proof Since every finitely generated k-algebra is a quotient of a polynomial ring, it is enough to prove the statement about $k[X_1, \ldots, X_n]$. Now, first note that an ideal of the form $I = (X_1 - v_1, \ldots, X_n - v_n)$ is maximal, since $k[X_1, \ldots, X_n]/I \cong k$. Conversely, let $I \subseteq k[X_1, \ldots, X_n]$ be any maximal ideal, and consider $A := k[X_1, \ldots, X_n]/I$. Then A is a finitely generated k-algebra which is in fact a field. Let $R = k[a_1, \ldots, a_d] \subseteq A$ be a Noether normalization, as in Theorem 2.1.4. Now Corollary 2.1.2(d) implies that R must also be a field. This is only possible if $d = 0$, that is, we have $R = k$. Consequently, A is a finite algebraic extension of k. Since k is algebraically closed, we have $A = k$. Thus, for each i, there exists some $v_i \in k$ such that $X_i - v_i \in I$, and so $I = (X_1 - v_1, \ldots, X_n - v_n)$. $\qquad\square$

We are now ready to give the abstract definition of affine varieties.

2.1.6 Definition Let X be any (non-empty) set and k be a field. We consider the k-algebra $\mathrm{Maps}(X, k)$ of all maps $X \to k$, with pointwise defined operations. For any $x \in X$, we have the *evaluation map* ε_x which sends a function $f \colon X \to k$ to its value $f(x)$. Now let A be a k-subalgebra of $\mathrm{Maps}(X, k)$. Then the pair (X, A) is called an *affine variety* (over k) if the following conditions hold.

(a) A is finitely generated as an algebra over k, and we have $1 \in A$.

(b) For $x \neq y$ in X, there exists some $f \in A$ such that $f(x) \neq f(y)$.

(c) For every k-algebra homomorphism $\lambda\colon A \to k$, there exists some $x \in X$ such that $\lambda = \varepsilon_x$.

(By convention, we also consider $(\varnothing, \{0\})$ as an affine variety.) We call A the *algebra of regular functions* on X and also write $A = A[X]$ to indicate the relation with X. Note that (b), (c) mean that the assignment $x \mapsto \varepsilon_x$ is a bijection between X and the set of k-algebra homomorphisms $A \to k$. The condition (a) means that A is a quotient of a polynomial ring in finitely many indeterminates over k. Hence, by Hilbert's basis theorem 1.1.1, A is a noetherian ring.

If (X, A) and (Y, B) are affine varieties over k, then a map $\varphi\colon X \to Y$ is called a *morphism of affine varieties* if $g \circ \varphi \in A$ for all $g \in B$. In this case, we have an induced k-algebra homomorphism $\varphi^*\colon B \to A$, $g \mapsto g \circ \varphi$. As in §1.3.3 and Proposition 1.3.4, one checks that the assignment $\varphi \mapsto \varphi^*$ is functorial and sets up a bijection between morphisms $X \to Y$ and k-algebra homomorphisms $B \to A$. Given a homomorphism $\alpha\colon B \to A$, the unique morphism $\varphi\colon X \to Y$ such that $\varphi^* = \alpha$ is determined by the condition that

$$\varepsilon_x \circ \alpha = \varepsilon_{\varphi(x)} \qquad \text{for any } x \in X.$$

Indeed, by Definition 2.1.6(c), there exists some $y \in Y$ such that $\varepsilon_x \circ \alpha = \varepsilon_y$. By Definition 2.1.6(b), y is uniquely determined, and so we can set $\varphi(x) := y$. In this way, we obtain a map $\varphi\colon X \to Y$ satisfying the condition $\varphi^* = \alpha$.

2.1.7 The abstract Zariski topology   Let (X, A) be an affine variety over k. For any subset $S \subseteq A$, we define

$$\mathbf{V}_X(S) := \{x \in X \mid f(x) = 0 \text{ for all } f \in S\}.$$

By exactly the same argument as in §1.1.6, we see that the sets $\mathbf{V}_X(S)$ $(S \subseteq A)$ form the closed sets of a topology on X which is called the *Zariski* topology. Furthermore, for any subset $Y \subseteq X$, we define

$$\mathbf{I}_A(Y) := \{f \in A \mid f(y) = 0 \text{ for all } y \in Y\}.$$

Then $\mathbf{I}_A(Y)$ is an ideal in A, the *vanishing* ideal of Y. As in §1.1.7, we see that

$$\mathbf{V}_X(\mathbf{I}_A(Y)) = \bar{Y} = \text{closure of } Y \text{ in the Zariski topology on } X.$$

Thus, the operators $\mathbf{V}_X$ and $\mathbf{I}_A$ have the same formal properties as the operators $\mathbf{V}$ and $\mathbf{I}$ introduced in Section 1.1.

Arguing as in the proof of Proposition 1.1.12(a), we see that X is a noetherian topological space. Furthermore, as in Proposition 1.1.12(b), we see that X is irreducible if and only if A is an integral domain.

Let $\varphi\colon X \to Y$ be a morphism of affine varieties. Then the following hold.

(a) φ is continuous.

(b) If X is irreducible, then $\overline{\varphi(X)} \subseteq Y$ is also irreducible.

This follows by exactly the same arguments as in Remark 1.3.2.

2.1.8 Example Let $V \subseteq k^n$ be a non-empty algebraic set and $A[V]$ its affine algebra, regarded as the algebra of regular functions on V, as in §1.3.3. Then $(V, A[V])$ is an affine variety in the sense of Definition 2.1.6.

Indeed, we have $A[V] = k[X_1, \ldots, X_n]/\mathbf{I}(V)$, and this is a finitely generated k-algebra. Furthermore, for $v = (v_1, \ldots, v_n) \in V$, we have $\bar{X}_i(v) = v_i$ for $1 \leqslant i \leqslant n$. Thus, given $v \neq w$ in V, there exists some i such that $\bar{X}_i(v) \neq \bar{X}_i(w)$. It remains to show that every k-algebra homomorphism $\lambda\colon A[V] \to k$ is an evaluation map. To prove this, set $v_i := \lambda(\bar{X}_i)$ for $1 \leqslant i \leqslant n$ and $v := (v_1, \ldots, v_n) \in k^n$. We claim that $v \in V$ and $\lambda = \varepsilon_v$. Indeed, consider the canonical map $\pi\colon k[X_1, \ldots, X_n] \to A[V]$. Then $\lambda \circ \pi\colon k[X_1, \ldots, X_n] \to k$ is a k-algebra homomorphism whose kernel contains the ideal $I := (X_1 - v_1, \ldots, X_n - v_n)$. This ideal is maximal (since $k[X_1, \ldots, X_n]/I \cong k$) and so $\ker(\lambda \circ \pi) = I$. We have $\mathbf{I}(V) \subseteq I$ and so $\{v\} = \mathbf{V}(I) \subseteq \mathbf{V}(\mathbf{I}(V)) = V$; see §1.1.7. Finally, we have $\lambda(\bar{X}_i) = v_i = \bar{X}_i(v) = \varepsilon_v(\bar{X}_i)$ for all i and so $\lambda = \varepsilon_v$, as required.

To state the next result, we need one more definition. Given any ideal I in a commutative ring A, the set $\sqrt{I} := \{a \in I \mid a^m \in I \text{ for some } m \geqslant 1\}$ is called the *radical* of I. Note that $\sqrt{I}$ itself is an ideal: if $a, b \in \sqrt{I}$, we have $a^r \in I$ and $b^s \in I$ for some $r, s \geqslant 1$ and so $(a + b)^{r+s} = \sum_{i=0}^{r+s} \binom{r+s}{i} a^i b^{(r-i)+s} \in I$.

2.1.9 Theorem (Hilbert's nullstellensatz, strong form) *Let k be an algebraically closed field and (X, A) be an affine variety over k. For any ideal $I \subseteq A$ and any closed subset $Y \subseteq X$, we have*

$$\mathbf{I}_A(\mathbf{V}_X(I)) = \sqrt{I} \quad and \quad \mathbf{V}_X(\mathbf{I}_A(Y)) = Y.$$

Furthermore, the maximal ideals in A are the ideals $\ker(\varepsilon_p)$ for $p \in X$, where $\varepsilon_p \colon A \to k$ denotes the evaluation homomorphism.

Proof By the same proof as in §1.1.7, we have $\mathbf{V}_X(\mathbf{I}_A(Y)) = Y$ for any closed set $Y \subseteq X$. Now let $I \subseteq A$ be any ideal. Then we certainly have $\sqrt{I} \subseteq \mathbf{I}_A(\mathbf{V}_X(I))$. To prove the reverse inclusion, let $f \in A$ be such that $f \notin \sqrt{I}$. We shall show that there exists some $x \in \mathbf{V}_X(I)$ such that $f(x) \neq 0$. This is done as follows. Consider the localization A_f and the natural map $\iota \colon A \to A_f$ as defined in Exercise 1.8.11. Now let $\sqrt{I}_f$ be the ideal generated by $\iota(\sqrt{I})$ in A_f. Since $f \notin \sqrt{I}$, the image of f in $A/\sqrt{I}$ is not nilpotent and so $A_f/\sqrt{I}_f \neq \{0\}$ by Exercise 1.8.11(c). Now, since $A_f/\sqrt{I}_f$ is a finitely generated k-algebra, the quotient of $A_f/\sqrt{I}_f$ by any maximal ideal is isomorphic to k, by the weak form of Hilbert's nullstellensatz. Hence, there exists a k-algebra homomorphism $\lambda \colon A_f \to k$ with $\sqrt{I}_f \subseteq \ker(\lambda)$. By Definition 2.1.6(c), we have $\varepsilon_x = \lambda \circ \iota \colon A \to k$ for some $x \in X$. Since $I \subseteq \ker(\varepsilon_x)$, we have $x \in \mathbf{V}_X(\ker(\varepsilon_x)) \subseteq \mathbf{V}_X(I)$. Furthermore, since $\iota(f) \in A_f$ is invertible, we have $\varepsilon_x(f) = \lambda(\iota(f)) \neq 0$, as required.

Finally, consider the statement concerning the maximal ideals in A. Every ideal $\ker(\varepsilon_p)$ certainly is maximal. Conversely, if $\mathfrak{m} \subseteq A$ is a maximal ideal, then using Theorem 2.1.5, we see that $A/\mathfrak{m} \cong k$. Thus, $\mathfrak{m}$ is the kernel of an evaluation homomorphism by Definition 2.1.6(c). $\square$

By Example 2.1.8, the above result applies to all algebraic sets in k^n. Next we show that every abstract affine variety is isomorphic to an algebraic set.

2.1.10 Proposition *Let (X, A) be an affine variety over k, where k is algebraically closed. Since A is finitely generated, there exists some $n \geqslant 1$ and a surjective algebra homomorphism $\pi \colon k[T_1, \ldots, T_n] \to A$ (where T_i are indeterminates). Then the following hold:*

(a) Define $\varphi \colon X \to k^n$ by $\varphi(x) = (\pi(T_1)(x), \ldots, \pi(T_n)(x))$. Then φ is a morphism of affine varieties and $\varphi(X) = \mathbf{V}(\ker(\pi))$ is closed.

(b) The restricted map $X \to \varphi(X)$, $x \mapsto \varphi(x)$, is an isomorphism where the closed subset $\varphi(X) \subseteq k^n$ is regarded as an affine variety as in Example 2.1.8.

Proof   Set $I := \ker(\pi) \subseteq k[T_1, \ldots, T_n]$ and $V := \mathbf{V}(I) \subseteq k^n$.

First note that φ is a morphism of affine varieties such that $\varphi^* = \pi$. (Indeed, given $f \in k[T_1, \ldots, T_n]$, we have $f \circ \varphi = f(\pi(T_1), \ldots, \pi(T_n)) = \pi(f) \in A$.) Next note that I certainly is a radical ideal, that is we have $\sqrt{I} = I$. (Indeed, if $f^m \in \ker(\pi)$ for some $f \in k[T_1, \ldots, T_n]$ and some $m \geqslant 1$, then $\pi(f)^m = 0$ in A. But A is an algebra of k-valued functions and so $\pi(f) = 0$.) Thus, Theorem 2.1.9 shows that $I = \mathbf{I}(V)$. Consequently, we have $A[V] = k[T_1, \ldots, T_n]/I$ and so π induces a k-algebra isomorphism $\pi_1 \colon A[V] \to A$. Hence, by the remarks following Definition 2.1.6, there exists an isomorphism of affine varieties $\varphi_1 \colon X \to V$ such that $\varphi_1^* = \pi_1$. Let $\iota \colon V \hookrightarrow k^n$ be the inclusion. Then ι^* is the natural map $k[T_1, \ldots, T_n] \to A[V]$ and so $\varphi^* = \pi = \pi_1 \circ \iota^* = \varphi_1^* \circ \iota^* = (\iota \circ \varphi_1)^*$. This implies $\varphi = \iota \circ \varphi_1$. Thus, we see that φ_1 is the restricted map $X \to V$, $x \mapsto \varphi(x)$. $\square$

Thus, if k is algebraically closed, then everything that we proved for algebraic sets in Chapter 1 will remain valid for an abstract affine variety (X, A) as above.

2.1.11 Definition   Let (X, A) be an affine variety over k. We define the dimension of X by $\dim X = \partial_k(A)$. Furthermore, for $p \in X$, we define the tangent space by $T_p(X) = \mathrm{Der}_k(A, k_p)$. If X is irreducible, we say that $p \in X$ is *non-singular* if $\dim X = \dim T_p(X)$.

We shall now show that the usual constructions involving algebraic sets all have a natural interpretation in the above abstract setting.

2.1.12 Closed subvarieties   Let (X, A) be an affine variety and $Y \subseteq X$ be a closed subset. Then the algebra homomorphism $A \to \mathrm{Maps}(Y, k)$ given by restricting functions from X to Y has kernel $I := \mathbf{I}_A(Y)$. Thus, we may regard A/I as a k-algebra of functions on Y. We claim that

$$(Y, A/I) \text{ is an affine variety}$$

and the inclusion $\iota \colon Y \hookrightarrow X$ is a morphism, where ι^* is the natural map $A \to A/I$.

Indeed, $B := A/I$ certainly is finitely generated. Furthermore, condition (b) in Definition 2.1.6 is clearly satisfied. Finally, if $\lambda \colon B \to k$ is a k-algebra homomorphism, then we can lift λ to A and obtain an element $x \in X$ such that $\lambda(f|_B) = f(x)$ for all $f \in A$.

Let $S \subseteq A$ be such that $Y = \mathbf{V}_X(S)$. Then, for any $f \in S$, we have $f|_B = 0$ and so $f(x) = 0$, that is, $x \in Y$.

2.1.13 Direct products of affine varieties Let (X, A) and (Y, B) be affine varieties over k. Then we have a natural map $A \otimes_k B \to \mathrm{Maps}(X \times Y, k)$ given by sending $f \otimes g$ to the function $X \times Y \mapsto k$, $(x, y) \mapsto f(x)g(y)$. As in Proposition 1.3.8(b), one sees that this map is in fact injective. Thus, we may regard $A \otimes_k B$ as a finitely generated subalgebra of $\mathrm{Maps}(X \times Y, k)$. It is easily checked that the remaining conditions in Definition 2.1.6 are satisfied, and so

$$(X \times Y, A \otimes_k B) \text{ is an affine variety.}$$

2.1.14 Affine open subvarieties Let (X, A) be an affine variety and $0 \neq f \in A$. Consider the open set $X_f := \{x \in X \mid f(x) \neq 0\}$, and let A_f be the k-subalgebra of $\mathrm{Maps}(X_f, k)$ generated by $1/f$ (restricted to X_f) and the functions in A (restricted to X_f). Then A_f is finitely generated, and it is easily checked that the remaining conditions in Definition 2.1.6 are also satisfied. Thus,

$$(X_f, A_f) \text{ is an affine variety.}$$

Using an argument similar to that in the proof of Lemma 1.1.15, we see that $A_f = A[X]/(fX - 1)$ is the localization of A in f; see Exercise 1.8.11. The embedding $\iota\colon X_f \to X$ is a morphism of affine varieties, where the corresponding algebra homomorphism $\iota^*\colon A \to A_f$ is the natural map.

2.2 Finite morphisms and Chevalley's theorem

Whenever we consider affine varieties, we will assume from now on that the ground field k is algebraically closed. The purpose of this section is to introduce an important class of morphisms: the so-called finite morphisms. We shall see (1) that these morphisms have many special properties and (2) that many questions about arbitrary morphisms can be reduced to questions about finite morphisms.

We will usually denote the algebra of regular functions on an affine variety X by $A[X]$ and just write X instead of $(X, A[X])$.

We begin by establishing some general results on morphisms. For this purpose, we introduce the following notation. Let $\varphi\colon X \to Y$ be a morphism between affine varieties. We say that φ is *dominant* if $\overline{\varphi(X)} = Y$. We say that φ is a *closed embedding* if the image $\varphi(X) \subseteq Y$ is a closed subset and the restricted map $\varphi_1\colon X \to \varphi(X)$, $x \mapsto \varphi(x)$, is an isomorphism of affine varieties. Here, we regard $\varphi(X) \subseteq Y$ as an affine variety as in §2.1.12.

2.2.1 Proposition *Let $\varphi\colon X \to Y$ be a morphism between affine varieties. Then the following hold.*

(a) φ is dominant if and only if $\varphi^*\colon A[Y] \to A[X]$ is injective.

(b) φ is a closed embedding if and only if $\varphi^*\colon A[Y] \to A[X]$ is surjective.

Proof (a) Let $Z := \overline{\varphi(X)} \subseteq Y$. Assume first that $Z \subsetneq Y$. Since the operator $\mathbf{I}_Y$ is injective, there exists some non-zero $g \in \mathbf{I}_Y(Z)$. Then we have $\varphi^*(g)(x) = g(\varphi(x)) = 0$ for all $x \in X$. This means $g \in \ker(\varphi^*)$, and so φ^* is not injective. Conversely, if φ^* is not injective, there exists some non-zero $g \in A[Y]$ such that $g(\varphi(x)) = \varphi^*(g)(x) = 0$ for all $x \in X$, and so $g \in \mathbf{I}_Y(\varphi(X))$. Consequently, we have $\varphi(X) \subseteq \mathbf{V}_Y(\{g\})$ and, since the latter set is closed, $\overline{\varphi(X)} \subseteq \mathbf{V}_Y(\{g\})$. Since $g \neq 0$, this shows $\mathbf{V}_Y(\{g\}) \subsetneq Y$ and so $\overline{\varphi(X)} \subsetneq Y$.

(b) If φ is a closed embedding, then we have a factorization $\varphi = \iota \circ \varphi_1$, where ι is the inclusion $\varphi(X) \subseteq Y$, and so $\varphi^* = \varphi_1^* \circ \iota^*$. Since φ_1^* is an isomorphism and ι^* is surjective, we conclude that φ^* is surjective. Conversely, assume that φ is a morphism such that φ^* is surjective. First we claim that $\varphi(X)$ is a closed subset of Y; more precisely:

$$\varphi(X) = \mathbf{V}_Y(\ker(\varphi^*)) \subseteq Y.$$

To see this, let $x \in X$ and $g \in \ker(\varphi^*)$. Then $g(\varphi(x)) = \varphi^*(g)(x) = 0$ and so $\varphi(x) \in \mathbf{V}_Y(\ker(\varphi^*))$. Conversely, let $y \in Y$ be such that $g(y) = 0$ for all $g \in \ker(\varphi^*)$. Consider the evaluation map $\varepsilon_y\colon A[Y] \to k$. Since φ^* is surjective and $g(y) = 0$ for all $g \in \ker(\varphi^*)$, we have a well-defined k-algebra homomorphism $\lambda\colon A[X] \to k$ such that $\varepsilon_y = \lambda \circ \varphi^*$. By Definition 2.1.6(c), there exists some $x \in X$ such that $\lambda = \varepsilon_x$ and so

$$\varepsilon_y(g) = \varepsilon_x(\varphi^*(g)) = \varphi^*(g)(x) = g(\varphi(x)) = \varepsilon_{\varphi(x)}(g) \quad \text{for all } g \in A[Y].$$

Thus, we have $y = \varphi(x) \in \varphi(X)$, as desired.

As in the proof of Proposition 2.1.10, one sees that $\sqrt{\ker(\varphi^*)} = \ker(\varphi^*)$. Hence, Theorem 2.1.9 implies that $\mathbf{I}_{A[Y]}(\varphi(X)) = \ker(\varphi^*)$. We can now argue as in the proof of Proposition 2.1.10 to conclude that the restricted map $\varphi_1 \colon X \to \varphi(X)$ is an isomorphism of affine varieties, as required. $\qquad\square$

2.2.2 Definition Let $\varphi \colon X \to Y$ be a morphism of affine varieties. We say that φ is a *finite* morphism if $A[X]$ is integral over $\varphi^*(A[Y])$.

The simplest possible example is given as follows. Consider k as an affine variety, where $k[X]$ is the algebra of regular functions on k. Let $e \geqslant 1$ and consider the morphism $\varphi_e \colon k \to k$, $x \mapsto x^e$. Then X certainly is integral over $\varphi_e^*(k[X]) = k[X^e]$, and so φ_e is a finite morphism.

A general class of examples is given by the Noether normalization theorem. Let X be any affine variety over k and $k[a_1, \ldots, a_d] \subseteq A[X]$ be as in Theorem 2.1.4. Since $a_1, \ldots, a_d$ are algebraically independent, we may regard $k[a_1, \ldots, a_d]$ as the algebra of regular functions on k^d. Then the embedding $k[a_1, \ldots, a_d] \subseteq A[X]$ is the algebra homomorphism corresponding to a dominant finite morphism $\varphi \colon X \to k^d$.

2.2.3 Lemma *Let $\varphi \colon X \to Y$ be a finite morphism of affine varieties over k. Then there exists a constant $c > 0$ such that $|\varphi^{-1}(q)| < c$ for all $q \in Y$.*

Proof By Proposition 2.1.10, we may assume without loss of generality that $X \subseteq k^n$ and $Y \subseteq k^m$ are algebraic sets and that $A[X] = k[x_1, \ldots, x_n]$ and $A[Y] = k[y_1, \ldots, y_m]$, where x_i and y_j are the corresponding coordinate functions.

Let $q \in \varphi(X)$ and $p \in X$ be such that $\varphi(p) = q$. Then we have $p = (x_1(p), \ldots, x_n(p))$, and it is enough to show that the number of possible values for $x_i(p)$ $(1 \leqslant i \leqslant n)$ is bounded independently of q. Now let us fix some i. Since $A[X]$ is integral over $\varphi^*(A[Y])$, we can write

$$x_i^l + \varphi^*(b_{l-1})x_i^{l-1} + \cdots + \varphi^*(b_1)x_i + \varphi^*(b_0) = 0,$$

where $l \geqslant 1$ and $b_j \in A[Y]$.

Evaluating at p and using $\varphi^*(b_j)(p) = b_j(q)$, we see that $x_i(p)$ is a zero of $T^l + b_{l-1}(q)T^{l-1} + \cdots + b_1(q)T + b_0(q) \in k[T]$ (where T is an indeterminate). So there are only finitely many possibilities for

$x_i(p)$. Since this works for all i, we also obtain a global bound $c > 0$ for the number of preimages of q. $\qquad\square$

More precise results on finite morphisms will be a consequence of the following basic result, which is in fact a special case of the 'going-up theorem' of Cohen–Seidenberg; see, for example, Corollary 5.8 and Theorem 5.10 of Atiyah and Macdonald (1969).

2.2.4 Lemma ('Going up', weak form)   *Assume that $A \subseteq B$ are commutative rings with 1 such that B is finitely generated and integral over A. Let $\mathfrak{m} \subsetneq A$ be a maximal ideal. Then there exists a maximal ideal $\mathfrak{q} \subsetneq B$ such that $\mathfrak{q} \cap A = \mathfrak{m}$.*

Proof   By Corollary 2.1.2(a), we can write $B = Ab_1 + \cdots + Ab_n$, where $b_i \in B$. Now consider the ideal $B\mathfrak{m} \subseteq B$ generated by $\mathfrak{m}$. Assume, if possible, that $B\mathfrak{m} = B$. Then, by Nakayama's lemma (see Exercise 2.7.3), there exists some $f \in 1+\mathfrak{m}$ such that $fB = \{0\}$. But we have $1 \in B$ and so $f = 0$, that is, $1 \in \mathfrak{m}$, a contradiction. Thus, we must have $B\mathfrak{m} \subsetneq B$. So then there exists a maximal ideal $\mathfrak{q} \subsetneq B$ such that $B\mathfrak{m} \subseteq \mathfrak{q}$. Then we have $\mathfrak{m} \subseteq \mathfrak{q} \cap A \subsetneq A$. Since $\mathfrak{m}$ is maximal, we deduce that $\mathfrak{m} = \mathfrak{q} \cap A$, as required. $\qquad\square$

2.2.5 Proposition   *Let $\varphi \colon X \to Y$ be a finite morphism of affine varieties.*

(a) Let $X' \subseteq X$ and $Y' \subseteq Y$ be closed subsets such that $\varphi(X') \subseteq Y'$. Then the restricted morphism $\varphi' \colon X' \to Y'$ is also finite.

(b) The map φ is closed (that is, maps closed subsets to closed subsets). In particular, if φ is dominant then φ is surjective.

Proof   We set $A = A[X]$ and $B = A[Y]$.

(a) Let $I := \mathbf{I}_A(X') \subseteq A$ and $J := \mathbf{I}_B(Y')$. By Example 2.1.12, the algebras of regular functions on X' and Y' can be identified with A/I and B/J, respectively. Then $\varphi' \colon X' \to Y'$ corresponds to the induced algebra homomorphism $(\varphi')^* \colon B/J \to A/I$. Now, let $a \in A$ and consider an equation of integral dependence for a over $\varphi^*(B)$. Then we also get an equation of integral dependence for the image of a in A/I over $(\varphi')^*(B/J)$, as desired.

(b) First we show that $\varphi(X) \subseteq Y$ is closed. For $q \in Y$, consider the evaluation homomorphism $\varepsilon_q \colon B \to k$ and set $I_q := \ker(\varepsilon_q)$. We claim that

$$\varphi(X) = \{q \in Y \mid \ker(\varphi^*) \subseteq I_q\}.$$

Indeed, first let $p \in X$ and $q = \varphi(p) \in Y$. Then, for any $g \in \ker(\varphi^*)$, we have $g(q) = g(\varphi(p)) = \varphi^*(g)(p) = 0$ and so $\ker(\varphi^*) \subseteq I_q$. Conversely, let $q \in Y$ be such that $\ker(\varphi^*) \subseteq I_q$. Then $\varphi^*(I_q) \subseteq \varphi^*(B)$ is a maximal ideal. So, by Lemma 2.2.4 ('going-up'), there exists a maximal ideal $\mathfrak{m} \subseteq A$ such that $\mathfrak{m} \cap \varphi^*(B) = \varphi^*(I_q)$. By Theorem 2.1.9, we have $\mathfrak{m} = \ker(\varepsilon_p)$ for some $p \in X$. Then $\varepsilon_p \circ \varphi^* = \varepsilon_q$ and so $\varphi(p) = q$. Thus, the claim is proved. Now note that, for any $q \in Y$, we have $f(q) = \varepsilon_q(f) = 0$ for all $f \in \ker(\varphi^*)$ if and only if $\ker(\varphi^*) \subseteq I_q$. Thus, $\mathbf{V}_Y(\ker(\varphi^*)) = \{q \in Y \mid \ker(\varphi^*) \subseteq I_q\}$ is a closed set.

Now consider any closed subset $Z \subseteq X$. By (a), the restriction $\varphi|_Z \colon Z \to Y$ is also finite, and so $\varphi(Z) \subseteq Y$ is closed. $\qquad\square$

2.2.6 Lemma *Let R and S be finitely generated k-algebras, where k is any field. Assume that R and S are integral domains and that $R \subseteq S$. Let K be the field of fractions of S and $F \subseteq K$ be the field of fractions of R. Then K is a finitely generated field extension of F, and we set $d := \partial_k(K) - \partial_k(F) \geqslant 0$. Then there exists some non-zero $f \in R$ and $s_1, \ldots, s_d \in S$ such that*

(a) S_f is integral over $R_f[s_1, \ldots, s_d]$,
(b) $s_1, \ldots, s_d \in S$ are algebraically independent over F.

Here, we have $R_f = \{a/f^n \mid a \in R,\ n \geqslant 0\}$ as in Exercise 1.8.11, where the quotients are taken in the field of fractions of R.

Proof We consider the subring $S^* := \{s/r \in K \mid s \in S,\ 0 \neq r \in R\} \subseteq K$. We have $F \subseteq S^* \subseteq K$, and S^* is a finitely generated F-algebra (since S is finitely generated over k); furthermore, K is also the field of fractions of S^*. So, using Lemma 1.2.18, we have $\partial_F(S^*) = \partial_F(K) = \partial_k(K) - \partial_k(F) = d$.

Now, by the Noether normalization theorem, there exist $y_1, \ldots, y_d \in S^*$ which are algebraically independent over F and such that S^* is integral over $F[y_1, \ldots, y_d]$. We can write $y_i = s_i/r$ with $s_i \in S$ and $0 \neq r \in R$. Then $s_1, \ldots, s_d$ are also algebraically independent over F, and so these elements satisfy the condition in (b). We now have a subring $R[s_1, \ldots, s_d] \subseteq S$ and, for any non-zero $f \in R$, we also have $R_f[s_1, \ldots, s_d] \subseteq S_f$.

We claim that there exists some f such that (a) holds. To see this, consider any $s \in S$. Then s is integral over $F[y_1, \ldots, y_d]$, and so we have $s^m + p_{m-1}s^{m-1} + \cdots + p_1 s + p_0 = 0$ for some $m \geqslant 1$ and $p_i \in F[y_1, \ldots, y_d]$. Writing each coefficient of each p_i as a quotient with denominator in R and using the expressions $y_i = s_i/r$ as above, there exists some $0 \neq g \in R$ such that $gp_i \in R[s_1, \ldots, s_d]$ for all i. Then the above relation shows that s is integral over $R_g[s_1, \ldots, s_d]$. We apply this construction to a finite set of k-algebra generators of S. For each of these generators, we obtain a corresponding element g. Taking the product of all these elements g, we see that there exists some non-zero $f \in R$ such that every algebra generator of S is integral over $R_f[s_1, \ldots, s_d]$. Thus, by Corollary 2.1.2, every element in S is integral over $R_f[s_1, \ldots, s_d]$. Finally, this implies that also every element in S_f is integral over that subring. Indeed, let $s/f^n \in S_f$ and $s^m + p_{m-1}s^{m-1} + \cdots + p_1 s + p_0 = 0$ be an equation of integral dependence for s, where $m \geqslant 1$ and $p_i \in R_f[s_1, \ldots, s_d]$. Dividing this equation by f^{nm}, we obtain an equation of integral dependence for s/f^n over $R_f[s_1, \ldots, s_d]$. $\qquad\square$

We now translate the above result to the context of affine varieties. For this purpose, recall from §2.1.14 that, if (X, A) is an affine variety and $0 \neq f \in A$, then we have an affine open subvariety (X_f, A_f), where $X_f = \{x \in X \mid f(x) \neq 0\}$ and the embedding $X_f \subseteq X$ is a morphism.

2.2.7 Theorem   *Let $\varphi \colon X \to Y$ be a dominant morphism of irreducible affine varieties. In particular, $\varphi^* \colon A[Y] \to A[X]$ is injective and $d := \dim X - \dim Y \geqslant 0$. Then we have a factorization*

$$\varphi|_{X_{\varphi^*(g)}} \colon X_{\varphi^*(g)} \xrightarrow{\ \tilde{\varphi}\ } Y_g \times k^d \xrightarrow{\ \mathrm{pr}_1\ } Y_g \qquad \textit{for some } 0 \neq g \in A[Y],$$

where $\tilde{\varphi}$ is a finite dominant morphism and pr_1 is the first projection.

Proof   To simplify notation, let us write $A = A[X]$ and $B = A[Y]$, and denote $b^* = \varphi^*(b)$ for $b \in B$. Since X and Y are irreducible, A and B are integral domains and we have $B^* \subseteq A$. Let K be the field of fractions of A, and $F \subseteq K$ be the field of fractions of B^*. Since K is finitely generated over F and φ^* is injective, we have

$$\partial_k(K) - \partial_k(F) = \partial_k(A) - \partial_k(B) = \dim X - \dim Y = d.$$

We now apply Lemma 2.2.6 to the rings $R = B^*$ and $S = A$. So there exist elements $a_1, \ldots, a_d \in A$ and a non-zero element $g \in B$

such that

(a) A_{g^*} is integral over $B_{g^*}^*[a_1, \ldots, a_d]$,
(b) $a_1, \ldots, a_d \in A$ are algebraically independent over F.

Now consider the homomorphism $\varphi^* \colon B \to B^* \subseteq A$. We can canonically extend it to a ring homomorphism $\psi \colon B_g \to B_{g^*}^* \subseteq A_{g^*}$, by setting $\psi(b/g^m) = b^*/(g^*)^m$ for any $b \in B$ and $m \geqslant 0$. (Check this!) Now, denoting by $T_1, \ldots, T_d$ independent indeterminates over B_g, we can extend ψ to a unique ring homomorphism $\tilde{\psi} \colon B_g[T_1, \ldots, T_d] \to B_{g^*}^*[a_1, \ldots, a_d] \subseteq A_{g^*}$ such that $T_i \mapsto a_i$. Since $a_1, \ldots, a_d$ are algebraically independent, $\tilde{\psi}$ is an isomorphism onto its image.

Now B_g and A_{g^*} can be identified with the algebras of regular functions on Y_g and X_{g^*}, respectively; see §2.1.14. Furthermore, $B_g[T_1, \ldots, T_d]$ can be identified with the algebra of regular functions on $Y_g \times k^d$; see §2.1.13. Then $\tilde{\psi}$ is the algebra homomorphism corresponding to a dominant regular map $\tilde{\varphi} \colon X_{g^*} \to Y_g \times k^d$; furthermore, by (a), we have that $\tilde{\varphi}$ is a finite morphism. Finally, the embedding $B_g \subseteq B_g[T_1, \ldots, T_d]$ corresponds to the projection map $\mathrm{pr}_1 \colon Y_g \times k^d \to Y_g$, and the composition of $B_g \subseteq B_g[T_1, \ldots, T_d]$ with $\tilde{\psi}$ is nothing but the canonical extension of φ^* to the homomorphism $B_g \to A_{g^*}$, $b/g^m \mapsto b^*/(g^*)^m$. The latter homomorphism corresponds to the restriction of φ to X_{g^*}; see once more §2.1.14. Thus, we have a factorization of that restriction as desired. $\qquad\square$

2.2.8 Corollary *Let $\varphi \colon X \to Y$ be a dominant morphism of irreducible affine varieties. Then the image of any non-empty open subset $U \subseteq X$ contains a non-empty open subset of Y.*

Proof First we show that $\varphi(X)$ itself contains some non-empty open subset of Y. For this purpose, consider a factorization of φ as in Theorem 2.2.7. Since $\tilde{\varphi}$ is a finite morphism, it is surjective by Proposition 2.2.5. Hence, since pr_1 also is surjective, we have $\varphi(X_{\varphi^*(g)}) = Y_g$. Thus, $\varphi(X)$ contains the open set Y_g.

Now let $U \subseteq X$ be any non-empty open subset. Since, by §1.1.14, U is a finite union of affine open sets, it is enough to prove the assertion in the case where $U = X_f$ for some non-zero $f \in A[X]$. Then the inclusion $X_f \hookrightarrow X$ corresponds to the embedding $A[X] \hookrightarrow A[X]_f$. Consequently, the restriction $\varphi|_{X_f} \colon X_f \to Y$ is still a dominant morphism between irreducible affine varieties, and so $\varphi(X_f)$ contains a non-empty open subset of Y. $\qquad\square$

2.2.9 Corollary   *Let $\varphi\colon X \to Y$ be a dominant morphism of irreducible affine varieties. Then there exists a non-empty open subset $U \subseteq Y$ with $U \subseteq \varphi(X)$ and such that $\dim \varphi^{-1}(y) = \dim X - \dim Y$ for all $y \in U$.*

In particular, if $|\varphi^{-1}(\varphi(x))| < \infty$ for all $x \in X$, then $\dim X = \dim Y$.

Proof   We set $d := \dim X - \dim Y$. Consider a factorization of φ as in Theorem 2.2.7 and set $U := Y_g$. Now let $y \in Y_g$. Since $\varphi^{-1}(Y_g) = X_{\varphi^*(g)}$, we also have $\varphi^{-1}(y) = \tilde{\varphi}^{-1}(\{y\} \times k^d)$. Now $\{y\} \times k^d$ is irreducible of dimension d; see Example 1.2.16(a). Let $Z_1, \ldots, Z_m$ be the irreducible components of $\varphi^{-1}(y)$. Then each restriction $\tilde{\varphi}|_{Z_i}\colon Z_i \to \{y\} \times k^d$ is also finite by Proposition 2.2.5(a), and so we have $\dim Z_i \leqslant d$ (since $A(Z_i)$ is algebraic over the field of fractions of the image of $A[\{y\} \times k^d]$ in $A[Z_i]$). On the other hand, $\tilde{\varphi}(Z_i) \subseteq \{y\} \times k^d$ is closed and irreducible for all i. Thus, since $\{y\} \times k^d$ is irreducible, there exists some i_0 such $\tilde{\varphi}(Z_{i_0}) = \{y\} \times k^d$, and so we have $\dim Z_{i_0} = d$. $\square$

2.2.10 Example   Consider the regular map $\varphi\colon k^2 \to k^2$, $(x, y) \mapsto (xy, y)$, introduced in Example 1.3.5(a). We have already seen that $\varphi(k^2) = \{(0,0)\} \cup \{(x,y) \in k^2 \mid y \neq 0\}$. Thus, $\varphi(k^2)$ is dense in k^2, and so φ is dominant but not surjective. In particular, φ cannot be a finite morphism; see Proposition 2.2.5. We have $\varphi^{-1}(\{0,0\}) = \{(x,0) \mid x \in k\}$, and this is 1-dimensional. On the other hand, for any $(x,y) \in k^2$ with $y \neq 0$, we have $\varphi^{-1}(x,y) = \{(x/y, y)\}$, and this is 0-dimensional. Furthermore, the set of all $(x,y) \in k^2$ with $y \neq 0$ is open in k^2.

To state the next result, we need the following definition. Let Z be any topological space. A subset $Y \subseteq Z$ is called *locally closed* if Y is the intersection of an open subset and a closed subset of Z. (For example, all open subsets are locally closed, and so are all closed subsets.) We say that $Y \subseteq Z$ is *constructible* if Y is a finite union of locally closed subsets. For some basic properties of constructible subsets, see Exercise 2.7.6.

2.2.11 Theorem (Chevalley)   *Let X and Y be affine varieties (not necessarily irreducible) and $\varphi\colon X \to Y$ be a morphism (not necessarily dominant). Then the image of any constructible subset of X is a constructible subset of Y.*

Proof We proceed in two steps.

Step 1. First assume that X is irreducible. Then we claim that $\varphi(X) \subseteq Y$ is constructible. We prove this by induction on $\dim Y$. If $\dim Y = 0$, then Y is a finite set and so every subset of Y is constructible. Now assume that $\dim Y > 0$. Since X is irreducible, $Y' = \overline{\varphi(X)}$ is a closed irreducible subset of Y and we may regard φ as a dominant morphism $\varphi \colon X \to Y'$. Furthermore, if $\varphi(X)$ is constructible in Y' then it is also constructible in Y. Thus, we may assume without loss of generality that $Y' = Y$. Applying Corollary 2.2.8, we obtain a non-empty open subset $U \subseteq Y$ such that $U \subseteq \varphi(X)$. Let Z' be an irreducible component of $Y \setminus U$ and X' be an irreducible component of $\varphi^{-1}(Z')$. Restricting φ to X', we obtain a morphism $\varphi' \colon X' \to Z'$. Since $U \neq \varnothing$, we have $\dim Z' < \dim Y$ by Proposition 1.2.20. So, by induction, $\varphi(X')$ is constructible in Z' and, hence, also in Y. We conclude that $U' := \varphi(\varphi^{-1}(Y \setminus U))$ is constructible in Y. Consequently, $\varphi(X) = U \cup U'$ is constructible in Y, as required.

Step 2. Now let X be arbitrary and $Z \subseteq X$ be any constructible subset. By definition, we can write $Z = \bigcup_{i=1}^{r}(O_i \cap C_i)$, where $O_i \subseteq X$ is open and $C_i \subseteq X$ is closed. Now each closed set in X can be written as a finite union of irreducible closed sets, and each open set in X can be written as a finite union of principal open sets; see §2.1.14. Thus, we may assume without loss of generality that each C_i is irreducible and each O_i is a principal open set. Then we have $\varphi(Z) = \bigcup_i \varphi(O_i \cap C_i)$, and so it is enough to consider the restriction of φ to $O_i \cap C_i$. Thus, we may assume without loss of generality that X is irreducible and $Z \subseteq X$ is a principal open subset, that is, we have $Z = X_f$ for some non-zero $f \in A[X]$. Since X_f also is an irreducible affine variety and the inclusion $X_f \subseteq X$ is a morphism, we can apply Step 1 to the restriction $\varphi|_{X_f} \colon X_f \to Y$ and conclude that its image is constructible, as desired. $\qquad\square$

In the case of homomorphisms between algebraic groups, the above results take a particularly simple form. Let us first give an 'abstract' definition of algebraic groups, in the spirit of Definition 2.1.6.

2.2.12 Affine algebraic groups Let (G, A) be an affine variety and $\mu \colon G \times G \to G$ be a morphism. We say that G is an *affine algebraic monoid* if the conditions in Definition 1.3.9 are satisfied, that is, μ is associative and there exists an identity element

$1 \in G$. If, moreover, every element $g \in G$ has an inverse and the map $\iota \colon G \to G$, $g \mapsto g^{-1}$, is a morphism, then G is called an *affine algebraic group*. A homomorphism of affine algebraic groups is a group homomorphism which is also a morphism of the underlying affine varieties.

By Example 2.1.8 and the characterization of morphisms in Proposition 1.3.4, it is clear that every linear algebraic group in the sense of Definition 1.3.9 also is an affine algebraic group in the above abstract sense. (The converse will be proved in Corollary 2.4.4.) We remark that the statement of Proposition 1.3.13 remains true, with the same proof, word by word. Thus, given an affine algebraic group G, there is a unique irreducible component G° containing the identity element of G. Furthermore, G° is a closed normal subgroup whose cosets are the irreducible components of G. In particular, we have $\dim G = \dim G^\circ$.

2.2.13 Lemma *Let G be an affine algebraic group. If $H \subseteq G$ is a subgroup (not necessarily closed), then $\overline{H} \subseteq G$ also is a subgroup. If H is any constructible subset of G such that $\overline{H}$ is a subgroup, then $\overline{H} = H \cdot H$. In particular, if H is a subgroup which is a constructible subset of G, then H is closed.*

Proof Let $H \subseteq G$ be a subgroup. Since inversion is a homeomorphism (with respect to the Zariski topology), we have $\overline{H}^{-1} = \overline{H^{-1}} = \overline{H}$. Similarly, left multiplication by any $x \in H$ is a homeomorphism, and so $x\overline{H} = \overline{xH} = \overline{H}$, that is, we have $H \cdot \overline{H} \subseteq \overline{H}$. Consequently, if $x \in \overline{H}$, we have $Hx \subseteq \overline{H}$ and so $\overline{H}x = \overline{Hx} \subseteq \overline{H}$. Thus, we see that $\overline{H}$ is a subgroup.

Now, if H is any constructible subset, there exists some dense open set $U \subseteq \overline{H}$ such that $U \subseteq H$. Now assume also that $\overline{H}$ is a subgroup. Since inversion is a homeomorphism, U^{-1} is also dense and open in $\overline{H}$. Since left multiplication by any $h \in \overline{H}$ is a homeomorphism, hU^{-1} is also dense and open in $\overline{H}$. Thus, we must have $U \cap hU^{-1} \neq \varnothing$ and so $h \in U \cdot U$. Hence, $\overline{H} = U \cdot U \subseteq H \cdot H = H$. $\square$

2.2.14 Proposition *Let $\varphi \colon \tilde{G} \to G$ be a homomorphism of affine algebraic groups.*

(a) *The kernel $\ker(\varphi) \subseteq \tilde{G}$ and the image $\varphi(\tilde{G}) \subseteq G$ are closed subgroups. We have $\dim \tilde{G} = \dim \varphi(\tilde{G}) + \dim \ker(\varphi)$.*
(b) *We have $\varphi(\tilde{G}^\circ) = \varphi(\tilde{G})^\circ$.*

Proof First note that $\ker(\varphi) = \varphi^{-1}(1)$ is a closed normal subgroup and that $\varphi(\tilde{G})$ is a subgroup. The fact that $\varphi(\tilde{G})$ is closed follows from Lemma 2.2.13 and Chevalley's Theorem 2.2.11. This proves (a) except for the statement concerning the dimensions. Next, consider (b). By Remark 1.3.2, we have $\varphi(\tilde{G}^\circ) \subseteq \varphi(\tilde{G})^\circ$. On the other hand, by the part of (a) already shown, $\varphi(\tilde{G}^\circ)$ is a closed subgroup of finite index in $\varphi(\tilde{G})$ and so it contains $\varphi(\tilde{G})^\circ$.

It remains to prove the statement in (a) concerning the dimensions. For this purpose, we may now assume that φ is surjective. Then, by (b), we also have $\varphi(\tilde{G}^\circ) = G^\circ$. Denote by $\varphi^\circ \colon \tilde{G}^\circ \to G^\circ$ the restriction of φ. Then φ° is a surjective morphism between irreducible affine varieties. The preimage of any $g \in G^\circ$ is a coset of $\ker(\varphi^\circ)$. Thus, all these preimages have the same dimension and so $\dim \tilde{G}^\circ = \dim \ker(\varphi^\circ) + \dim G^\circ$, by Corollary 2.2.9. But all irreducible components of G (or of $\tilde{G}$) have the same dimension, and so $\dim G = \dim G^\circ$ and $\dim \tilde{G} = \dim \tilde{G}^\circ$. Finally, $\ker(\varphi^\circ) = \ker(\varphi) \cap \tilde{G}^\circ$ is a closed subgroup of finite index in $\ker(\varphi)$ and so contains $\ker(\varphi)^\circ$. Hence $\dim \ker(\varphi^\circ) = \dim \ker(\varphi)$. $\qquad\square$

2.3 Birational equivalences and normal varieties

In this section, we study *bijective* morphisms between irreducible affine varieties. We have already encountered examples which show that such a morphism need not be an isomorphism. We would like to know: In addition to being bijective, what is required for a morphism to be an isomorphism? We shall see that there is a satisfactory answer to this question. This will involve two basic notions: birational equivalences and normal varieties. Again, finite morphisms will play a key role. Let us first consider the following example.

2.3.1 Example Let $C := \{(x, y) \in k^2 \mid x^3 = y^2\}$. Then $C \subseteq k^2$ is irreducible and we have a bijective morphism $\varphi \colon k \to C, t \mapsto (t^2, t^3)$. (Check this!) We have $\varphi^*(A[C]) = k[T^2, T^3] \subsetneqq k[T]$, and so φ is not an isomorphism.

On the other hand, note the following three facts: (1) The field of fractions of $k[T^2, T^3]$ certainly equals $k(T)$. Thus φ at least induces an isomorphism on the level of fields of fractions. (2) It is not difficult to check that $\mathbf{I}(C) = (X^3 - Y^2) \subseteq k[X, Y]$. This yields $T_{(0,0)}(C) = k^2$ and $T_{(x,y)}(C) \cong k$ for all $(x, y) \in C \setminus \{(0, 0)\}$. Thus,

$(0,0)$ is the unique singular point of C. (3) The variable T is integral over $\varphi^*(A[C])$ and so $A[C]$ is not integrally closed in its field of fractions. We shall see that the failure of being an isomorphism is related to facts (2) and (3).

2.3.2 Definition Let X and Y be irreducible affine varieties and $\varphi\colon X \to Y$ be a dominant morphism. Since $A[X]$ and $A[Y]$ are integral domains, we have corresponding fields of fractions denoted by $A(X)$ and $A(Y)$, respectively. Then φ^* induces a k-algebra homomorphism $A(Y) \to A(X)$ which we denote by the same symbol. We say that φ is a *birational equivalence* if φ^* induces a field isomorphism $A(Y) \cong A(X)$.

Clearly, if $\varphi\colon X \to Y$ is an isomorphism, then φ is a birational equivalence. The above example shows that the converse need not be true. In Theorem 2.3.10 below, we shall establish a differential criterion for birational equivalence.

2.3.3 Definition Let X be an irreducible affine variety. We say that X is *normal* if $A[X]$ is integrally closed in its field of fractions.

In Proposition 2.3.11, we will see that there always exists an affine open subvariety in X which is normal. Furthermore, in Example 2.3.14 below, we will see that all connected affine algebraic groups are normal.

2.3.4 Remark Let us give an illustration of how the above notions can be combined to yield a criterion for a morphism to be an isomorphism. So, let $\varphi\colon X \to Y$ be a morphism between irreducible affine varieties and assume that

   (a) φ is a finite morphism;
   (b) φ is a birational equivalence;
   (c) Y is normal.

Then φ is an isomorphism. Indeed, by (a), $\varphi^*(A[Y]) \subseteq A[X]$ is an integral extension. By (b), $\varphi^*(A[Y])$ and $A[X]$ have the same field of fractions. Hence, by (c), we must have $\varphi^*(A[Y]) = A[X]$. Thus, φ is an isomorphism as claimed.

This is the simplest possible case of a much deeper result: *Zariski's main theorem*. In one formulation, this asserts that condition (a) can be replaced by the substantially weaker condition that φ is surjective and all fibres of φ are finite—but we will not prove this here. See §5 of Dieudonné(1974), §5.2 of Springer (1998), and §III.9 of Mumford (1988), for further details.

In any case, as far as applications are concerned, the trouble is that the assumptions (b) and (c) are rather strong and may be difficult to verify. Our next aim is to develop critera for checking these conditions. This will involve some properties of the local ring of an affine variety at a point.

2.3.5 The local ring of a point　Let X be an irreducible affine variety over k, and fix a point $p \in X$. We consider the evaluation homomorphism

$$\varepsilon_p \colon A[X] \to k, \quad f \mapsto f(p).$$

Since X is irreducible, $A[X]$ is an integral domain. Let $A(X)$ be the field of fractions of $A[X]$. Then it is easily checked that

$$\mathcal{O}_p := \{x \in A(X) \mid x = a/b \text{ with } a, b \in A[X] \text{ and } \varepsilon_p(b) \neq 0\} \subseteq A(X)$$

is a subring of $A(X)$ containing $A[X]$. We can extend ε_p canonically to a ring homomorphism $\tilde{\varepsilon}_p \colon \mathcal{O}_p \to k$ by setting $\tilde{\varepsilon}_p(x) := \varepsilon_p(a)/\varepsilon_p(b)$ for $x = a/b \in \mathcal{O}_p$. The ring $\mathcal{O}_p$ is local, that is, it has a unique maximal ideal which is given by

$$\mathfrak{m}_p := \ker(\tilde{\varepsilon}_p) \subset \mathcal{O}_p.$$

(Indeed, if $x \in \mathcal{O}_p$ is such that $\tilde{\varepsilon}_p(x) \neq 0$, then, writing $x = a/b$, we see that $\varepsilon_p(a) \neq 0$ and $\varepsilon_p(b) \neq 0$ and so $x^{-1} \in \mathcal{O}_p$. Thus, every non-invertible element of $\mathcal{O}_p$ lies in $\mathfrak{m}_p$.) We call $\mathcal{O}_p$ the *local ring of the point* $p \in X$.

2.3.6 Lemma　*In the above set-up, the following hold.*

(a) We have $A[X] = \bigcap_{p \in X} \mathcal{O}_p$, where the intersection is taken inside $A(X)$.

(b) The ring $\mathcal{O}_p$ is noetherian.

Proof　(a) Let $f \in \bigcap_{p \in X} \mathcal{O}_p$ and consider the ideal $I = \{a \in A[X] \mid af \in A[X]\}$. Suppose that $I \neq A[X]$. Then there exists some $p \in X$ such that $I \subseteq \ker(\varepsilon_p)$. On the other hand, since $f \in \mathcal{O}_p$, there exists some $b \in I$ such that $\varepsilon_p(b) \neq 0$, a contradiction. So we have $1 \in I$ and hence $f \in A[X]$.

(b) Let $I \subseteq \mathcal{O}_p$ be an ideal and set $J := I \cap A[X]$. Assume that J is generated by $a_1, \ldots, a_d \in A[X]$ as an ideal in $A[X]$. Let $f \in I$ and $g \in A[X]$ be such that $g(p) \neq 0$ and $gf \in A[X]$. Then $gf \in A[X] \cap I = J$ and so $gf = \sum_{i=1}^{r} r_i a_i$ with $r_i \in A[X]$. Consequently, I is generated by $a_1, \ldots, a_d$ as an ideal in $\mathcal{O}_p$. $\qquad\square$

2.3.7 Lemma *Let $\varphi\colon X \to Y$ be a surjective morphism of irreducible affine varieties such that $\varphi^*(\mathcal{O}_{\varphi(p)}) = \mathcal{O}_p$ for all $p \in X$. Then φ is an isomorphism.*

Proof Since φ is dominant, the map $\varphi^*\colon A[Y] \to A[X]$ is injective. We must show that it is also surjective. So let $f \in A[X]$ and consider the ideal

$$I = \{g \in A[Y] \mid \varphi^*(g)f \in \varphi^*(A[Y])\}.$$

Assume, if possible, that $I \neq A[Y]$. Then there exists some $q \in Y$ such that $I \subseteq \ker(\varepsilon_q)$. Since φ is surjective, there exists some $p \in X$ such that $\varphi(p) = q$. Now, by assumption, we have $f \in \mathcal{O}_p = \varphi^*(\mathcal{O}_q)$. So there exist some $g, h \in A[Y]$ such that $h(q) \neq 0$ and $\varphi^*(g/h) = f$. This implies that $\varphi^*(h)f \in \varphi^*(A[Y])$ and so $h \in I \subseteq \ker(\varepsilon_q)$, a contradiction. Thus, $I = A[Y]$ and so $f \in \varphi^*(A[Y])$. $\qquad\square$

Now we can give a new characterization of tangent spaces and non-singular points, as promised in Section 1.4. In the following statement, note that $\mathfrak{m}_p/\mathfrak{m}_p^2$ is naturally a vector space over k, via the identification $k = \mathcal{O}_p/\mathfrak{m}_p$. (See also Exercise 2.7.4.)

2.3.8 Proposition *Let X be an irreducible affine variety and $p \in X$.*

(a) There is a well-defined k-linear isomorphism

$$\delta_p\colon \operatorname{Hom}_k(\mathfrak{m}_p/\mathfrak{m}_p^2, k) \to \operatorname{Der}_k(A[X], k_p), \quad \mu \mapsto D_\mu,$$

such that $D_\mu(a) = \mu(a + \mathfrak{m}_p^2)$ for all $a \in A[X]$ with $a(p) = 0$.

(b) $p \in X$ is non-singular if and only if $\mathfrak{m}_p$ is generated by $\dim X$ elements.

Proof (a) Let $\mu\colon \mathfrak{m}_p/\mathfrak{m}_p^2 \to k$ be k-linear and define a map $D_\mu\colon A[X] \to k$ by $D_\mu(a) = \mu((a - a(p)) + \mathfrak{m}_p^2)$ for $a \in A[X]$. We check that $D_\mu \in \operatorname{Der}_k(A[X], k_p)$. We have $(a - a(p))(b - b(p)) \in \mathfrak{m}_p^2$

and so

$$ab - a(p)b(p) \equiv a(p)(b - b(p)) + b(p)(a - a(p)) \bmod \mathfrak{m}_p^2.$$

Applying μ to this equation yields that D_μ is a derivation as required. Thus, we obtain a k-linear map $\delta_p \colon \operatorname{Hom}_k(\mathfrak{m}_p/\mathfrak{m}_p^2, k) \to \operatorname{Der}_k(A[X], k_p)$, where $\delta_p(\mu) = D_\mu$. Since μ is k-linear, the map D_μ is uniquely determined by its values on functions $a \in A[X]$ such that $a(p) = 0$. To show that δ_p is an isomorphism, we construct an inverse map. To do this, first note that we can extend uniquely any $D \in \operatorname{Der}_k(A[X], k_p)$ to a derivation $\tilde{D} \in \operatorname{Der}_k(\mathcal{O}_p, k_p)$, by setting

$$\tilde{D}(f) = \frac{1}{b(p)^2}(a(p)D(b) - b(p)D(a)) \qquad \text{for } f = a/b \in \mathcal{O}_p.$$

(This is just analogous to the rule for the derivative of the quotient of two functions.) Then we have $\tilde{D}(ab) = a(p)\tilde{D}(b) + b(p)\tilde{D}(a) = 0$ for all $a, b \in \mathfrak{m}_p$, and so D induces a k-linear function $\lambda_D \colon \mathfrak{m}_p/\mathfrak{m}_p^2 \to k$, $a + \mathfrak{m}_p^2 \mapsto \tilde{D}(a)$. Thus, we obtain a k-linear map

$$\lambda_p \colon \operatorname{Der}_k(A[X], k_p) \to \operatorname{Hom}_k(\mathfrak{m}_p/\mathfrak{m}_p^2, k), \quad D \mapsto \lambda_D.$$

It is readily checked that λ_p and δ_p are inverse to each other.

(b) By Remark 1.4.9, we can identify $T_p(X) = \operatorname{Der}_k(A[X], k_p)$. Thus, by (a), we have $\dim_k T_p(X) = \dim_k(\mathfrak{m}_p/\mathfrak{m}_p^2)$. Hence the assertion follows by combining this with Exercise 2.7.4 and Theorem 1.4.11. $\square$

2.3.9 Remark The isomorphism in Proposition 2.3.8(a) is functorial, in the following sense. Let $\varphi \colon X \to Y$ be a dominant morphism between irreducible affine varieties. Let $p \in X$ and $q := \varphi(p) \in Y$. Then the corresponding map $\varphi^* \colon A[Y] \to A[X]$ induces a homomorphism $A(Y) \to A(X)$ which we denote by the same symbol. With this convention, we can apply φ^* to $\mathcal{O}_q \subset A(Y)$ and obtain a subring $\varphi^*(\mathcal{O}_q) \subseteq A(X)$. It is readily checked that

$$\varphi^*(\mathcal{O}_q) \subseteq \mathcal{O}_p \subseteq A(X).$$

Consequently, we also have $\varphi^*(\mathfrak{m}_q) \subseteq \mathfrak{m}_p$, and so φ induces a k-linear map

$$\bar{\varphi}^* \colon \mathfrak{m}_q/\mathfrak{m}_q^2 \to \mathfrak{m}_p/\mathfrak{m}_p^2, \qquad a + \mathfrak{m}_q^2 \mapsto \varphi^*(a) + \mathfrak{m}_p^2.$$

Then we get an induced k-linear map on the dual spaces, and this makes the following diagram commutative:

$$
\begin{array}{ccc}
\mathrm{Hom}_k(\mathfrak{m}_p/\mathfrak{m}_p^2, k) & \xrightarrow{\ \mu \,\mapsto\, \mu \circ \bar{\varphi}^*\ } & \mathrm{Hom}_k(\mathfrak{m}_q/\mathfrak{m}_q^2, k) \\[4pt]
\Big\downarrow{\scriptstyle \delta_p} & & \Big\downarrow{\scriptstyle \delta_q.} \\[4pt]
\mathrm{Der}_k(A[X], k_p) & \xrightarrow{\ \ \ d_p\varphi\ \ \ } & \mathrm{Der}_k(A[Y], k_q).
\end{array}
$$

With the above preparations at hand, we can now establish the following basic result which provides a criterion for checking condition (b) in Remark 2.3.4.

2.3.10 Theorem (differential criterion for birational equivalence) *Let X and Y be irreducible affine varieties. Assume that $\varphi \colon X \to Y$ is a dominant injective morphism such that*

$$d_p\varphi \colon T_p(X) \to T_{\varphi(p)}(Y) \text{ is surjective for some non-singular } p \in X.$$
$$(*)$$

Then φ is a birational equivalence. More precisely, there exists some $x \in X$ such that $\varphi^*(\mathcal{O}_{\varphi(x)}) = \mathcal{O}_x$, where φ^* also denotes the induced map $A(Y) \to A(X)$.

Proof First note that, since φ is injective, we have $\dim X = \dim Y$; see Corollary 2.2.9. Hence, $(*)$ implies that $\dim T_{\varphi(p)}(Y) \leqslant \dim T_p(X) = \dim X = \dim Y$. Thus, by Theorem 1.4.11, $\varphi(p) \in Y$ is non-singular and so $d_p\varphi$ is an isomorphism. We now proceed in two steps.

Step 1. First we prove the desired assertion under the additional assumption that φ is a finite morphism. As in Remark 2.3.9, φ^* induces a homomorphism $A(Y) \to A(X)$ which we denote by the same symbol. We want to show that this induced map is an isomorphism. For this purpose, consider the local rings $\mathcal{O}_p \subset A(X)$ and $\mathcal{O}_{\varphi(p)} \subset A(Y)$. As in Remark 2.3.9, we have $\varphi^*(\mathcal{O}_{\varphi(p)}) \subseteq \mathcal{O}_p$. Then it will be enough to prove that we have in fact

$$\varphi^*(\mathcal{O}_q) = \mathcal{O}_p, \qquad \text{where } q := \varphi(p).$$
$$(\dagger)$$

To prove this, we consider $\mathcal{O}_p$ as a $\varphi^*(\mathcal{O}_q)$-module. By the discussion in Remark 2.3.9, we conclude that the k-linear map

$$\bar{\varphi}^* \colon \mathfrak{m}_q/\mathfrak{m}_q^2 \to \mathfrak{m}_p/\mathfrak{m}_p^2, \qquad a + \mathfrak{m}_q^2 \mapsto \varphi^*(a) + \mathfrak{m}_p^2,$$

is an isomorphism. Thus, if $\mathfrak{m}_q$ is generated by $a_1, \ldots, a_d$, then $\mathfrak{m}_p/\mathfrak{m}_p^2$ is spanned by the images of $\varphi^*(a_1), \ldots, \varphi^*(a_d)$. So Exercise 2.7.4 shows that $\mathfrak{m}_p$ is generated by $\varphi^*(a_1), \ldots, \varphi^*(a_d)$, that is, we have $\mathfrak{m}_p = \varphi^*(\mathfrak{m}_q)\mathcal{O}_p$. Since, clearly, we have $\mathcal{O}_p = \mathfrak{m}_p + \varphi^*(\mathcal{O}_q)$ (note that $1 \notin \mathfrak{m}_p$), we can now deduce that

$$\mathcal{O}_p = \varphi^*(\mathfrak{m}_q)\mathcal{O}_p + \varphi^*(\mathcal{O}_q).$$

Hence (†) would follow if we could apply *Nakayama's lemma* (see Exercise 2.7.3). Now, by Lemma 2.3.6, $\varphi^*(\mathcal{O}_q)$ is noetherian. So it will be enough to show that $\mathcal{O}_p$ is contained in a submodule of a finitely generated $\varphi^*(\mathcal{O}_q)$-module (see Exercise 1.8.1). To prove this, let $x \in \mathcal{O}_p$ and write $x = a/b$, where $a, b \in A[X]$ and $b(p) \neq 0$. Then consider the set $Y' := \varphi(\mathbf{V}_X(\{b\})) \subseteq Y$. Since φ is a finite morphism, Y' is a closed subset of Y. Furthermore, we necessarily have $q \notin Y'$. (Indeed, if $q \in Y'$, then $q = \varphi(v)$ for some $v \in X$ such that $b(v) = 0$. Since φ is injective, this would imply $p = v$ and so $b(p) = 0$, a contradiction.) So there exists some $g \in A[Y]$ such that $g(q) \neq 0$ and $g(y) = 0$ for all $y \in Y'$. Then $\varphi^*(g)(x) = g(\varphi(x)) = 0$ for all $x \in \mathbf{V}_X(\{b\})$, and so $\varphi^*(g) \in \mathbf{I}_{A[X]}(\mathbf{V}_X(\{b\})) = \sqrt{(b)}$, where the last equality holds by Hilbert's nullstellensatz. Hence we can write $\varphi^*(g^m) = \varphi^*(g)^m = cb$ for some $m \geqslant 1$ and some $c \in A[X]$. It follows that $x = (ca)/(cb) = ca/\varphi^*(g^m) \in A[X]\varphi^*(\mathcal{O}_q)$. Thus, we have shown that $\mathcal{O}_p \subseteq A[X]\varphi^*(\mathcal{O}_q)$. Now, since φ is a finite morphism, we have that $\varphi^*(A[Y]) \subseteq A[X]$ is an integral extension. So, since $A[X]$ is a finitely generated k-algebra, there exist $u_1, \ldots, u_l \in A[X]$ which generate $A[X]$ as a $\varphi^*(A[Y])$-module. Then we also have

$$\mathcal{O}_p \subseteq \sum_{i=1}^{l} \varphi^*(\mathcal{O}_q)u_i.$$

Thus, $\mathcal{O}_p$ is a submodule of a finitely generated $\varphi^*(\mathcal{O}_q)$-module, as required.

Step 2. Now we show that the general case can be reduced to the special case considered in Step 1. So let $\varphi \colon X \to Y$ be a dominant injective morphism between irreducible affine varieties and assume that (∗) holds. Now, as in the proof of Proposition 1.4.15, one shows that there is an open subset $U \subseteq X$ containing p such that the above conditions are satisfied for all $x \in U$, that is, x is non-singular, $\varphi(x)$ is non-singular, and $d_x\varphi$ is bijective.

We can now proceed as follows. Since $\dim X = \dim Y$, there exists some non-zero $g \in A[Y]$ such that φ restricts to a finite bijective morphism $\varphi_g \colon X_{\varphi^*(g)} \to Y_g$; see Theorem 2.2.7. Now let us choose $x \in U \cap X_{\varphi^*(g)}$. By Lemma 1.1.14, we may identify $T_x(X_{\varphi^*(g)})$ with $T_x(X)$ and $T_{\varphi(x)}(Y_g)$ with $T_{\varphi(x)}(Y)$. Then $(*)$ also holds for the restricted morphism φ_g. So, by Step 1, φ_g is a birational equivalence. It remains to note that the algebra homomorphism φ_g^* is the natural map $A[Y]_g \to A[X]_{\varphi^*(g)}$ induced by φ^*. Hence φ is a birational equivalence. $\qquad\square$

2.3.11 Proposition *Let X be an irreducible affine variety. Then there is a non-empty open set $U \subseteq X$ such that $\mathcal{O}_p$ is integrally closed in $A(X)$ for all $p \in U$.*

Proof Let K be the field of fractions of $A[X]$. We begin by showing that there exists some $0 \neq f \in A[X]$ and a finitely generated k-subalgebra $B \subseteq K$ such that the following conditions hold.

    (a) B is integrally closed in its field of fractions $F \subseteq K$.
    (b) We have $B \subseteq A[X]_f$, and $A[X]_f$ is integral over B.
    (c) K is a finite separable extension of F.

This is proved as follows. By Exercise 1.8.15, there exist algebraically independent elements $z_1, \ldots, z_d \in K$ such that K is a finite separable algebraic extension of $F := k(z_1, \ldots, z_d)$. We will construct B as a certain localization of the ring $R := k[z_1, \ldots, z_d]$. Since K is the field of fractions of $A[X]$, we can write $z_i = a_i/h$, where $0 \neq h \in A[X]$ and $a_i \in A[X]$ for $1 \leqslant i \leqslant d$. Thus, we have $R \subseteq A[X]_h$. We claim that there exists some non-zero $g \in R$ such that every element in $A[X]_h$ is integral over R_g. To see this, let $a \in A[X]_h$. Then a is algebraic over F. Since F is the field of fractions of R, this implies that a is integral over R_s for some non-zero $s \in R$. Applying this to a finite set of algebra generators of $A[X]_h$, we see that there exists some non-zero $g \in R$ such that all elements of $A[X]_h$ are integral over R_g, as claimed. Then we have $R_g \subseteq A[X]_{gh}$, and $A[X]_{gh}$ will be integral over R_g (see the argument in the proof of Lemma 2.2.6). Now set $B := R_g$ and $f := gh$. Then (b) holds. By construction, (c) also holds. Finally, since R is a polynomial ring, R is integrally closed in F. Then R_g will also be integrally closed in F (see Exercise 2.7.7) and so (a) holds.

Now let S be the integral closure of B in K. We claim that K is the field of fractions of S; more precisely:

there exists an F-basis $\{u_1, \ldots, u_n\}$ of K such that all u_i lie in S. (1)

Indeed, let $u \in K$. Then u is algebraic over F and, hence, satisfies an equation of the form $b_m u^m + b_{m-1} u^{m-1} + \cdots + b_0 = 0$, where $m \geqslant 1$, $b_i \in B$, and $b_m \neq 0$. Multiplying this equation by b_m^{m-1}, we see that $b_m u$ is integral over B. Thus, given any F-basis of K, we may multiply the basis elements by suitable elements of B to get an F-basis consisting of elements in S. Next we claim that

there exist $z_1, \ldots, z_n$ in K such that $S \subseteq Bz_1 + \cdots + Bz_n$. (2)

To prove this, we use the trace map $T_{K/F}$ introduced in Exercise 2.7.5. Since $K \supseteq F$ is separable, the symmetric F-bilinear form $K \times K \rightarrow F$, $(x, y) \mapsto T_{K/F}(xy)$, is non-degenerate; see Exercise 2.7.5(d). Let $\{z_1, \ldots, z_n\}$ be the F-basis of K dual to an F-basis $\{u_1, \ldots, u_n\} \subseteq S$ as in (1), that is, we have $T_{K/F}(z_j u_i) = \delta_{ij}$ for all i, j. Now let $s \in S$. Then we can write $s = \sum_{j=1}^{n} x_j z_j$ with $x_j \in F$. Since $u_i \in S$, we have $su_i \in S$ for all i and so $T_{K/F}(su_i) \in B$, see Exercise 2.7.5(c). It follows that $T_{K/F}(su_i) = \sum_{j=1}^{n} x_j T_{K/F}(z_j u_i) = x_i \in B$. Thus, every element of S is contained $Bz_1 + \cdots + Bz_n$, as claimed.

Now, by (2), S is contained in a finitely generated B-module. Since B is noetherian, S is finitely generated as a B-module (see Exercise 1.8.1) and so

S is a finitely generated k-algebra. (3)

Let us write $S = k[s_1, \ldots, s_l]$, where $s_i \in S$. Since $S \subseteq K$ and K is the field of fractions of $A[X]$, there exists some non-zero $c \in A[X]$ and $a_i \in A[X]$ such that $s_i = a_i/c$ for $1 \leqslant i \leqslant l$. Thus, we have $S \subseteq A[X]_c$. We claim that this implies that $\mathcal{O}_p$ is integrally closed in K for all $p \in U := X_c$. Indeed, let $x \in K$ be integral over $\mathcal{O}_p$ where $p \in U$. Then we have an equation of the form

$$x^m + (a_{m-1}/b_{m-1})x^{m-1} + \cdots + (a_1/b_1)x + a_0/b_0 = 0,$$

where $m \geqslant 1$ and $a_i, b_i \in A[X]$ are such that $b_i(p) \neq 0$. Now set $d := b_0 b_1 \cdots b_{m-1}$ and multiply the above equation by d^m. Then one obtains an equation of integral dependence of dx over $A[X]$.

Hence dx is integral over $A[X]_f \supseteq A[X]$ and, hence, also over B. Thus, we have $dx \in S \subseteq A[X]_c$ and so we can write $dx = a/c^l$, where $a \in A[X]$ and $l \geqslant 0$. This yields that $x = a/(dc^l) \in \mathcal{O}_p$, as required. $\qquad\qquad\qquad\qquad\qquad\qquad\qquad\qquad\qquad\qquad\qquad\qquad$ $\square$

2.3.12 Remark   Let X be an irreducible affine variety. Then one can show that the following implication holds:

$$p \in X \text{ is non-singular} \quad \Rightarrow \quad \mathcal{O}_p \text{ is a factorial ring};$$

in particular, if $p \in X$ is non-singular, then $\mathcal{O}_p$ is integrally closed in $A(X)$. Here, we will not give the proof of this result; for further details see §III.7 of Mumford (1988) and §II.3 of Shafarevich (1994). The above Proposition 2.3.11 in combination with Theorem 1.4.11 shows that, at least, there exist non-singular points $p \in X$ such that $\mathcal{O}_p$ is integrally closed in $A(X)$.

As in the previous section, one may expect that the above results admit a simpler formulation when we consider algebraic groups. This is indeed the case.

2.3.13 Algebraic group actions   Let G be an affine algebraic group and X be an affine variety. We say that X is a *G-variety* if there exists a morphism of affine varieties

$$\alpha \colon G \times X \to X, \qquad (g, x) \mapsto g.x,$$

which defines an abstract operation of G on the set X; that is, we have $1.x = x$ and $(gh).x = g.(h.x)$ for all $g, h \in G$ and $x \in X$. We say that X is a *homogeneous G-variety* if the action of G on X is transitive. Note the following simple facts.

For each $g \in G$, the map $\pi_g \colon X \to X$, $x \mapsto g.x$, is obtained by composing α with the morphism $X \to G \times X$, $x \mapsto (g, x)$. Hence π_g is a morphism. Furthermore, we have $\pi_1 = \mathrm{id}_X$ and $\pi_g \circ \pi_h = \pi_{gh}$ for all $g, h \in G$. Consequently, each π_g is an isomorphism of affine varieties of X, with inverse $\pi_{g^{-1}}$.

2.3.14 Example   Let X be an irreducible affine variety which is a homogeneous G-variety for some affine algebraic group G. Then X is normal. Indeed, let K be the field of fractions of $A[X]$. By Proposition 2.3.11, there exists some $p \in X$ such that $\mathcal{O}_p$ is integrally closed in K. Now let $q \in X$ and choose $g \in G$ such that $p = g.q$. Then consider the isomorphism $\pi_g \colon X \to X$, $x \mapsto g.x$. The corresponding algebra homomorphism $\pi_g^* \colon A[X] \to A[X]$ also is an isomorphism, and this induces a field automorphism of K. It follows that $\mathcal{O}_q = \pi_g^*(\mathcal{O}_{g.q})$ will also be integrally closed in K. Since this holds for all $q \in X$, we see that all local rings are integrally closed in K. Hence X is normal (see Exercise 2.7.7).

The above discussion applies, in particular, to the case where G is a connected affine algebraic group. Then we may regard G as a homogeneous G-variety via left multiplication. Hence, G is normal.

2.3.15 Proposition *Let G be an affine algebraic group and X, Y be irreducible affine varieties which are homogeneous G-varieties. Let $\varphi \colon X \to Y$ be a G-equivariant bijective morphism such that*

$$d_p\varphi \colon T_p(X) \to T_{\varphi(p)}(Y) \text{ is surjective for some non-singular } p \in X.$$

Then φ is an isomorphism.

Proof By Theorem 2.3.10, there exists some $x \in X$ such that $\varphi^*(\mathcal{O}_{\varphi(x)}) = \mathcal{O}_x$. Since φ is G-equivariant, this implies that

$$\varphi^*(\mathcal{O}_{\varphi(g^{-1}.x)}) = \varphi^*(\mathcal{O}_{g^{-1}.\varphi(x)}) = \varphi^*(\pi_g^*(\mathcal{O}_{\varphi(x)}))$$
$$= \pi_g^*(\varphi^*(\mathcal{O}_{\varphi(x)})) = \pi_g^*(\mathcal{O}_x) = \mathcal{O}_{g^{-1}.x} \quad \text{for all } g \in G.$$

Thus, since X is a homogeneous G-variety, we have $\varphi^*(\mathcal{O}_{\varphi(x')}) = \mathcal{O}_{x'}$ for all $x' \in X$. It remains to apply Lemma 2.3.7. $\qquad\square$

2.3.16 Example Let $\varphi \colon \tilde{G} \to G$ be a bijective homomorphism between connected affine algebraic groups. Assume that $d_1\varphi \colon T_1(\tilde{G}) \to T_1(G)$ is surjective. Then φ is an isomorphism. Indeed, $\tilde{G}$ and G are homogeneous $\tilde{G}$-varieties, where $\tilde{G}$ acts on itself by left multiplication and on G via $\tilde{G} \times G \to G$, $(\tilde{g}, h) \mapsto \varphi(\tilde{g})h$. Then φ is G-equivariant and the assertion follows from Theorem 2.3.15.

We remark that, if k has characteristic 0, then it can be shown that the assumption on $d_1\varphi$ is in fact unnecessary; see Corollary 11.1.3 of Goodman and Wallach (1998). However, in characteristic $p > 0$, that assumption cannot be omitted. Take, for example, the additive group $G = \mathbb{G}_a(k)$, where k is an algebraic closure of the field with p elements, and let $\varphi \colon G \to G$, $x \mapsto x^p$. Then φ is a bijective homomorphism, but φ^{-1} certainly is not a morphism of affine varieties; see Example 1.3.5(c).

2.4 Linearization and generation of algebraic groups

Our first aim in this section is to show that every 'abstract' affine algebraic group, as defined in §2.2.12, is isomorphic to a linear algebraic group in the sense of Definition 1.3.9. Then we establish a basic connectedness criterion and give some applications. Throughout, k is an algebraically closed field.

2.4.1 Example Consider the general linear monoid $M_n(k)$ with algebra of regular functions given by $\mathcal{A} = k[X_{ij} \mid 1 \leqslant i, j \leqslant n]$; see Definition 1.3.9. Let

$$\det := \sum_{\sigma \in \mathfrak{S}_n} \operatorname{sgn}(\sigma)\, X_{1\sigma(1)}\, X_{2\sigma(2)} \cdots X_{n\sigma(n)} \in \mathcal{A}.$$

Since $\det(I_n) = 1$, we certainly have that det is not nilpotent in $\mathcal{A}$. Thus, using the notation in §2.1.14 (see also Exercise 1.8.11) we have an affine variety $(\mathrm{GL}_n(k), \mathcal{A}_{\det})$. The matrix multiplication in $M_n(k)$ defines a multiplication map $\mu\colon \mathrm{GL}_n(k) \times \mathrm{GL}_n(k) \to \mathrm{GL}_n(k)$ and an inversion map $\iota\colon \mathrm{GL}_n(k) \to \mathrm{GL}_n(k)$. We claim that $(\mathrm{GL}_n(k), \mathcal{A}_{\det})$ is an affine algebraic group.

So, we must check that μ and ι are morphisms. First, using the formula for the inverse of a matrix, we see that ι is a morphism. Next, we have

$$\mu^*(X_{ij}) = \sum_{l=1}^{n} X_{il} \otimes X_{lj} \in \mathcal{A}_{\det} \otimes \mathcal{A}_{\det} \qquad \text{for all } i, j \in \{1, \ldots, n\}.$$

So it remains to consider $\mu^*(1/\det)$. For $x, y \in G$, we have $(1/\det \circ \mu)(x, y) = \mu(\iota^*(\det)(x), \iota^*(\det)(y))$. Now ι is a morphism and so $\iota^*(\det) \in \mathcal{A}_{\det}$. Thus, we also see that $\mu^*(1/\det) \in \mathcal{A}_{\det} \otimes \mathcal{A}_{\det}$, as required.

Note also that the group $\mathrm{GL}_1(k)$ is nothing but the *multiplicative group* of k, which we denote by $\mathbb{G}_m(k)$.

2.4.2 Lemma *Let (G, A) be an affine algebraic group over k, and assume that $\varphi\colon G \to \mathrm{GL}_n(k)$ is a homomorphism of abstract groups. For each $x \in G$, let us write $\varphi(x) = (a_{ij}(x))$, where $a_{ij}(x) \in k$ for all i, j. Then φ is a homomorphism of algebraic groups if all the maps $a_{ij}\colon G \to k$ belong to A.*

Proof We use the notation in Example 2.4.1. For $1 \leqslant i, j \leqslant n$, we have $X_{ij} \circ \varphi = a_{ij} \in A$ by assumption. Next, we have $(1/\det \circ \varphi)(x) = \det(\varphi(x))^{-1} = \det(\varphi(x^{-1})) = \det(\varphi(\iota(x)))$ for all $x \in G$. Thus, we obtain $1/\det \circ \varphi = \det \circ \varphi \circ \iota$. We certainly have $\det \circ \varphi \in A$. Finally, since ι is a morphism, we also have $(\det \circ \varphi) \circ \iota \in A$, as required. $\qquad \square$

As a first application, we shall prove that an affine algebraic group (G, A) can always be embedded in some general linear group. We first need the following result which is a basic tool in the study of algebraic groups. Let $\mu \colon G \times G \to G$ be the multiplication map. For any $x \in G$, we define

$$\lambda_x \colon G \to G, \quad y \mapsto xy, \quad \text{and} \quad \rho_x \colon G \to G, \quad y \mapsto yx.$$

As in §1.3.12, we see that λ_x and ρ_x are isomorphisms of affine varieties. Since ρ_x is an isomorphism, the corresponding k-algebra homomorphism $\rho_x^* \colon A[G] \to A[G]$ also is an isomorphism, where $\rho_x^* \circ \rho_y^* = \rho_{xy}^*$. Hence, we have a group homomorphism $\rho^* \colon G \to \mathrm{Aut}_k(A[G])$, where $\mathrm{Aut}_k(A[G])$ denotes the group of all k-algebra automorphisms of $A[G]$. Note that the analogous construction with λ_x yields an anti-homomorphism $\lambda^* \colon G \to \mathrm{Aut}_k(A[G])$, $x \mapsto \lambda_x^*$.

2.4.3 Theorem *Let (G, A) be an affine algebraic group, with multiplication map $\mu \colon G \times G \to G$. Let $F \subset A$ be any finite subset. Then the subspace*

$$E := \langle \rho_x^*(f) \mid f \in F, x \in G \rangle_k \subseteq A$$

has finite dimension. If $\{e_1, \ldots, e_n\}$ is a basis of E, then we have

$$\mu^*(e_j) = \sum_{i=1}^{n} e_i \otimes a_{ij} \qquad \textit{for all } j \in \{1, \ldots, n\}, \textit{ where } a_{ij} \in A.$$

Furthermore, the map $\varphi \colon G \to \mathrm{GL}_n(k)$, $x \mapsto (a_{ij}(x))$, is a homomorphism of affine algebraic groups.

Proof Fix some non-zero $f \in F$ and consider the subspace $E_f = \langle \rho_x^*(f) \mid x \in G \rangle_k \subseteq A$. Then choose $m \geqslant 1$ minimal such that $\mu^*(f) = \sum_{i=1}^{m} g_i \otimes h_i$ where $g_i, h_i \in A$. Now, for $x, y \in G$, we have

$\rho_y^*(f)(x) = f(xy) = \mu^*(f)(x,y) = \sum_{i=1}^m g_i(x)h_i(y)$ and so

$$\rho_y^*(f) = \sum_{i=1}^m h_i(y)g_i \qquad \text{for all } y \in G. \tag{1}$$

Similarly, we have $\mu^*(\rho_x^*(f))(y,z) = f(yzx) = f(y\rho_x(z)) = \mu^*(f)(y, \rho_x(z))$ for any $y, z \in G$, and so

$$\mu^*(\rho_x^*(f)) = \sum_{i=1}^m g_i \otimes \rho_x^*(h_i) \qquad \text{for all } x \in G. \tag{2}$$

Now, the minimality of m shows that $\{h_1, \ldots, h_m\}$ is linearly independent. Hence there exist elements $y_1, \ldots, y_m \in G$ such that the matrix $(h_i(y_j))_{1 \leqslant i,j \leqslant m}$ is invertible. Consequently, we can 'invert' the equations $\rho_{y_j}^*(f) = \sum_{i=1}^m h_i(y_j)g_i$ and find that each g_i is a linear combination of $\rho_{y_j}^*(f)$ $(1 \leqslant j \leqslant m)$. We conclude that $g_i \in E_f$ for all i. On the other hand, (1) shows that any $\rho_y^*(f)$ is a linear combination of $g_1, \ldots, g_m$. Hence $\{g_1, \ldots, g_m\}$ spans E_f. Applying this to each $f \in F$, we see that $E = \sum_{f \in F} E_f$ has finite dimension.

Now let $\{e_1, \ldots, e_n\}$ be a basis of E. By writing all of the above expressions in terms of this basis, we see that (2) yields relations of the form

$$\mu^*(e_j) = \sum_{i=1}^n e_i \otimes a_{ij} \qquad \text{for all } j \in \{1, \ldots, n\}, \text{ where } a_{ij} \in A. \tag{3}$$

For any $x, y \in G$ and $f \in F$, we have $\rho_y^*(\rho_x^*(f)) = \rho_{yx}^*(f) \in E$, and so E is invariant under all ρ_y^* $(y \in G)$. Thus, for any $y \in G$, we have an element $\varphi_y \in \mathrm{GL}(E)$ given by $\varphi_y(f) := \rho_y^*(f)$ $(f \in E)$. Then the map

$$\varphi \colon G \to \mathrm{GL}_n(k), \quad y \mapsto M_y = \text{matrix of } \varphi_y \text{ with respect to } \{e_1, \ldots, e_n\},$$

is a group homomorphism. Furthermore, for any $x \in G$, we have

$$\varphi_y(e_j)(x) = \rho_y^*(e_j)(x) = e_j(\mu(x,y)) = \mu^*(e_j)(x,y) = \sum_{i=1}^m a_{ij}(y)e_i(x)$$

for all $x \in G$ and so $\varphi_y(e_j) = \sum_{i=1}^n a_{ij}(y)e_i$. Thus, the coefficients of M_y are given by the regular functions $a_{ij} \in A$ evaluated at y. Hence φ is a homomorphism of affine algebraic groups by Lemma 2.4.2. $\qquad\square$

2.4.4 Corollary *Let (G, A) be an affine algebraic group over k. Then there exists some $n \geqslant 1$ and a closed embedding of algebraic groups $\varphi\colon G \to \mathrm{GL}_n(k)$; that is, $\varphi(G) \subseteq \mathrm{GL}_n(k)$ is a closed subgroup and the restricted map $G \to \varphi(G)$, $g \mapsto \varphi(g)$, is an isomorphism of algebraic groups.*

Proof Since A is a finitely generated k-algebra, we have $A = k[f_1, \ldots, f_m]$, where $f_j \in A$. We set $F = \{f_1, \ldots, f_m\}$ and apply Theorem 2.4.3. This yields a finite-dimensional subspace $E \subset A$ containing F with a basis $\{e_1, \ldots, e_n\}$ such that

$$\mu^*(e_j) = \sum_{i=1}^{n} e_i \otimes a_{ij} \qquad \text{for all } 1 \leqslant j \leqslant n, \text{ where } a_{ij} \in A.$$

Furthermore, $\varphi\colon G \to \mathrm{GL}_n(k)$, $x \mapsto (a_{ij}(x))$, is a homomorphism of affine algebraic groups. By Proposition 2.2.1, it remains to show that $\varphi^*\colon \mathcal{A}_{\mathrm{det}} \to A$ is surjective. Now, E contains all f_i and, hence, we have $A = k[e_1, \ldots, e_n]$. So it is enough to show that all e_j lie in the image of φ^*. First note that all a_{ij} lie in that image since $\varphi^*(X_{ij}) = a_{ij}$. Now, using the above relation, we obtain

$$e_j(y) = e_j(\mu(1, y)) = \mu^*(e_j)(1, y) = \sum_{i=1}^{n} e_i(1)a_{ij}(y) \quad \text{for all } y \in G$$

and so $e_j = \sum_{i=1}^{n} e_i(1)a_{ij}$. Hence we have $e_j \in \varphi^*(\mathcal{A}_{\mathrm{det}})$, as required. $\qquad\square$

2.4.5 Remark Let $\varphi\colon G \to \mathrm{GL}_n(k)$ be a closed embedding as above. Furthermore, note that we also have a closed embedding

$$\iota \ :\ \mathrm{GL}_n(k) \to \mathrm{SL}_{n+1}(k), \qquad A \mapsto \left[\begin{array}{c|c} A & 0 \\ \hline 0 & \det(A)^{-1} \end{array} \right].$$

Thus, the composition $\iota \circ \varphi\colon G \to \mathrm{SL}_{n+1}(k)$ yields an isomorphism of G with a closed subgroup of $\mathrm{SL}_{n+1}(k)$. So everything that we proved for a linear algebraic group in the sense of Definition 1.3.9 remains valid for an abstract affine group (G, A) as defined above. For example, Proposition 1.3.13 (concerning the irreducible components of an algebraic group) remains true without any change. Furthermore, we have $\dim T_g(G) = \dim T_1(G) = \dim G$ for all $g \in G$ (see Proposition 1.5.2) and the tangent space $T_1(G)$ carries in a natural way the structure of a Lie algebra, where the Lie product is given by the formula in Remark 1.5.4.

2.4.6 Theorem *Let G be an affine algebraic group and $\{(Y_\lambda, B_\lambda)\}_{\lambda \in \Lambda}$ be a family of irreducible affine varieties. We assume that there exist morphisms $\varphi_\lambda \colon Y_\lambda \to G$ such that $1 \in X_\lambda := \varphi_\lambda(Y_\lambda)$ for all $\lambda \in \Lambda$. Let H be the subgroup of G generated by all X_λ. Then H is closed and irreducible and we have*

$$H = X_{\lambda_1}^{\varepsilon_1} \cdots X_{\lambda_n}^{\varepsilon_n} \quad \text{for some } n \geqslant 0, \text{ where } \lambda_i \in \Lambda \text{ and } \varepsilon_i \in \{\pm 1\}.$$

In particular, a collection of closed irreducible subgroups in G generates a subgroup which is itself closed and irreducible.

Proof We may assume that, for each λ, there exists some λ' such that $X_{\lambda'} = X_\lambda^{-1}$. Now let $n \geqslant 0$ and $\underline{\lambda} = (\lambda_1, \ldots, \lambda_n)$ with $\lambda_i \in \Lambda$, and consider the morphism

$$\varphi_{\underline{\lambda}} \colon Y_{\lambda_1} \times \cdots \times Y_{\lambda_n} \to G, \quad (y_1, \ldots, y_n) \mapsto \varphi_{\lambda_1}(y_1) \cdots \varphi_{\lambda_n}(y_n).$$

Denote by $X_{\underline{\lambda}} = X_{\lambda_1} \cdots X_{\lambda_n}$ the image of $\varphi_{\underline{\lambda}}$. Since each Y_λ is irreducible, the above direct product is irreducible and, hence, $\overline{X}_{\underline{\lambda}}$ is an irreducible closed subset of G. Now let $\underline{\nu} = (\nu_1, \ldots, \nu_m)$ for some $m \geqslant 0$, where $\nu_i \in \Lambda$. Denote by $(\underline{\lambda}, \underline{\nu})$ the concatenation of $\underline{\lambda}$ and $\underline{\nu}$. Then $X_{\underline{\lambda}} \cdot X_{\underline{\nu}} = X_{(\underline{\lambda}, \underline{\nu})}$ and we claim that

$$\overline{X}_{\underline{\lambda}} \cdot \overline{X}_{\underline{\nu}} \subseteq \overline{X}_{(\underline{\lambda}, \underline{\nu})}. \tag{a}$$

Indeed, for a fixed $x \in X_{\underline{\nu}}$, the map $X_{\underline{\lambda}} \to X_{(\underline{\lambda}, \underline{\nu})}$, $y \mapsto yx$, is continuous, and so $\overline{X}_{\underline{\lambda}} x \subseteq \overline{X}_{(\underline{\lambda}, \underline{\nu})}$. Thus, we have $\overline{X}_{\underline{\lambda}} \cdot X_{\underline{\nu}} \subseteq \overline{X}_{(\underline{\lambda}, \underline{\nu})}$. On the other hand, for a fixed $y \in \overline{X}_{\underline{\lambda}}$, the map $X_{\underline{\nu}} \to \overline{X}_{(\underline{\lambda}, \underline{\nu})}$, $x \mapsto yx$, is continuous, and so $y\overline{X}_{\underline{\nu}} \subseteq \overline{X}_{(\underline{\lambda}, \underline{\nu})}$. Thus, (a) is proved.

Now choose $\underline{\lambda}$ as above such that $\dim \overline{X}_{\underline{\lambda}}$ is maximal. Then, for any $\underline{\nu}$, we have $\overline{X}_{\underline{\lambda}} \subseteq \overline{X}_{\underline{\lambda}} \cdot 1 \subseteq \overline{X}_{\underline{\lambda}} \cdot \overline{X}_{\underline{\nu}} \subseteq \overline{X}_{(\underline{\lambda}, \underline{\nu})}$ (since $1 \in X_\mu$ for $\mu \in \Lambda$). By the maximality of $\underline{\lambda}$ and Proposition 1.2.20, we have $\overline{X}_{\underline{\lambda}} = \overline{X}_{(\underline{\lambda}, \underline{\nu})}$ and, hence, using a similar argument, also $\overline{X}_{\underline{\nu}} \subseteq \overline{X}_{(\underline{\lambda}, \underline{\nu})} = \overline{X}_{\underline{\lambda}}$. In particular, this implies that

$$\overline{X}_{\underline{\lambda}} \cdot \overline{X}_{\underline{\lambda}} \subseteq \overline{X}_{\underline{\lambda}} \quad \text{and} \quad \overline{X}_{\underline{\lambda}}^{-1} \subseteq \overline{X}_{\underline{\lambda}}, \tag{b}$$

and so $\overline{X}_{\underline{\lambda}} \subseteq G$ is a subgroup. Now, $X_{\underline{\lambda}}$ is constructible by Theorem 2.2.11, and so $\overline{X}_{\underline{\lambda}} = X_{\underline{\lambda}} \cdot X_{\underline{\lambda}}$ by Lemma 2.2.13. Finally, since $\overline{X}_\nu \subseteq \overline{X}_\lambda$ for all ν, we have $H = \overline{X}_\lambda$, and the sequence $(\underline{\lambda}, \underline{\lambda})$ has the required properties. $\qquad\square$

2.4.7 Example Let G be an affine algebraic group. For $g, h \in G$, the *commutator* is defined by $[g, h] := g^{-1}h^{-1}gh \in G$. If $U, H \subseteq G$ are subgroups, the corresponding commutator subgroup is defined to be the subgroup

$$[U, H] := \langle [u, h] \mid u \in U, h \in H \rangle \subseteq G.$$

Now, if U and H are closed and U or H is connected, then $[U, H]$ is closed and connected. Indeed, assume for example that H is connected. For $u \in U$, we consider the morphism $\varphi_u \colon H \to G,\ h \mapsto [u, h]$. We certainly have $1 \in \varphi_u(H)$. Thus, all the conditions in Theorem 2.4.6 are satisfied, and so $[U, H] = \langle \varphi_u(H) \mid u \in U \rangle$ is a closed connected subgroup, as claimed.

2.4.8 Definition Let G be an affine algebraic group. We define a chain of subgroups $G \supseteq G^{(1)} \supseteq G^{(2)} \supseteq \cdots$ as follows. We set

$$G^{(0)} := G \quad \text{and} \quad G^{(i+1)} := [G^{(i)}, G^{(i)}] \quad \text{for } i \geqslant 0.$$

It is well known (and easy to check) that the subgroups $G^{(i)}$ are normal and that the quotients $G^{(i)}/G^{(i+1)}$ are abelian. The group G is called *solvable* if $G^{(r)} = \{1\}$ for some $r \geqslant 0$. Now assume that G is connected. Then, by Example 2.4.7, all subgroups $G^{(i)}$ are closed and connected.

Similarly, we can define a descending chain of normal subgroups by

$$K_0(G) := G \quad \text{and} \quad K_{i+1}(G) := [G, K_i(G)] \quad \text{for } i \geqslant 0.$$

We say that G is *nilpotent* if $K_r(G) = \{1\}$ for some $r \geqslant 1$. Clearly, nilpotent groups are solvable. As before, we see that, if G is connected, then the groups $K_i(G)$ are closed and connected.

2.4.9 Example Let $B_n(k) \subseteq \mathrm{GL}_n(k)$ be the closed subgroup consisting of all upper triangular invertible matrices, that is, all matrices $A = (a_{ij})_{1 \leqslant i,j \leqslant n}$ such that $a_{ij} = 0$ for $1 \leqslant j < i \leqslant n$. As abstract groups, we have $B_n(k) = U_n(k)T_n(k)$, where $U_n(k)$ is the group of all upper triangular matrices with 1 on the diagonal (see Example 1.3.11) and $T_n(k)$ is the group of all invertible diagonal matrices. Since $U_n(k)$ and $T_n(k)$ are connected (see Example 1.5.5) and the natural map given by multiplication $U_n(k) \times T_n(k) \to B_n(k)$ is surjective, we conclude that $B_n(k)$ is connected.

Furthermore, $U_n(k)$ is normal in $B_n(k)$, and we have $B_n(k)/U_n(k) \cong T_n(k)$ (as abstract groups). Hence, since $T_n(k)$ is abelian, we have $[B_n(k), B_n(k)] \subseteq U_n(k)$. Now it is easily checked that there exists an element $t \in T_n(k)$ whose centralizer in $B_n(k)$ is just $T_n(k)$. So, by Exercise 2.7.13, we can even conclude that

$$[B_n(k), B_n(k)] = U_n(k).$$

Now consider the group $U_n(k)$. A matrix calculation shows that

$$K_r(U_n(k)) = \{(a_{ij}) \in U_n(k) \mid a_{ij} = 0 \text{ if } 1 \leqslant j - i \leqslant r\} \quad \text{for } r \geqslant 1;$$

(see, for example, Propostion III.16.3 of Huppert (1983). Thus, $K_n(U_n(k)) = \{1\}$, and so $U_n(k)$ is nilpotent; consequently, $B_n(k)^{(n)} \subseteq K_n(U_n(k)) = \{1\}$, and so $B_n(k)$ is solvable. The group $B_n(k)$ provides a model for the further study of connected solvable groups. In Section 3.5, we will obtain a generalization of the decomposition $B_n(k) = U_n(k)T_n(k)$ for an arbitrary connected solvable group.

2.4.10 Example Consider the algebraic group $\mathrm{SL}_n(k)$. Let $U_n(k)$ and $U_n'(k)$ be the subgroups of $\mathrm{SL}_n(k)$ consisting of all upper unitriangular matrices and all lower unitriangular matrices, respectively, as in Example 1.5.5. These are closed connected subgroups in $\mathrm{SL}_n(k)$. We claim that

$$\mathrm{SL}_n(k) = [\mathrm{SL}_n(k), \mathrm{SL}_n(k)] = \langle U_n(k), U_n'(k)\rangle.$$

This is seen as follows. By the Gauss elimination algorithm, we know that $\mathrm{SL}_n(k)$ is generated by $U_n(k)$ and $U_n'(k)$, together with the set of all monomial matrices in $\mathrm{SL}_n(k)$. (Recall that a matrix is monomial if there is precisely one non-zero coefficient in each row and each column.) Let us now check that every monomial matrix in $\mathrm{SL}_n(k)$ lies in $\langle U_n(k), U_n'(k)\rangle$. For this purpose, we use the following computations in $\mathrm{SL}_2(k)$:

$$n_0(a) := \begin{bmatrix} 0 & a \\ -a^{-1} & 0 \end{bmatrix} = \begin{bmatrix} 1 & a \\ 0 & 1 \end{bmatrix} \cdot \begin{bmatrix} 1 & 0 \\ -a^{-1} & 1 \end{bmatrix} \cdot \begin{bmatrix} 1 & a \\ 0 & 1 \end{bmatrix}$$

for any non-zero $a \in k$. This also shows that the diagonal matrix

$$\begin{bmatrix} a & 0 \\ 0 & a^{-1} \end{bmatrix} = n_0(a)n_0(-1)$$

can be written as a product of upper and lower unitriangular matrices in $\mathrm{SL}_2(k)$. Now let $1 \leqslant i \leqslant n-1$. Then we have an embedding

$$\varphi_i \colon \mathrm{SL}_2(k) \hookrightarrow \mathrm{SL}_n(k), \qquad A \mapsto \left[\begin{array}{c|c|c} I_{i-1} & 0 & 0 \\ \hline 0 & A & 0 \\ \hline 0 & 0 & I_{n-i-1} \end{array}\right].$$

(Thus, the four coefficients of A are placed at positions (i,i), $(i,i+1)$, $(i+1,i)$, and $(i+1,i+1)$ in the bigger matrix.) It is then easy to show (using the fact that every permutation of $1,\dots,n$ can be written as a product of transpositions permuting two consecutive numbers) that every monomial matrix in $\mathrm{SL}_n(k)$ can be expressed as a product $\varphi_{i_1}(A_1) \cdots \varphi_{i_r}(A_r)$, for various $i_j \in \{1,\dots,n-1\}$ and various monomial matrices $A_j \in \mathrm{SL}_2(k)$. Thus, we have shown that $\mathrm{SL}_n(k) = \langle U_n(k), U_n'(k) \rangle$. Furthermore, by Example 2.4.9, we know that $U_n(k) = [B_n(k), B_n(k)] \subseteq [\mathrm{SL}_n(k), \mathrm{SL}_n(k)]$. A similar argument also applies to $U_n'(k)$. Thus, we conclude that $\mathrm{SL}_n(k) = [\mathrm{SL}_n(k), \mathrm{SL}_n(k)]$.

2.4.11 Example Let G be one of the classical groups defined in §1.3.15 and §1.3.16; thus, we have

$$G = \begin{cases} \mathrm{SO}_{2m+1}(k), & \text{any } m \geqslant 1,\ \mathrm{char}(k) \neq 2, \\ \mathrm{Sp}_{2m}(k), & \text{any } m \geqslant 1,\ \text{any characteristic}, \\ \mathrm{SO}_{2m}^{+}(k), & \text{any } m \geqslant 2,\ \text{any characteristic}. \end{cases}$$

Let us set $n = 2m+1$ if $G = \mathrm{SO}_{2m+1}(k)$, and $n = 2m$ otherwise. Then G is a closed subgroup of $\mathrm{GL}_n(k)$. Furthermore, by Theorems 1.7.4 and 1.7.8, we know that G is connected and that there is a split BN-pair in G. We claim that

$$G = [G,G] = \langle U, U' \rangle \qquad \text{and} \qquad Z(G) = \{\pm \mathrm{id}\},$$

where U and U' are the subgroups of all upper unitriangular matrices and all lower unitriangular matrices in G, respectively.

We sketch the proof but leave some details as an exercise to the reader. Let T be the subgroup of all invertible diagonal matrices in G. By Remark 1.5.12, the groups U, U', and T are connected. Using Exercise 1.8.27 and Lemma 2.2.13, we conclude that $G = \langle U, T, U' \rangle$. Now recall from Lemmas 1.5.9–11 the explicit description of the matrices in T. Using that explicit description, one finds some $t \in T$ such that $\mathrm{C}_G(t) = T$. Then, by Exercise 2.7.13, we may conclude that

$U, U' \subseteq [G, G]$. So it will be enough to show that $T \subseteq \langle U, U' \rangle$. Now, in each case, we have an embedding $\mathrm{SL}_m(k) \to G$, $A \mapsto A$, defined similarly as in the proof of Lemma 1.7.6. (If $G = \mathrm{SO}_{2m+1}(k)$, one places 1 at position $(m+1, m+1)$.) Thus, since $\mathrm{SL}_m(k) = \langle U_m(k), U'_m(k) \rangle$, one already finds a big portion of T in $\langle U, U' \rangle$. In fact, it remains to show that the diagonal matrices with entries $1, \ldots, 1, a, a^{-1}, 1, \ldots, 1$ lie in $\langle U, U' \rangle$ for all non-zero $a \in k$ (where a and a^{-1} are at the positions m and $m+1$, respectively; if $G = \mathrm{SO}_{2m+1}(k)$, one has to insert a coefficient 1 between a and a^{-1}). Thus, the problem can be reduced to an explicit computation in $\mathrm{Sp}_2(k) = \mathrm{SL}_2(k)$, $\mathrm{SO}_3(k)$, or $\mathrm{SO}_4^+(k)$; to deal with these groups, use Exercises 1.8.19 and 1.8.20.

To determine $Z(G)$, we argue as follows. Since $Z(G)$ certainly normalizes $B = UT$, we have $Z(G) \subseteq B$ by Lemma 1.6.4, and hence $Z(G) \subseteq \bigcap_{n \in N} nBn^{-1} = T$. Thus, it remains to find all diagonal matrices which commute with all elements of U and U'. It is easily checked that $\pm\,\mathrm{id}$ are the only such matrices.

2.4.12 Remark Let G be a group with a split BN-pair. Assume that

$$G = [G, G], \quad B \text{ is solvable}, \quad \text{and} \quad W \text{ is irreducible}.$$

Then it can be shown that $Z := \bigcap_{g \in G} gBg^{-1} \subseteq H$ is the centre of G and that G/Z is a simple group. See Chapter IV, §2, no. 7 of Bourbaki (1968) of §11.1 of Carter (1972). This criterion can be applied to show that G/Z is simple, where $G = \mathrm{SL}_n(k)$ or G is one of the classical groups as in Example 2.4.11. More generally, this also works for groups of Lie type as in §1.5.15; see §11.1 of Carter (1972) and §4 of Steinberg (1967).

2.5 Group actions on affine varieties

The aim of this section is to establish some basic results concerning algebraic group actions on affine varieties. The two main results are Proposition 2.5.2, which shows that closed orbits always exist, and Theorem 2.5.5, which shows that orbit maps are open maps. At the end of this section, we will also discuss the problem of defining 'affine quotients'. In general, this is a very difficult problem, but we will at least show that quotients by finite groups always exist.

Let G be an affine algebraic group and X be an affine variety (both over an algebraically closed field k). Recall from §2.3.13 that we call X is a *G-variety* if there exists a morphism of affine varieties

$$\alpha\colon G \times X \to X, \qquad (g, x) \mapsto g.x,$$

which defines an abstract operation of G on the set X. We have already noted that, for each $g \in G$, the map $\pi_g\colon X \to X$, $x \mapsto g.x$, is an isomorphism of affine varieties. Similarly, for each $x \in X$, the orbit map $\varphi_x\colon G \to X$, $g \mapsto g.x$, is obtained by composing α with the morphism $G \to G \times X$, $g \mapsto (g, x)$. Hence φ_x is a morphism. The orbit of x is the set $O_x := \{g.x \mid g \in G\}$. For any subsets $H \subseteq G$ and $Y \subseteq X$, we write $H.Y := \{h.y \mid h \in H, y \in Y\}$.

2.5.1 Lemma *Let X be a G-variety. Then the following hold.*

(a) Let Y, $Z \subseteq X$, with Z closed. Then the transporter $\mathrm{Tran}_G(Y, Z) := \{g \in G \mid g.Y \subseteq Z\}$ is a closed subset of G. In particular, for $x \in X$, the stabilizer $\mathrm{Stab}_G(x) = \mathrm{Tran}_G(\{x\}, \{x\})$ is a closed subgroup.

(b) The identity component G° leaves each irreducible component of X invariant.

Proof (a) Let $y \in Y$ and consider the orbit map $\varphi_y\colon G \to X$. Then $\varphi_y^{-1}(Z) \subseteq G$ is closed. But then $\mathrm{Tran}_G(Y, Z) = \bigcap_{y \in Y} \varphi_y^{-1}(Z)$ also is closed.

(b) Let $X' \subseteq X$ be an irreducible component and $G' = \mathrm{Tran}_G(X', X')$ be its stabilizer in G. By (a), G' is a closed subset in G, and it is a subgroup. Since G permutes the finitely many irreducible components of X, we have that G' is a closed subgroup of finite index in G, and so $G^\circ \subseteq G'$ by Proposition 1.3.13. $\qquad\square$

2.5.2 Proposition *Let X be a G-variety.*

(a) Let O be a G-orbit. Then O is open in $\overline{O}$, and $\overline{O}$ is a union of orbits.

(b) For G-orbits $O, O' \subseteq X$, we write $O \preceq O'$ if $O \subseteq \overline{O}'(= Zariski closure of O'). This defines a partial order on the set of G-orbits in X.

(c) There exist minimal elements in the partial order defined in (b) and each of these is a closed orbit.

Proof (a) First we show that $\overline{O}$ is a union of orbits. For $g \in G$, consider the map $\pi_g\colon X \to X$, $y \mapsto g.y$. Since π_g is continuous for each $g \in G$, we actually have that π_g is a homeomorphism (with inverse given by $\pi_{g^{-1}}$). Thus, we have $\pi_g(\overline{O}) \subseteq \overline{O}$, and so $\overline{O}$ is a union of orbits. Now let $O = O_x$, where $x \in X$. Since O is the image of the morphism $\varphi_x \colon G \to X$, $g \mapsto g.x$, we know by Theorem 2.2.11 that O is constructible. So there exists a subset $U \subseteq O$ such that U is dense open in $\overline{O}$ (see Exercise 2.7.6). Let $g_0 \in G$ be such that $g_0.x \in U$. Then $x \in g_0^{-1}.U$ and so $g.x \in gg_0^{-1}.U \subseteq O$ for all $g \in G$. It follows that $O = \bigcup_{g \in G} g.U$, which is open in $\overline{O}$ (since $g.U = \pi_g(U)$ and π_g is a homeomorphism).

(b) To show that $\preceq$ is a partial order, it only remains to check that if O, O' are G-orbits such that $\overline{O} = \overline{O}'$, then $O = O'$. Now, by (a), O is open in $\overline{O}$. Hence, if $O' \neq O$ is a G-orbit contained in $\overline{O}$, then O' is contained in the closed set $\overline{O} \setminus O$ and so $\overline{O}' \subseteq \overline{O} \setminus O$.

(c) For $x \in X$ we set $S_x := \overline{O}_x \setminus O_x$. By (a), this is a closed subset of X. We can now construct a decreasing chain of closed subsets in X as follows. Let $x_1 \in X$ and assume that $S_{x_1} \neq \varnothing$. Then let $x_2 \in S_{x_1}$. By (a), the orbit O_{x_2} is still contained in S_{x_1}, and we have $S_{x_2} \subsetneq S_{x_1}$. If $S_{x_2} \neq \varnothing$, we choose $x_3 \in S_{x_2}$ and consider O_{x_3}. Thus, we obtain a chain $S_{x_1} \supsetneq S_{x_2} \supsetneq S_{x_3} \supsetneq \cdots$ of closed subvarieties. Since X is noetherian, there exists some $n \geqslant 1$ such that $S_{x_n} = \varnothing$. This means that O_{x_n} is a closed subset of X. $\qquad\square$

2.5.3 Proposition *Let X be a G-variety and $x \in X$. Denote by O_x° the orbit of x under the action of the identity component G°. Then we have $\dim \overline{O}_x = \dim \overline{O}_x^\circ$, $\dim \operatorname{Stab}_G(x) = \dim \operatorname{Stab}_{G^\circ}(x)$, and*

$$\dim G = \dim \overline{O}_x + \dim \operatorname{Stab}_G(x).$$

Proof Write $G = \coprod_{i=1}^t g_i G^\circ$, where $g_i \in G$. Then we have $O_x = \bigcup_{i=1}^t g_i O_x^\circ$ and so $\overline{O}_x = \bigcup_{i=1}^t \overline{g_i O_x^\circ}$. Since left multiplication by any element of G is an isomorphism of affine varieties, $\overline{g_i O_x^\circ}$ is a closed irreducible subset, and we have $\dim \overline{g_i O_x^\circ} = \dim \overline{O}_x^\circ$ for all i. This shows that $\dim \overline{O}_x = \dim \overline{O}_x^\circ$. Furthermore, since G° is a normal subgroup, the product $G^\circ \operatorname{Stab}_G(x)$ is a subgroup of G, and we have the following isomorphism of abstract groups:

$$\operatorname{Stab}_G(x)/\operatorname{Stab}_{G^\circ}(x) = \operatorname{Stab}_G(x)/(G^\circ \cap \operatorname{Stab}_G(x)) \cong G^\circ \operatorname{Stab}_G(x)/G^\circ.$$

Hence, since G° has finite index in G, we conclude that $\operatorname{Stab}_{G^\circ}(x)$ has finite index in $\operatorname{Stab}_G(x)$, and so $\dim \operatorname{Stab}_G(x) = \dim \operatorname{Stab}_{G^\circ}(x)$. To prove the remaining statement, we may now assume without loss of generality that G is connected.

Consider the orbit map $\varphi_x \colon G \to \overline{O}_x \subseteq X$. Then φ_x is a dominant morphism between irreducible affine varieties. By Corollary 2.2.9, there exists a non-empty open subset $U \subseteq O_x$ such that $\dim \varphi_x^{-1}(y) = \dim G - \dim \overline{O}_x$ for all $y \in U$. $\overline{O}_x$ by Proposition 2.5.2, we conclude that $O_x \cap U \neq \varnothing$. Now let $y \in O_x \cap U$ and $g \in G$ be such that $y = g.x$. Then we have

$$
\begin{aligned}
\varphi_x^{-1}(y) &= \{ h \in G \mid h.x = y \} \\
&= \{ h \in G \mid g^{-1}h \in \mathrm{Stab}_G(x) \} \\
&= g\,\mathrm{Stab}_G(x).
\end{aligned}
$$

Since left multiplication with g is an isomorphism of affine varieties, we have $\dim \varphi_x^{-1}(y) = \dim \mathrm{Stab}_G(x)$, and so $\dim G = \dim \overline{O}_x + \dim \mathrm{Stab}_G(x)$. $\qquad\square$

2.5.4 Example   Let G be an affine algebraic group. Then there are several ways to see G itself as a G-variety. For example, we may consider the action given by left multiplication. Another example is the *conjugation action* $G \times G \to G$, $(g, x) \mapsto gxg^{-1}$. For $x \in G$, the orbit $O_x = \{ gxg^{-1} \mid g \in G \}$ is nothing but the *conjugacy class* of x in G, and the stabilizer of $x \in G$ is the centralizer $\mathrm{C}_G(x) := \{ g \in G \mid gx = xg \}$. Thus, Proposition 2.5.3 shows that

$$
\dim G = \dim \overline{O}_x + \dim \mathrm{C}_G(x).
$$

See Section 2.6 for worked examples in the case where $G = \mathrm{SL}_n(k)$; in particular, Theorem 2.6.5 gives an explicit combinatorial characterization of the partial order on conjugacy classes of unipotent elements.

Furthermore, let $S \subseteq G$ be a closed subset. Then

$$
\mathrm{N}_G(S) := \{ g \in G \mid gsg^{-1} \in S \text{ for all } s \in S \}
$$

is a closed subgroup in G, which is called the *normalizer* of S. To prove this, just note that $\mathrm{N}_G(S) = \mathrm{Tran}_G(S, S)$ and use Lemma 2.5.1.

2.5.5 Theorem (Steinberg)   *Let X_1 and X_2 be G-varieties and $\varphi \colon X_1 \to X_2$ be a dominant G-equivariant morphism. Assume that X_1 consists of a single G-orbit. Then φ is an open map, that is, it sends open sets to open sets.*

Proof We begin with some reductions. First we show that it is enough to consider the special case where $X_1 = G$ and G acts by left multiplication. Indeed, let us fix $x_1 \in X_1$ and set $x_2 = \varphi(x_1) \in X_2$. Then we have the orbit maps $\varphi_i \colon G \to X_i$, $g \mapsto g.x_i$ $(i = 1, 2)$. Since $\varphi_2(g) = g.x_2 = g.\varphi(x_1) = \varphi(g.x_1) = \varphi(\varphi_1(g))$ for all $g \in G$, we have $\varphi_2 = \varphi \circ \varphi_1$. Now, since φ_1 is assumed to be surjective, we see that the openness of φ would follow if we knew that φ_2 was open. So let us now change the notation and consider a G-variety X, let $x \in X$ and consider the orbit map $\varphi_x \colon G \to X$. Assume that this map is dominant (that is, the G-orbit of x is dense in X). Then we must show that φ_x is an open map. We claim that it is enough to show that

$$\varphi_x|_U \colon U \to X \text{ is an open map for some non-empty open } U \subseteq G.$$
$$(*)$$

Indeed, if we have such an open subset U, we can proceed as follows. Since left multiplication by any $g \in G$ is a homeomorphism, the map $\varphi_x|_{gU} \colon gU \to X$ is also open. Since $G = \bigcup_{g \in G} gU$, this implies that φ_x is open.

Let us now prove $(*)$ in the special case where G is connected. Since φ_x is assumed to be dominant, this also forces X to be irreducible. Let $d := \dim G - \dim X$. By Theorem 2.2.7, there exists a non-empty affine open subset $U \subseteq G$ and a non-empty affine open subset $V \subseteq X$ such that we have a factorization

$$\varphi_x|_U \colon U \xrightarrow{\ \tilde{\varphi}\ } V \times k^d \xrightarrow{\ \mathrm{pr}_1\ } V,$$

where $\tilde{\varphi}$ is a dominant finite morphism; in particular, we have $\varphi_x(U) = V$. We claim that U has the desired property. Since $V \subseteq X$ is open, it is enough to show that the restricted map $\varphi_x|_U \colon U \to V$ (which is still a dominant morphism between irreducible affine varieties) is open.

To prove this, we argue as follows. By Lemma 2.2.5, $\tilde{\varphi}$ is a closed map. Hence, by taking complements, one sees that $\tilde{\varphi}(U_1)$ is open for any open subset $U_1 \subseteq U$ which consists of complete fibres of $\tilde{\varphi}$, that is, we have $U_1 = \tilde{\varphi}^{-1}(\tilde{\varphi}(U_1))$. Since the projection map pr_1 is open by Exercise 1.8.8, we conclude that

$$\varphi_x(U_1) \subseteq V \text{ is open for any open } U_1 \subseteq U \text{ such that } U_1 = \tilde{\varphi}^{-1}(\tilde{\varphi}(U_1)).$$

Now, given any open subset $U_1 \subseteq U$, we set

$$\tilde{U}_1 := U_1 H \cap U \supseteq U_1, \qquad \text{where } H := \mathrm{Stab}_G(x).$$

Then $\tilde{U}_1$ is open since $U_1 H = \bigcup_{h \in H} U_1 h$ is open. Furthermore, we have $\varphi_x(\tilde{U}_1) = \tilde{U}_1.x = U_1.x = \varphi_x(U_1)$. We claim that $\tilde{U}_1 = \tilde{\varphi}^{-1}(\tilde{\varphi}(\tilde{U}_1))$. The inclusion '$\subseteq$' is clear. Now let $u \in U$ be such that $\tilde{\varphi}(u) = \tilde{\varphi}(u_1)$ for some $u_1 \in \tilde{U}_1$. Composing with pr_1, we obtain $u.x = \varphi_x(u) = \varphi_x(u_1) = u_1.x$ and so $u_1^{-1} u.x = x$. This implies that $u \in u_1 H \cap U \subseteq \tilde{U}_1$, as required. Thus, we conclude that $\varphi_x(U_1) = \varphi_x(\tilde{U}_1)$ is open, as claimed.

So $(*)$ is proved in the special case where G is irreducible. Now consider the general case. Then $(*)$ holds for the restricted orbit map $\varphi_x^\circ \colon G^\circ \to X^\circ$, where $X^\circ = \overline{O}_x^\circ$. So there is a non-empty open set $U \subseteq G^\circ$ such that $\varphi_x|_U \colon U \to X^\circ$ is open. Now, by Exercise 2.8.10, O_x° is open in O_x and, hence, open in $X = \overline{O}_x$. Consequently, $\varphi_x|_U \colon U \to X$ is an open map. $\qquad\square$

2.5.6 Example   (a) Let X be a G-variety and O_x be the G-orbit of some $x \in X$. Then the orbit map $\varphi_x \colon G \to \overline{O}_x$, $g \mapsto g.x$, is an open map. (Apply Theorem 2.5.5 where $X_1 = G$ and G acts by left multiplication.)

(b) Let $\varphi \colon G_1 \to G_2$ be a surjective homomorphism of affine algebraic groups. Then φ is an open map. (Apply Theorem 2.5.5, where $G = G_1$ acts on G_1 by left multiplication and on G_2 by $g.g_2 = \varphi(g)(g_2)$.) In particular, a bijective homomorphism between affine algebraic groups is a homeomorphism.

2.5.7 Corollary   *With the same assumptions as in Theorem 2.5.5, let Y be any affine variety. Then the map $\varphi \times \mathrm{id} \colon X_1 \times Y \to X_2 \times Y$ is open.*

Proof   Let us regard Y as an algebraic set in some affine space k^n. Then $X_i \times Y$ is a closed subset of $X_i \times k^n$ for $i = 1, 2$. Hence, if $U \subseteq X_1 \times Y$ is open, we have $U = (X_1 \times Y) \cap U'$ for some open set $U' \subseteq X_1 \times k^n$ and

$$(\varphi \times \mathrm{id})(U) = (X_2 \times Y) \cap (\varphi \times \mathrm{id})(U').$$

Thus, it will be enough to prove the assertion in the special case where $Y = k^n$. In this case, we use the fact that k^n is a k^n-variety, where we regard k^n as an affine algebraic group under addition. Then $X_i \times k^n$ are $(G \times k^n)$-varieties, with action given by $(g, v).(x_i, v') := (g.x_i, v + v')$ (where $g \in G$, $x_i \in X_i$, and $v, v' \in k^n$). Furthermore, $X_1 \times k^n$ is a single orbit and $\varphi \times \mathrm{id}$ is a $(G \times k^n)$-equivariant dominant morphism. Thus, $\varphi \times \mathrm{id}$ is open by Theorem 2.5.5. $\qquad\square$

Given a G-variety X, we can consider the set of orbits of G on X. A natural question is whether this set of orbits might be itself regarded as an affine variety. This leads us to the following notion.

2.5.8 Affine quotients Let G be an affine algebraic group and X be a G-variety, as in §2.3.13. For any $g \in G$, consider the corresponding morphism $\pi_g \colon X \to X$. This morphism is in fact an isomorphism, with inverse being given by $\pi_{g^{-1}}$. Hence $\pi_g^* \colon A[X] \to A[X]$ is a k-algebra isomorphism. Explicitly, π_g^* is given by the formula $\pi_g^*(f)(x) = f(g.x)$, where $f \in A[X]$ and $x \in X$. We set

$$A[X]^G := \{ f \in A[X] \mid f(g.x) = f(x) \text{ for all } g \in G,\, x \in X \}.$$

Then $A[X]^G$ is a k-subalgebra of $A[X]$. By definition, an *affine quotient* of X by G is a pair (X_0, φ) consisting of an affine variety X_0 over k and a morphism of affine varieties $\varphi \colon X \to X_0$ such that the following conditions are satisfied:

(a) For each $x_0 \in X_0$, the set $\varphi^{-1}(x_0)$ is a G-orbit. In particular, as a set, X_0 is in bijection with the G-orbits on X.

(b) The map φ is open, that is, $\pi(U) \subseteq X_0$ is open for any open $U \subseteq X$.

(c) We have $\varphi^*(A[X_0]) = A[X]^G$.

If it exists, an affine quotient is unique up to unique isomorphism. This follows from the following universal property. Assume that (X_0, φ) is an affine quotient, and let $\psi \colon X \to Y$ be any morphism of affine varieties which is constant on the orbits of G. Then there exists a unique morphism of affine varieties $\bar{\psi} \colon X_0 \to Y$ such that $\psi = \bar{\psi} \circ \varphi$.

Indeed, define $\bar{\psi}(\varphi(x)) := \psi(x)$ for $x \in X$. (Note that this is well-defined and the unique possibility.) We must check that $\bar{\psi}$ is a morphism of affine varieties. So let $f \in A[Y]$. Since ψ is constant on the orbits of G, we have $\psi^*(f) \in A[X]^G$. By condition (c), there exists a unique $f_0 \in A[X_0]$ such that $\psi^*(f) = \varphi^*(f_0)$. Now let $x_0 \in X_0$ and choose $x \in X$ with $\varphi(x) = x_0$. Then we have

$$(f \circ \bar{\psi})(x_0) = f(\psi(x)) = \psi^*(f)(x) = \varphi^*(f_0)(x) = f_0(\varphi(x)) = f_0(x_0),$$

that is, $f \circ \bar{\psi} = f_0 \in A[X_0]$. So $\bar{\psi}$ is a morphism.

2.5.9 Example Let X be a G-variety, where X is irreducible. Assume that an affine quotient (X_0, φ) exists. Then, since φ is surjective, X_0 must be irreducible, too. The following example shows that affine quotients do not always exist. Let $G = \mathrm{GL}_n(k)$ and $X = k^n$, where G acts on X in the natural way. Then there are only two orbits: $\{0\}$ and $k^n \setminus \{0\}$. If an affine quotient existed, it would have only two elements and, hence, would not be irreducible.

In fact, the question of whether an affine quotient exists or not is rather subtle. With the methods that we have at hand right now, we can give a complete answer in the special case where G is a finite group. This will be used, for example, in Chapter 4 where we study the finite fixed-point set of an algebraic group under a Frobenius map. Note that a finite group always is an affine algebraic group in a natural way. (The algebra of regular functions is given by the algebra of all k-valued functions on the group.)

2.5.10 Theorem (quotients by finite groups) *Let X be a G-variety, where G is a finite group. Let X/G be the set of G-orbits on X and $\pi\colon X \to X/G$ be the natural map. We set*

$$A[X/G] = \{\bar{f}\colon X/G \to k \mid \bar{f} \circ \pi \in A[X]\}.$$

Then $(X/G, A[X/G])$ is an affine variety, and the pair $(X/G, \pi)$ is an affine quotient; furthermore, $\pi\colon X \to X/G$ is a finite morphism.

Proof First note that $A[X/G]$ is a k-subalgebra of $\mathrm{Maps}(X/G, k)$, and the map $\pi^*\colon A[X/G] \to A[X]$, $f \mapsto f \circ \pi$, is an injective k-algebra homomorphism; furthermore, $\pi^*(A[X/G]) = A[X]^G$. We now proceed in several steps.

Step 1. First we show that $A[X]^G$ is a finitely generated k-algebra and that $A[X] \supseteq A[X]^G$ is an integral extension. This is done as follows. Since $A[X]$ is finitely generated, we can write $A[X] = k[f_1, \ldots, f_r]$, where $f_i \in A[X]$. Now set

$$F_i := \prod_{g \in G}\left(T - \pi_g^*(f_i)\right) \qquad \text{for } 1 \leqslant i \leqslant r,$$

where T is an indeterminate. Thus, F_i is a monic polynomial with coefficients in $A[X]$. By construction, since the product runs over all elements of G, the coefficients are even invariant under π_g^* for all $g \in G$. So the coefficients of F_i lie in $A[X]^G$. Thus, since $F_i(f_i) = 0$, we have found an equation of integral dependence of f_i over $A[X]^G$.

This already shows that $A[X]$ is integral over $A[X]^G$. Now write $F_i = \sum_{j=0}^{|G|} a_{ij} T^j$ ($a_{ij} \in A[X]^G$) and consider the k-algebra

$$A_0 := k[a_{ij} \mid 1 \leqslant i \leqslant r,\ 0 \leqslant j \leqslant |G|] \subseteq A[X]^G.$$

Then A_0 is finitely generated, and every f_i is integral over A_0. Consequently, by Corollary 2.1.2(a), $A[X]$ is a finitely generated A_0-module. So $A[X]^G$ will also be a finitely generated A_0-module; see Exercise 1.8.1. This implies that $A[X]^G$ is finitely generated as a k-algebra.

Step 2. We now show that $(X/G, A[X/G])$ is an affine variety. This means that we have to verify the conditions in Definition 2.1.6. By Step 1 and the fact that $\pi^* \colon A[X/G] \to A[X]^G$ is an isomorphism, we know that $A[X/G]$ is a finitely generated k-algebra. Now let $x, y \in X$ be such that the G-orbits O_x and O_y are disjoint. We must show that there exists some $\bar{f} \in A[X/G]$ such that $\bar{f}(O_x) \neq \bar{f}(O_y)$. In other words, we must show that there exists some $f \in A[X]^G$ such that $f(x) \neq f(y)$. Now, since G is a finite group, O_x is a finite set and, hence, closed. So there exists some $a \in A[X]$ such that $a(g.x) = 0$ for all $g \in G$ and $a(y) = 1$. Hence, setting $a' := a - 1 \in A[X]$, we have $a'(g.x) = -1$ for all $g \in G$, and $a'(y) = 0$. Now set $f := \prod_{g \in G} \pi_g^*(a') \in A[X]^G$. Then we have $f(x) = \prod_{g \in G} \pi_g^*(a')(x) = \prod_{g \in G} a'(g.x) = \pm 1$ and $f(y) = \prod_{g \in G} \pi_g^*(a')(y) = \prod_{g \in G} a'(g.y) = 0$, as required. Finally, let $\lambda \colon A[X/G] \to k$ be any k-algebra homomorphism. We must show that λ is given by evaluation at a point in X/G. Now, $\ker(\lambda)$ is a maximal ideal in $A[X/G]$. Since $A[X] \supseteq A[X]^G = \pi^*(A[X/G])$ is an integral extension, Lemma 2.2.4 ('going up') shows that there exists a maximal ideal $\mathfrak{m} \subseteq A[X]$ such that $\pi^*(\ker(\lambda)) = A[X]^G \cap \mathfrak{m}$. Now, by Theorem 2.1.9, we have $\mathfrak{m} = \ker(\varepsilon_x)$ for some $x \in X$. Let O_x be the G-orbit of x. Then it is readily checked that λ is the evaluation at O_x.

Step 3. It remains to check that the three conditions in §2.5.8 are satisfied. This is clear for (a) and (c). To verify (b), let $U \subseteq X$ be an open subset. We must show that $\pi(U) \subseteq X/G$ is open. Now $U' = \bigcup_{g \in G} g.U$ also is open, and we have $\pi(U) = \pi(U')$. Thus, we may assume without loss generality that $U = \pi^{-1}(\pi(U))$. But then we have $\pi(X \setminus U) = (X/G) \setminus \pi(U)$. Since π is a closed map (see Lemma 2.2.5), this shows that $\pi(U)$ is open, as desired. $\square$

2.5.11 Example Let $G \subseteq \mathrm{GL}_n(k)$ be a finite subgroup. Then we may regard k^n as a G-variety in a natural way; the algebra of regular functions on k^n is just given by the polynomial ring $R = k[X_1, \ldots, X_n]$. By Theorem 2.5.10, the ring of invariants R^G is finitely generated; let us write $R^G = k[f_1, \ldots, f_m]$, where $F := \{f_1, \ldots, f_m\} \subseteq R^G$. Now consider the ideal

$$I_F := \{h \in k[Y_1, \ldots, Y_m] \mid h(f_1, \ldots, f_m) = 0\} \subseteq k[Y_1, \ldots, Y_m]$$

and set $V_F := \mathbf{V}(I_F) \subseteq k^m$. We have a corresponding morphism $\varphi \colon k^n \to V_F$, $x \mapsto (f_1(x), \ldots, f_m(x))$, which is constant on the G-orbits on k^n. Hence, by the universal property in §2.5.8, we obtain an induced morphism

$$\bar{\varphi} \colon k^n/G \to V_F.$$

We claim that this map is an isomorphism. Indeed, the map $k[Y_1, \ldots, Y_m] \to R^G$, $h \mapsto h(f_1, \ldots, f_m)$, is a surjective k-algebra homomorphism with kernel I_F. Since R^G is the algebra of regular functions on k^n/G, it has no nilpotent elements except 0. Thus, we have $I_F = \sqrt{I_F}$ and so $I_F = \mathbf{I}(V_F)$ by Hilbert's nullstellensatz (Theorem 2.1.9). Hence, we obtain an induced isomorphism $A[V_F] \to R^G$, which is easily seen to be equal to $\bar{\varphi}^*$, as required. Now, assuming that we have found $F = \{f_1, \ldots, f_m\}$, we can explicitly find generators for I_F. Indeed, by Example 1.3.6, we have $I_F = J_F \cap k[Y_1, \ldots, Y_m]$, where

$$J_F := (f_1 - Y_1, \ldots, f_m - Y_m) \subseteq k[X_1, \ldots, X_n, Y_1, \ldots, Y_m].$$

Hence, choosing the **LEX** order with $X_1 \succ \cdots \succ X_n \succ Y_1 \succ \cdots \succ Y_m$, we have $I_F = (G \cap k[Y_1, \ldots, Y_m])$, where G is a *Groebner basis* for J_F.

2.5.12 Quotients by finite normal subgroups Let G be an affine algebraic group and Z be a finite closed normal subgroup of G. We consider G as a Z-variety, where $z \in Z$ acts by left multiplication on G. Then $G/Z = \{Zg \mid g \in G\}$ is the set of Z-orbits on G, and this is an abstract group. Since Z is finite, we also have a structure of G/Z as an affine variety by Theorem 2.5.10. We claim that G/Z is an affine algebraic group. To see this, we must check that multiplication and inversion in G/Z are given by morphisms of affine varieties.

Let $\pi \colon G \to G/Z$, $g \mapsto Zg$, be the natural map and $\iota \colon G \to G$ be inversion in G. Then $\pi \circ \iota \colon G \to G/Z$ is a morphism of affine varieties which is constant on the cosets of Z. Hence, by the universal property of the affine quotient, we obtain an induced map $\bar{\iota}\colon G/Z \to G/Z$ which is a morphism of affine varieties and which gives exactly the inversion in G/Z. Now consider the multiplication map $\mu\colon G \times G \to G$. The map $\pi \circ \mu\colon G \times G \to G/Z$ is a morphism of affine varieties which is constant on the cosets of the finite closed normal subgroup $Z \times Z \subseteq G \times G$. Hence, applying the universal property of the affine quotient of $G \times G$ by $Z \times Z$ and using Exercise 2.7.11, we obtain an induced map $\bar{\mu}\colon (G/Z) \times (G/Z) \to G/Z$ which is a morphism of affine varieties and which gives exactly the multiplication map for G/Z.

2.5.13 Example Consider the group $G = \mathrm{SL}_n(k)$, and let $Z \subseteq G$ be the centre of G. Then $Z = \{aI_n \mid a^n = 1\}$ is a finite group. So, by §2.5.12, the factor group $\mathrm{PSL}_n(k) = \mathrm{SL}_n(k)/Z$ can be naturally regarded as an affine algebraic group.

Similarly, let G be one of the classical groups introduced in §1.3.15 and §1.3.16, that is, G equals $\mathrm{SO}_{2m+1}(k)$ (any $m \geqslant 1$, $\mathrm{char}(k) \neq 2$), $\mathrm{Sp}_{2m}(k)$, or $\mathrm{SO}_{2m}^+(k)$ (any $m \geqslant 1$, any characteristic). If $G = \mathrm{SO}_{2m}^+(k)$, we assume that $m \geqslant 2$. Then, by Example 2.4.11, we have $Z(G) = \{\pm 1\}$. Hence, we obtain affine algebraic groups $G/Z(G)$ which we denote by $\mathrm{PSO}_{2m+1}(k)$, $\mathrm{PSp}_{2m}(k)$, and $\mathrm{PSO}_{2m}^+(k)$, respectively.

We close this section with the following result which provides a criterion for checking if a morphism of affine varieties is a quotient morphism as in §2.5.8. It relies on the results on birational equivalences and normal varieties in Section 2.3. (If we had shown the general form of *Zariski's main theorem*, then the assumption that φ be finite in the following result would be unnecessary.)

2.5.14 Proposition *Let X be an irreducible G-variety, where G is a finite group. Let $\varphi\colon X \to X_0$ be a dominant finite morphism such that the following conditions are satisfied.*

 (a) X_0 is normal.
 (b) For each $x_0 \in X_0$, the set $\varphi^{-1}(x_0) \subseteq X$ is a G-orbit.
 (c) $d_p\varphi\colon T_p(X) \to T_{\varphi(p)}(X_0)$ is surjective for some $p \in X$.

Then the pair (X_0, φ) is an affine quotient of X by G.

Proof Let $(X/G, \pi)$ be the affine quotient constructed in Theorem 2.5.8. Since φ is constant on the G-orbits on X, the universal property in 2.5.8 implies that there exists a unique morphism $\bar{\varphi} \colon X/G \to X_0$ such that $\varphi = \bar{\varphi} \circ \pi$. We shall show that $\bar{\varphi}$ is an isomorphism. To prove this, we use the criterion in Remark 2.3.4 (the very weak form of Zariski's main theorem). First note that $\bar{\varphi}$ is injective by (b). Next, X_0 is normal by (a). So it only remains to show that $\bar{\varphi}$ is a birational equivalence. By Theorem 2.3.10, it will be enough to find some non-singular $q \in X/G$ such that $d_q\bar{\varphi}$ is surjective.

To show the existence of q, first note that $\dim X = \dim X_0$. Hence (c) implies that $\dim T_{\varphi(p)}(X) \leqslant \dim T_p(X) = \dim X = \dim X_0$. Thus, by Theorem 1.4.11, $\varphi(p) \in X_0$ is non-singular and so $d_p\varphi$ is an isomorphism. As in the proof of Theorem 1.4.15, there exists an open set $U \subseteq X$ containing p such that the above properties hold for all $x \in U$, that is, x is non-singular, $\varphi(x)$ is non-singular, and $d_x\varphi$ is an isomorphism. Now, by §2.5.8(b), $\pi(U) \subseteq X/G$ is a non-empty open set. Hence, by Theorem 1.4.11, there exist some non-singular point $x \in U$ such that $q := \pi(x) \in \pi(U)$ is non-singular. Now consider the factorization of differentials $d_x\varphi = d_q\bar{\varphi} \circ d_x\pi$. Since $d_x\varphi$ is an isomorphism, we conclude that $d_q\bar{\varphi}$ is surjective, as required. $\square$

2.6 The unipotent variety of the special linear groups

This section contains a detailed study of the conjugation action of $\mathrm{SL}_n(k)$ on the variety of unipotent matrices. This will illustrate many of the ideas in the previous sections. We assume that the reader is familiar with the theory of Jordan normal forms for matrices. Throughout, let

$$G = \mathrm{SL}_n(k), \qquad \text{where } k \text{ is an algebraically closed field.}$$

A matrix $g \in G$ is called *unipotent* if all eigenvalues of g are 1. Note that, since k is algebraically closed, we have $\mathrm{GL}_n(k) = ZG$, where Z is the group of all scalar matrices in $\mathrm{GL}_n(k)$. Thus, two elements of G are conjugate in G if and only if they are conjugate in $\mathrm{GL}_n(k)$, and the latter condition holds if and only if the two elements have the same Jordan normal form. In particular, an element $g \in G$ is unipotent if and only if g is conjugate to a matrix in $U_n(k)$, where $U_n(k)$ is the group of all upper triangular matrices with 1 on the diagonal; see Example 1.3.11(a). Recall that the conjugation action of G on itself is given by the regular map

$$G \times G \to G, \qquad (g, x) \mapsto gxg^{-1}.$$

The *unipotent variety* is defined by $\mathbb{U}_n := \{g \in G \mid g \text{ unipotent}\}$. We claim that

$\mathbb{U}_n$ *is a closed irreducible subset of* G
(invariant v under conjugation).

This is seen as follows. A matrix $g \in G$ is unipotent if and only if its characteristitic polynomial is of the form $(X - 1)^n$ (where X is an indeterminate). Hence, using the Cayley–Hamilton theorem, we have $\mathbb{U}_n = \{g \in G \mid (g - I_n)^n = 0\}$, and so we see that $\mathbb{U}_n$ is closed. Now we consider the morphism $\varphi \colon G \times U_n(k) \to \mathbb{U}_n$, $(g, u) \mapsto gug^{-1}$. By the above remarks, φ is surjective. Furthermore, G and $U_n(k)$ are irreducible and so $G \times U_n(k)$ is irreducible by Proposition 1.3.8. Hence, Remark 1.3.2 implies that $\mathbb{U}_n$ is irreducible, as claimed.

We want to understand how $\mathbb{U}_n$ decomposes into conjugacy classes. The Jordan normal form of an element in $\mathbb{U}_n$ consists of Jordan blocks corresponding to the eigenvalue 1, and such a block is given by

$$J_m = \begin{bmatrix} 1 & 1 & 0 & \cdots & 0 \\ 0 & 1 & 1 & \ddots & \vdots \\ \vdots & \ddots & \ddots & \ddots & 0 \\ 0 & & 0 & 1 & 1 \\ 0 & \cdots & 0 & 0 & 1 \end{bmatrix} \in M_m(k).$$

Thus, we only need to specify how many blocks of a given size there are. Hence,

the conjugacy classes in $\mathbb{U}_n$ *are parametrized by*
the partitions of n.

Recall that a partition of n is a finite and weakly decreasing sequence of non-negative integers $\lambda = (\lambda_1 \geqslant \lambda_2 \geqslant \cdots \geqslant \lambda_r \geqslant 0)$ such that $\sum_{i=1}^r \lambda_i = n$. We will not distinguish between partitions which have different length but the same non-zero parts. We write $\lambda \vdash n$ if λ is a partition of n and denote by O_λ the corresponding conjugacy class in $\mathbb{U}_n$. A representative $u_\lambda \in O_\lambda$ is given by the block diagonal matrix with r Jordan blocks of sizes $\lambda_1, \ldots, \lambda_r$ on the diagonal. The following result gives some information on the centralizer of a unipotent element. For any $x \in G$, the centralizer is defined by $C_G(x) = \{g \in G \mid gxg^{-1} = x\}$.

2.6.1 Proposition *Let $\lambda \vdash n$. Then $\mathrm{C}_G(u_\lambda)$ is a closed subgroup and we have*

$$\dim \mathrm{C}_G(u_\lambda) = n - 1 + 2n(\lambda) \quad \text{where} \quad n(\lambda) := \sum_{i=1}^{r} (i-1)\lambda_i.$$

Proof We set $\mathcal{C}_\lambda := \{A \in M_n(k) \mid Au_\lambda = u_\lambda A\}$. Then $\mathcal{C}_\lambda$ is a k-subspace of $M_n(k)$ and, hence, a closed irreducible subset of $M_n(k)$; see Exercise 1.8.6. Furthermore, the dimension of $\mathcal{C}_\lambda$ as an algebraic set is the same as the dimension as a k-subspace of $M_n(k)$. Now note that $\mathrm{C}_G(u_\lambda)$ is the intersection of $\mathcal{C}_\lambda$ with the zero set of the polynomial given by the determinant. Thus, $\mathrm{C}_G(u_\lambda)$ is a hypersurface in $\mathcal{C}_\lambda$, and so $\dim \mathrm{C}_G(u_\lambda) = \dim \mathcal{C}_\lambda - 1$ by Example 1.2.16(b).

It remains to determine $\dim_k \mathcal{C}_\lambda$. For this purpose, write $\lambda = (\lambda_1 \geqslant \cdots \geqslant \lambda_r > 0)$. Let $A \in M_n(k)$ and partition A into blocks as follows:

$$A = \begin{bmatrix} A_{11} & A_{12} & \cdots & A_{1r} \\ A_{21} & A_{22} & \cdots & A_{2r} \\ \vdots & \vdots & & \vdots \\ A_{r1} & A_{r2} & \cdots & A_{rr} \end{bmatrix}, \quad \text{where} \quad A_{ij} \in k^{\lambda_i \times \lambda_j}.$$

Then the condition that $u_\lambda A = A u_\lambda$ is equivalent to the condition

$$J_{\lambda_i} A_{ij} = A_{ij} J_{\lambda_j} \quad \text{for all } i, j \in \{1, \ldots, r\}.$$

Hence our task is reduced to finding the set of matrices commuting with two Jordan blocks of possibly different size. A direct computation shows that these are precisely the matrices of the form

$$A_{ij} = \begin{bmatrix} 0 & \cdots & 0 & a_1 & a_2 & \cdots & & a_{\lambda_j} \\ 0 & \cdots & 0 & 0 & a_1 & a_2 & \cdots & a_{\lambda_j - 1} \\ \vdots & & \vdots & \vdots & \ddots & \ddots & \ddots & \\ 0 & \cdots & 0 & 0 & \cdots & 0 & a_1 & a_2 \\ 0 & \cdots & 0 & 0 & \cdots & 0 & 0 & a_1 \end{bmatrix} \quad \text{or} \quad \begin{bmatrix} a_1 & a_2 & \cdots & & a_{\lambda_j} \\ 0 & a_1 & a_2 & \cdots & a_{\lambda_j - 1} \\ \vdots & \ddots & \ddots & \ddots & \\ 0 & \cdots & 0 & a_1 & a_2 \\ 0 & \cdots & 0 & 0 & a_1 \\ 0 & \cdots & 0 & 0 & 0 \\ \vdots & & \vdots & \vdots & \vdots \\ 0 & \cdots & 0 & 0 & 0 \end{bmatrix}$$

according to whether $\lambda_i \leqslant \lambda_j$ or $\lambda_i > \lambda_j$. Thus, the dimension of the space of all such matrices A_{ij} is just $\min\{\lambda_i, \lambda_j\}$ and so

$\dim_k C_\lambda = \sum_{i,j} \min\{\lambda_i, \lambda_j\}$. Since $\lambda_i \geqslant \lambda_j$ if $i \leqslant j$, we can evaluate this sum as follows.

$$\dim_k C_\lambda = \sum_{i,j=1}^{r} \min\{\lambda_i, \lambda_j\} = n + 2 \sum_{1 \leqslant i < j \leqslant r} \lambda_j = n + 2n(\lambda),$$

as required. $\qquad\qquad\square$

2.6.2 Remark The above result does not decide whether $C_G(u_\lambda)$ is connected. In general, this is not the case. To formulate the issue, let $p := \operatorname{char}(k)$ if $\operatorname{char}(k) > 0$, and $p := 1$ if $\operatorname{char}(k) = 0$. Then $C_G(u_\lambda)/C_G(u_\lambda)^\circ$ is a cyclic group of order $\gcd(n', \lambda_1, \lambda_2, \ldots)$, where $\lambda_1, \lambda_2, \ldots$ are the non-zero parts of λ and n' is the largest divisor of n which is prime to p. In fact, we have $C_G(u_\lambda) = Z(G)\, C_G(u_\lambda)^\circ$, where $Z(G)$ is the centre of G, and so $C_G(u_\lambda)/C_G(u_\lambda)^\circ$ is seen to be a quotient of $Z(G)$. We leave the proof of these assertions as an exercise to the reader.

In order to study some geometric properties of the conjugacy classes of unipotent elements in G, we need a convenient characterization of the Jordan normal form of a unipotent matrix. This is provided by the following result.

2.6.3 Lemma *Let $A \in \mathbb{U}_n$ be any unipotent matrix and $\lambda \vdash n$. Then we have*

$$A \in O_\lambda \quad\Leftrightarrow\quad \sum_{i=1}^{t} \lambda_i^* = n - r_t(A) \qquad for\ 1 \leqslant t \leqslant n,$$

where we set $r_t(A) = \operatorname{rank}_k(A - I_n)^t$ for all t.

Here, the numbers λ_i^* are defined as follows. If λ has parts $\lambda_1 \geqslant \lambda_2 \geqslant \cdots$, we set $\lambda_i^* = |\{j \mid \lambda_j \geqslant i\}|$ for $i = 1, 2, \ldots$. Then $\lambda^* := (\lambda_1^* \geqslant \lambda_2^* \geqslant \cdots \lambda_n^* \geqslant 0)$ is called the *conjugate partition* (which also is a partition of n). Note that λ and λ^* determine each other, since $(\lambda^*)^* = \lambda$.

Proof Assume that $A \in O_\lambda$. For $1 \leqslant i \leqslant n$, let $l_i \geqslant 0$ be the number of parts of λ which are equal to i. (In other words, l_i is the number of Jordan blocks of size i in the Jordan normal form of A.)

With this notation, we have

$$\lambda_i^* = |\{j \mid \lambda_j \geq i\}| = l_i + l_{i+1} + \cdots + l_n \quad \text{for } 1 \leq i \leq n.$$

Now the numbers l_i satisfy the following relations:

$$
\begin{aligned}
l_1 + 2l_2 + 3l_3 \quad \cdots \quad +(n-1)l_{n-1} \quad\;\; +nl_n &= \operatorname{rank}_k(A-I_n)^0 = n \\
l_2 + 2l_3 \quad \cdots \quad +(n-2)l_{n-1} +(n-1)l_n &= \operatorname{rank}_k(A-I_n)^1 \\
&\;\;\vdots \\
l_{n-1} \quad\quad\quad +2l_n &= \operatorname{rank}_k(A-I_n)^{n-2} \\
l_n &= \operatorname{rank}_k(A-I_n)^{n-1}.
\end{aligned}
$$

(This easily follows from the fact that $\operatorname{rank}_k(J_l - I_l)^i = l - i$ for $0 \leq i \leq l$.) The above relations form a system of linear equations for the quantities l_i. Since the matrix of that system is triangular with an all-unity diagonal, we see that the numbers l_i are uniquely determined by the vector $(r_1(A), \ldots, r_n(A))$.

Furthermore, using the above expression for λ_i^*, we see that $r_t(A) = \lambda_{t+1}^* + \lambda_{t+2}^* + \cdots + \lambda_n^*$, and so $n - r_t(A) = \lambda_1^* + \cdots + \lambda_t^*$, as required. $\qquad\square$

To state the following result, we introduce an order relation on the partitions of n. Assume that $\lambda = (\lambda_1 \geq \cdots \geq \lambda_r \geq 0)$ and $\mu = (\mu_1 \geq \cdots \geq \mu_r \geq 0)$ are partitions of n. Then we write

$$\lambda \overset{\text{def}}{\trianglelefteq} \mu \quad \text{if and only if} \quad \sum_{i=1}^{t} \lambda_i \leq \sum_{i=1}^{t} \mu_i \quad \text{for } 1 \leq t \leq r.$$

Then $\trianglelefteq$ is a partial, but in general not total, order. It is called the *dominance order* on the set of partitions of n. We have $\lambda \trianglelefteq \mu$ if and only if $\mu^* \trianglelefteq \lambda^*$; see §I.1.11 of Macdonald (1995).

2.6.4 Lemma   *For any partition $\lambda \vdash n$, the set $\tilde{C}_\lambda := \bigcup_{\mu \trianglelefteq \lambda} O_\mu$ is closed in G.*

Proof   Let $\mu \vdash n$ and take $A, B \in \mathbb{U}_n$ with $A \in O_\lambda$ and $B \in O_\mu$. We have already remarked above that $\mu \trianglelefteq \lambda \Leftrightarrow \lambda^* \trianglelefteq \mu^*$. Furthermore, by Lemma 2.6.3, the latter condition is equivalent to $r_t(B) \leq r_t(A)$ for all t. Thus, we have shown that

$$\tilde{C}_\lambda = \{B \in \mathbb{U}_n \mid r_t(B) \leq r_t(A) \text{ for } 1 \leq t \leq n\} \quad \text{(where } A \in O_\lambda\text{).}$$
$$(*)$$

Now, the condition $r_t(B) \leq r_t(A)$ means that the determinants of all square submatrices of B of size strictly bigger than $r_t(A)$ are 0.

Since these determinants are given by polynomials in the coefficients of B, we see that $\tilde{C}_\lambda$ is closed. $\qquad\qquad\square$

As in Proposition 2.5.2, given partitions $\lambda, \mu \vdash n$, we write $O_\mu \preceq O_\lambda$ if $O_\mu \subseteq \overline{O}_\lambda$. Thus, by Lemma 2.6.4, we already know that $O_\mu \preceq O_\lambda \Rightarrow \mu \trianglelefteq \lambda$. The following result shows that the reverse implication also holds.

2.6.5 Theorem   *Let $\lambda, \mu \vdash n$. Then we have $O_\mu \subseteq O_\lambda$ if and only if $\mu \trianglelefteq \lambda$.*

Proof   By Lemma 2.6.4 we already know that $O_\mu \preceq O_\lambda \Rightarrow \mu \trianglelefteq \lambda$. To prove the reverse implication, let $\lambda_1 \geqslant \cdots \geqslant \lambda_r > 0$ be the non-zero parts of λ, and consider the subgroup $U_\lambda(k) \subseteq G$ consisting of all block matrices of the form

$$
B = \begin{bmatrix}
I_{\lambda_1} & B_{12} & \cdots & B_{1r} \\
0 & I_{\lambda_2} & \ddots & \vdots \\
\vdots & \ddots & \ddots & B_{r-1,r} \\
0 & \cdots & 0 & I_{\lambda_r}
\end{bmatrix}
\qquad \text{with } B_{ij} \in k^{\lambda_i \times \lambda_j},
$$

where the diagonal blocks are the identity matrices of sizes $\lambda_1, \ldots, \lambda_r$. It is readily checked that, for any matrix B as above, we have

$$
r_t(B) = \operatorname{rank}_k (B - I_n)^t \leqslant n - \sum_{i=1}^{t} \lambda_i \qquad \text{for } 1 \leqslant t \leqslant n. \qquad (\dagger)
$$

Thus, working with the conjugate partition λ^* and using Lemma 2.6.3, we have $r_t(B) \leqslant r_t(A')$ for all t, where $A' \in O_{\lambda^*}$. So, by relation $(*)$ in the proof of Lemma 2.6.4, we have

$$
U_\lambda(k) \subseteq \tilde{C}_{\lambda^*}. \qquad (\dagger')
$$

We can in fact find some element in $U_\lambda(k)$ for which equality holds in $(\dagger)$ for all $t = 1, \ldots, n$. Indeed, let $u'_\lambda \in U_\lambda(k)$ be the matrix

$$
u'_\lambda = \begin{bmatrix}
I_{\lambda_1} & E_{12} & 0 & \cdots & & 0 \\
0 & I_{\lambda_2} & E_{23} & \ddots & & \vdots \\
\vdots & \ddots & \ddots & \ddots & & 0 \\
0 & \cdots & 0 & I_{\lambda_{r-1}} & E_{r-1,r} \\
0 & \cdots & 0 & 0 & I_{\lambda_r}
\end{bmatrix}
\quad \text{where } E_{i,i+1} = \begin{bmatrix}
1 & 0 & \cdots & 0 \\
0 & 1 & \ddots & \vdots \\
\vdots & \ddots & \ddots & 0 \\
0 & \cdots & 0 & 1 \\
0 & \cdots & 0 & 0 \\
\vdots & & \vdots & \vdots \\
0 & \cdots & 0 & 0
\end{bmatrix}.
$$

(Note that $E_{i,i+1}$ has λ_i rows and λ_{i+1} columns, where $\lambda_i \geqslant \lambda_{i+1}$.) Then a direct computation shows that $r_t(u'_\lambda) = n - \sum_{i=1}^t \lambda_i$ for all t, and so $u'_\lambda \in O_{\lambda^*}$ by Lemma 2.6.3. Now we claim that

$$U_\lambda(k) \subseteq \overline{O}_{\lambda^*}.$$

For this purpose, we consider the morphism $\varphi_\lambda \colon G \times U_\lambda(k) \to G$, $(g, u) \mapsto gug^{-1}$. We denote by $\mathbb{U}_\lambda$ the image of φ_λ. For example, if $\lambda = (1, 1, \ldots, 1)$, then $U_\lambda(k) = U_n(k)$ and $\mathbb{U}_\lambda = \mathbb{U}_n$ is the unipotent variety of G. We will now try to prove results about $\mathbb{U}_\lambda$ which are similar to those obtained for $\mathbb{U}_n$. The subgroup $U_\lambda(k)$ certainly is closed and irreducible. (Indeed, since we may place any element of k in the blocks above the diagonal, we have that $U_\lambda(k)$ is isomorphic to some affine space k^N as an affine variety.) Consequently, $\mathbb{U}_\lambda$ is irreducible and, hence, so is $\overline{\mathbb{U}}_\lambda$. Furthermore, $\mathbb{U}_\lambda$ is invariant under conjugation by elements of G, that is, $\mathbb{U}_\lambda$ is a union of unipotent classes of G. Hence the same also holds for $\overline{\mathbb{U}}_\lambda$. Writing the irreducible closed set $\overline{\mathbb{U}}_\lambda$ as a union of finitely many conjugacy classes (and then taking the closure of these classes), we see that there exists some $\lambda_0 \vdash n$ such that $\overline{\mathbb{U}}_\lambda = \overline{O}_{\lambda_0}$. Hence it will be enough to show that $\lambda_0 = \lambda^*$. Since $u'_\lambda \in U_\lambda(k) \cap O_{\lambda^*}$, we have $O_{\lambda^*} \subseteq \overline{\mathbb{U}}_\lambda = \overline{O}_{\lambda_0}$ and so $O_{\lambda^*} \preceq O_{\lambda_0}$. We have already seen that this implies $\lambda^* \trianglelefteq \lambda_0$. On the other hand, we have $\mathbb{U}_\lambda \subseteq \tilde{C}_{\lambda^*}$ by $(\dagger')$ and so $O_{\lambda_0} \subseteq \tilde{C}_{\lambda^*}$. By definition, this means that $\lambda_0 \trianglelefteq \lambda^*$. Thus, we must have $\lambda_0 = \lambda^*$, as claimed.

So we have reached the conclusion that $U_\lambda(k) \subseteq \overline{O}_{\lambda^*}$. Since this holds for all λ, we also have $U_{\lambda^*}(k) \subseteq \overline{O}_\lambda$. Now it will be enough to show that, for each $\mu \vdash n$ with $\mu \trianglelefteq \lambda$, we can find a representative of O_μ in the subgroup $U_{\lambda^*}(k)$. We will first show how this works in the special case where

$$\lambda = (r \geqslant s \geqslant 0) \quad \text{and} \quad \mu = (r - 1 \geqslant s + 1 \geqslant 0), \quad \text{where } r + s = n.$$

Now the conjugate partition λ^* has s parts equal to 2 and the remaining non-zero parts are equal to 1. Similarly, μ^* has $s + 1$ parts equal to 2 and the remaining non-zero parts are equal to 1. Then it is obvious that $U_{\mu^*}(k) \subseteq U_{\lambda^*}(k)$. In particular, the representative $u'_{\mu^*} \in O_\mu$ defined above lies in $U_{\lambda^*}(k)$. Thus, the implication $\mu \trianglelefteq \lambda \Rightarrow O_\mu \preceq O_\lambda$ is proved in the above special case.

Finally, we show that the general case can be reduced to this special case. Indeed, in order to show that $O_\mu \preceq O_\lambda$ for all $\mu \vdash n$ with $\mu \neq \lambda$ and $\mu \trianglelefteq \lambda$, it is clearly enough to do this for those partitions μ which are maximal with respect to these properties. It is readily checked that such a partition μ has the following form:

$$\mu = (\lambda_1, \ldots, \lambda_{i-1}, \lambda_i - 1, \lambda_{i+1}, \ldots, \lambda_{j-1}, \lambda_j + 1, \lambda_{j+1}, \ldots, \lambda_r \geqslant 0),$$

where $1 \leqslant i < j \leqslant r$ and $\lambda_i - 1 \geqslant \lambda_{i+1}$; see §I.1.16 of Macdonald (1995). Now we can find representatives $u'_\lambda \in O_\lambda$ and $u'_\mu \in O_\mu$ such that

$$u'_\lambda = \begin{bmatrix} J_{\lambda_i} & 0 & 0 & \cdots & 0 \\ 0 & J_{\lambda_j} & 0 & \cdots & 0 \\ \hline 0 & 0 & & & \\ \vdots & \vdots & & C & \\ 0 & 0 & & & \end{bmatrix}, \quad u'_\mu = \begin{bmatrix} J_{\lambda_i - 1} & 0 & 0 & \cdots & 0 \\ 0 & J_{\lambda_j + 1} & 0 & \cdots & 0 \\ \hline 0 & 0 & & & \\ \vdots & \vdots & & C & \\ 0 & 0 & & & \end{bmatrix},$$

where C is a unipotent matrix. Setting $m = \lambda_i + \lambda_j$, we consider $\mathrm{SL}_m(k)$ as a subgroup of $\mathrm{SL}_n(k)$, by sending a matrix A in $\mathrm{SL}_m(k)$ to the block diagonal matrix in $\mathrm{SL}_n(k)$ with diagonal blocks A and I_{n-m}. Then it will be enough to show that the conjugacy class in $\mathrm{SL}_m(k)$ which is labelled by the partition $(\lambda_i \geqslant \lambda_j)$ is contained in the Zariski closure of the class labelled by $(\lambda_i - 1 \geqslant \lambda_j + 1)$. Thus, we are reduced to the special case considered above. $\qquad\square$

2.6.6 Example   There is a unique maximal partition with respect to the dominance order; this is the partition (n). The corresponding unipotent element $u_{(n)}$ is just the Jordan block of size n. By Theorem 2.6.5, all unipotent classes are contained in the closure of $O_{(n)}$. Hence, $O_{(n)}$ is dense and open in $\mathbb{U}_n$. For this reason, the elements in $O_{(n)}$ are called *regular unipotent* elements. We claim

$$\dim \mathbb{U}_n = \dim \overline{O}_{(n)} = n(n-1).$$

Indeed, by Proposition 2.6.1, we have $\dim C_G(u_{(n)}) = n - 1$. Thus, the above assertion follows from Proposition 2.5.3, using that $\dim G = n^2 - 1$.

2.7   Bibliographic remarks and exercises

For the proof of the strong form of Hilbert's nullstellensatz, we followed Chapter 7, Exercise 14, of Atiyah and Macdonald (1969). The abstract definition of affine varieties is taken from Steinberg (1974). The results on the images of a morphism can all be found in §I.8

of Mumford (1988). A classical application of Corollary 2.2.8 in combination with the differential criterion for dominance (see Proposition 1.4.15) is the proof that any two Cartan subalgebras of a finite-dimensional complex semisimple Lie algebra are conjugate by an inner automorphism; see §D.22 of Fulton and Harris (1991).

The discussion of birational equivalences and normal varieties partly follows Springer (1998) and Shafarevich (1994). The proof of Theorem 2.3.10 essentially appears in §II.5.4 of Shafarevich (1994). The proof of Proposition 2.3.11 follows §5.17 of Atiyah and Macdonald (1969) and §5.2.11 of Springer (1998). In fact, the argument in that proof can be generalized to show that, given an irreducible affine variety X over k, then the integral closure of $A[X]$ in its field of fractions is a finitely generated k-algebra; see §II.5.2, Theorem 4, of Shafarevich (1994). The examples in Section 2.4 are taken from p. 31 and p. 277 of Mumford (1988). For a thorough discussion of the geometric content of Zariski's main theorem, see §III.9 of Mumford (1988). See also §III.11 of Hartshorne (1977) for a 'higher point of view', involving the cohomology theory of schemes.

The connectedness criterion Theorem 2.4.6 can be found in any book on algebraic groups; see, for example, §7.5 of Humphreys (1991).

The proof of Theorem 2.5.5 and its corollary is adapted from the Appendix to §2.11 of Steinberg (1974). For a different argument, see 5.3.2 of Springer (1998). The construction of the affine quotient by a finite group in Theorem 2.5.10 appears in §5.52 of Fogarty (1969). More generally, it is known that if H is closed subgroup of an affine algebraic group G, then one can define on G/H a natural structure as a 'quasi-projective' variety. This fact can be proved using the general form of Zariski's main theorem (for this approach, see §5.5, of Springer (1998)), or one can proceed as in Chapter 12 of Humphreys (1991), using a purely algebraic result due to Chevalley; see §5.3A of Humphreys (1991). See also Chapter 2 of Borel (1991) for a more general discussion of quotients.

For more on invariants of finite groups as in Example 2.5.11, see Chapter 7 of Cox *et al.* (1992) and Sturmfels (1993). The results on unipotent classes in $\mathrm{SL}_n(k)$ in Section 2.6 have been generalized to symplectic and orthogonal groups and, more generally, to groups of Lie type; see Spaltenstein (1982).

In the exercises, all affine varieties and algebraic groups are defined over an algebraically closed ground field k.

2.7.1 Exercise Let V be a finite dimensional k-vector space. Let $A[V]$ be the k-algebra generated by all elements of the dual space $\mathrm{Hom}_k(V, k)$:

$$A[V] := k[\lambda \mid \lambda \in \mathrm{Hom}(V, k)] \subseteq \mathrm{Maps}(V, k).$$

Show that $(V, A[V])$ is an affine variety. Furthermore, if $\{e_1, \ldots, e_n\}$ is a basis of V and $\{\lambda_1, \ldots, \lambda_n\}$ is the dual basis of $\mathrm{Hom}_k(V, k)$, then the map $V \to k^n$, $v \mapsto (\lambda_1(v), \ldots, \lambda_n(v))$, is an isomorphism of affine varieties.

2.7.2 Exercise Let $A \neq \{0\}$ be a finitely generated k-algebra with 1 which has no nilpotent elements except 0. Let X be the set of all maximal ideals in A. For every $x \in X$, there is a unique k-algebra homomorphism $\lambda_x \colon A \to k$ with kernel x. For $f \in A$, we define a function $\dot{f} \colon X \to k$ by $\dot{f}(x) := \lambda_x(f)$ for $x \in X$.

Show that the assignment $f \mapsto \dot{f}$ is injective. Thus, we may regard A as a k-subalgebra of $\mathrm{Maps}(X, k)$. Then show that (X, A) is an affine variety.

[*Hint.* If $\lambda_x(f) = 0$ for all $x \in X$, then f lies in the intersection of all maximal ideals of A. To see that this intersection is 0, one can argue as follows. Write $A = k[X_1, \ldots, X_n]/I$ for some ideal I and define $V := \mathbf{V}(I) \subseteq k^n$. Then, by Hilbert's nullstellensatz, we have $\mathbf{I}(V) = \sqrt{I} = I$, thanks to the assumption on A. Hence (V, A) is an affine variety. Now use the injectivity of the operator $\mathbf{I}$.]

2.7.3 Exercise Let A be a commutative ring with 1, and let $I \subseteq A$ be an ideal and M be a finitely generated A-module. Prove the following version of *Nakayama's lemma*: If $M = I.M$, then there exists some $f \in 1 + I$ such that $f.M = \{0\}$.

Furthermore, assume that A is local and $I \subset A$ is the unique maximal ideal. Show that, if $N \subseteq M$ is a submodule with $M = I.M + N$, then $M = N$.

[*Hint.* Use a 'determinantal trick' as in the proof of Lemma 2.1.1.]

2.7.4 Exercise Let A be a commutative local ring (with 1). Let $\mathfrak{m}$ be the unique maximal ideal of A, and denote by $F = A/\mathfrak{m}$ the residue field. Then we may consider $\mathfrak{m}/\mathfrak{m}^2$ as a F-vector space where the operation of F is given by

$$(x + \mathfrak{m}).(a + \mathfrak{m}^2) := xa + \mathfrak{m}^2 \quad \text{for } x \in A \text{ and } a \in \mathfrak{m}.$$

Assume that $\mathfrak{m}$ is finitely generated, and let $d \geqslant 0$ be minimal such that $\mathfrak{m}$ can be generated by d elements. Show that $d = \dim_F(\mathfrak{m}/\mathfrak{m}^2)$.

[*Hint.* First check that, if $\mathfrak{m}$ is generated as an ideal by $a_1, \ldots, a_d$, then the images of $a_1, \ldots, a_d$ generate $\mathfrak{m}/\mathfrak{m}^2$ as a F-vector space. Conversely, assume that $\dim_F(\mathfrak{m}/\mathfrak{m}^2) = l$, and let $\{b_1 + \mathfrak{m}^2, \ldots, b_l + \mathfrak{m}^2\}$ be an F-basis, where $b_i \in \mathfrak{m}$. To show that $\mathfrak{m}$ is generated by $b_1, \ldots, b_l$, consider the left A-module $M := \mathfrak{m}/(b_1, \ldots, b_l)$ and check that $\mathfrak{m}M = M$. Then apply Exercise 2.7.3.]

2.7.5 Exercise Let $R \subseteq A$ be integral domains such that A is integral over R. Let K be the field of fractions of A and $F \subseteq K$ be the field of fractions of R. Assume that K is a finite-dimensional F-vector space. The purpose of this exercise is to establish some basic properties of the *trace map*. (This is used in the proof of Proposition 2.3.11.) For any $x \in K$, we define $\alpha_x \colon K \to K$ by $\alpha_x(y) = xy$. Then α_x is F-linear, and the trace of x is defined as

$$T_{K/F}(x) := \text{Trace}(\alpha_x) \in F.$$

Prove the following statements.

(a) The map $\alpha \colon K \to \text{End}_F(K)$, $x \mapsto \alpha_x$, is an F-algebra homomorphism. For $x \in K$, the minimal polynomial of x over F equals the minimal polynomial of α_x.

(b) $T_{K/F}(ax + by) = aT_{K/F}(x) + bT_{K/F}(y)$ for all $x, y \in K$ and $a, b \in F$.

(c) Assume that R is integrally closed in F. Then we have $T_{F/K}(x) \in R$ for all $x \in K$ which are integral over A.

(d) If $K \supseteq F$ is separable, there exists some $x \in K$ such that $T_{K/F}(x) \neq 0$. Deduce that the bilinear form $K \times K \to F$, $(x, y) \mapsto T_{K/F}(xy)$, is non-degenerate.

[*Hint.* (c) Let $x \in K$ be integral over A. Then x is integral over R, and so there exists a monic $g \in R[X]$ such that $g(x) = 0$. Let $f \in F[X]$ be the minimal polynomial of x over F. Then f divides g, and so all eigenvalues of α_x are integral over R, in some finite extension of K. Hence $\text{Trace}(\alpha_x) \in F$ is integral over R and so $\text{Trace}(\alpha_x) \in R$ since R is integrally closed in F. (d) This is a standard result on separable field extensions; see §VIII.5 of Lang (1984), for example.]

2.7.6 Exercise   This exercise is concerned with some properties of constructible sets. Let X be any topological space.

(a) Show that finite unions and finite intersections of constructible subsets are constructible. Show that the complement of a constructible subset is constructible.

(b) Assume that X is noetherian and that $Y \subseteq X$ is a constructible subset. Show that there exists a subset $U \subseteq Y$ which is dense and open in $\overline{Y}$.

[*Hint.* (b) First assume that $\overline{Y}$ is irreducible. Then we can write $Y = \bigcup_{i=1}^{n}(O_i \cap F_i)$, where each $O_i \subseteq X$ is open, each $F_i \subseteq X$ is closed, and $F_i \subsetneq F_1 = \overline{Y}$ for $i > 1$. Now we can take $U := O_1 \cap F_1$. In the general case, apply this argument to the irreducible components of $\overline{Y}$.]

2.7.7 Exercise   Let X be an irreducible affine variety. Let K be the field of fractions of $A[X]$. Show that the following conditions are equivalent:

(a) $A[X]$ is integrally closed in K;
(b) $A[X]_f$ is integrally closed in K for all non-zero $f \in A[X]$;
(c) $\mathcal{O}_p$ is integrally closed in K for all $p \in X$.

[*Hint.* '(a) $\Rightarrow$ (b)' Let $x \in K$ be integral over $A[X]_f$. Write down an equation for integral dependence of x over $A[X]_f$. Multiplying that equation by a sufficiently large power of f yields an equation of integral dependence of $x f^m$ over $A[X]$, for some $m \geqslant 0$. Hence, we have $x f^m \in A[X]$ and so $x \in A[X]_f$. '(b) $\Rightarrow$ (c)' Similarly. '(c) $\Rightarrow$ (a)' Use Lemma 2.3.6.]

2.7.8 Exercise   The purpose of this exercise is to show that there exist normal varieties which have singular points. Let V be one of the following algebraic sets.

(a) $V = \{(x, y, z) \in k^3 \mid x^2 + y^2 = z^2\}$, where $\operatorname{char}(k) \neq 2$.
(b) $V = \{(x, y, z) \in k^3 \mid xy = z^2\}$.

Show that V is irreducible with $\dim V = 2$. Furthermore, V is normal and $(0,0,0)$ is a singular point of V. The first example is taken from p. 277 of Mumford (1988); the second one appears (with hints) in §II.5.1 of Shafarevich (1994).

2.7.9 Exercise Show that the set of all diagonalizable matrices in $M_n(k)$ is dense.

[*Hint.* Let $S_n(k) \subseteq M_n(k)$ be the set of diagonal matrices, and consider the morphism $\varphi \colon \mathrm{GL}_n(k) \times S_n(k) \to M_n(k)$, $(g,t) \mapsto g^{-1}tg$. Determine the dimension of the closure of the image of φ by looking at the fibres $\varphi^{-1}(y)$ for suitable diagonal matrices and using Corollary 2.2.9.]

2.7.10 Exercise Let G be an affine algebraic group and X be a G-variety. Let $x \in X$ and O_x be the corresponding G-orbit. Let G° be the connected component of $1 \in G$ and O_x° be the G°-orbit. (By Proposition 1.3.13, G° is a closed subgroup of finite index in G.) Show that O_x° is open and closed in O_x.

[*Hint.* Since $[G : G^\circ] < \infty$, it is clear that O_x is a finite union of G°-orbits, and if O_y° is such an orbit, then there exists some $g \in G$ such that $g.x = y$. So there exist $g_1, \ldots, g_r \in G$ such that $O_x = \coprod_{i=1}^{r} O_{g_i.x}^\circ$, where $g_1 = 1$. This implies $\overline{O}_x = \bigcup_{i=1}^{r} \overline{O}_{g_i.x}^\circ$. Since all $\overline{O}_{g_i.x}^\circ$ have the same dimension, we have $O_{g_i.x}^\circ \not\subseteq \overline{O}_{g_j.x}^\circ$ for $i \neq j$, and so $\overline{O}_{g_i.x}^\circ \cap O_x = O_{g_i.x}^\circ$. Thus, $O_{g_i.x}^\circ$ is closed in O_x.]

2.7.11 Exercise The purpose of this exercise is to show that the notion of affine quotients in §2.5.8 is compatible with direct products. For $i = 1, 2$, let X_i be a G_i-variety and assume that an affine quotient (X_{0i}, π_i) exists.

(a) Show that $X_1 \times X_2$ is a $(G_1 \times G_2)$-variety, where the operation is given by the formula $(g_1, g_2).(x_1, x_2) = (g_1.x_1, g_2.x_2)$.

(b) Consider the morphism $\varphi \colon X_1 \times X_2 \to X_{01} \times X_{02}$ which sends (x_1, x_2) to $(\pi_1(x_1), \pi_2(x_2))$. Show that the pair $(X_{01} \times X_{02}, \varphi)$ is an affine quotient of $X_1 \times X_2$ by $G_1 \times G_2$.

[*Hint.* Consider the three conditions in §2.5.8. To check (b), note that it is enough to consider a subset $U \subseteq X_{01} \times X_{02}$ which is of the form $U = U_1 \times U_2$ with $U_i \subseteq X_{0i}$. Furthermore, U is open if and only if U_1 and U_2 are open.]

2.7.12 Exercise Let $G = \mathrm{SL}_2(k)$. Show that every element of G is conjugate to one of the following matrices:

$$\pm \begin{bmatrix} 1 & 0 \\ 0 & 1 \end{bmatrix}, \qquad \pm \begin{bmatrix} 1 & 1 \\ 0 & 1 \end{bmatrix}, \qquad \begin{bmatrix} x & 0 \\ 0 & x^{-1} \end{bmatrix} \ (x \neq \pm 1).$$

Show that the centralizers of these matrices are respectively G, ZU, and H, where Z is the centre of G, U is the group of all upper triangular matrices with an all-unity diagonal, and H is the group of diagonal matrices in G. Determine explicitly which of these centralizers are connected.

2.7.13 Exercise Let B be a connected affine algebraic group. Assume that there exist closed connected subgroups $U, T \subseteq B$ such that U is normal in B, T is abelian, $B = U.T$, and $U \cap T = \{1\}$. Furthermore, assume that $C_B(t) = T$ for some $t \in T$. Then show that $U = [B, B]$.

[*Hint.* First note that $[B, B] \subseteq U$. Now, since $C_B(t) = T$, we have $t^{-1}ut \neq u$ for all non-unit $u \in U$, and so the morphism $\varphi \colon U \to U$, $u \mapsto [t, u] = t^{-1}u^{-1}tu$, is injective. Hence Corollary 2.2.9 implies that φ is dominant. Now, we have $\varphi(U) \subseteq [B, B] \subseteq U$. Finally, use the fact that $[B, B]$ is closed by Example 2.4.7.]

3
Algebraic representations and Borel subgroups

One of the principal aims of this chapter is to prove the conjugacy (and further basic properties) of Borel subgroups, that is, maximal closed connected solvable subgroups in an affine algebraic group.

This will require various preparations concerning algebraic representations, which certainly is an interesting topic in its own right. To begin with, in Section 3.1, we study linear actions of algebraic groups on vector spaces. We prove the Lie–Kolchin theorem which shows that a solvable closed connected group of invertible matrices is always conjugate to a group of triangular matrices. At the end of this section, we give the general definition of semisimple and reductive algebraic groups. Typical examples are $\mathrm{SL}_n(k)$ and $\mathrm{GL}_n(k)$, respectively.

The linear action of an algebraic group on a vector space V will lead us to consider the induced action on projective space $\mathbb{P}(V)$. This is no longer an affine variety, but it carries a topology defined by the vanishing of *homogeneous* polynomials. One of the main properties of that topology is the so-called 'completeness of projective space'; this will be discussed in Section 3.2. In Section 3.3, we introduce the related grassmannian and flag varieties.

In Section 3.4, by combining practically everything that we did so far, we establish the main properties of Borel subgroups in algebraic groups. As a highlight, we obtain the combinatorial characterization of the *Bruhat–Chevalley order* in groups with a *BN*-pair in Theorem 3.4.8. This will be completed in Section 3.5, where we obtain some detailed structure results on connected solvable groups.

3.1 Algebraic representations, solvable groups, and tori

In this section, we obtain the first substantial results on solvable algebraic groups. All this will be based on some notions related to algebraic representations. Throughout, k is algebraically closed.

3.1.1 Definition   Let V be a finite-dimensional k-vector space and G be an affine algebraic group over k. Let $\rho\colon G \to \mathrm{GL}(V)$ be a homomorphism of abstract groups. Fixing a basis $\{e_1,\ldots,e_n\}$ of V, we have equations

$$\rho(g)(e_j) = \sum_{i=1}^{n} a_{ij}(g)\, e_i, \qquad \text{where } a_{ij}(g) \in k.$$

Let $\mathrm{CF}(V,\rho)$ be the subspace of $\mathrm{Maps}(G,k)$ which is spanned by the set of functions $a_{ij}\colon G \to k$ ($1 \leqslant i,j \leqslant n$). Since any two bases of V can be transformed into each other by a linear transformation, $\mathrm{CF}(V,\rho)$ is independent of the choice of the basis; it is called the *coefficient space* of (V,ρ). We say that V is an *algebraic G-module*, or that ρ is an *algebraic representation* of G, if $\mathrm{CF}(V,\rho) \subseteq A[G]$.

By Lemma 2.4.2, the map $G \to \mathrm{GL}_n(k)$, $g \mapsto (a_{ij}(g))$, is a homomorphism of affine algebraic groups. Conversely, given a homomorphism $\rho\colon G \to \mathrm{GL}_n(k)$ of affine algebraic groups, then $V = k^n$ is an algebraic G-module.

3.1.2 Invariant subspaces   Assume that V is an algebraic G-module, and let $U \subseteq V$ be a G-invariant subspace, that is, we have $\rho(g)(U) \subseteq U$ for all $g \in G$. Then U and V/U (with the induced actions of G) also are algebraic G-modules. This is seen as follows. Assume that $0 < m = \dim U < \dim V = n$ and let $\{e_1,\ldots,e_n\}$ be a basis of V such that $\{e_1,\ldots,e_m\}$ is a basis of U. Then the matrices $A(g) := (a_{ij}(g)) \in \mathrm{GL}_n(k)$ have a block-triangular shape

$$A(g) = \left[\begin{array}{c|c} A_1(g) & * \\ \hline 0 & A_2(g) \end{array}\right],$$

where $A_1(g) \in \mathrm{GL}_m(k)$ describes the representation of G on U (with respect to the basis $\{e_1,\ldots,e_m\}$) and $A_2(g) \in \mathrm{GL}_{n-m}(k)$ describes the representation of G on V/U (with respect to the basis given by the images of $\{e_{m+1},\ldots,e_n\}$ in V/U). Now it may happen that $V \neq \{0\}$

does not have any non-trivial proper G-invariant subspaces at all. In this case, we say that V is *irreducible*. Otherwise, we can always find a chain of G-invariant subspaces $\{0\} = V_0 \subsetneqq V_1 \subsetneqq V_2 \subsetneqq \cdots \subsetneqq V_r = V$ such that V_i/V_{i-1} is irreducible for all i. Choosing a basis of V which is adapted to this chain, we have

$$A(g) = \begin{bmatrix} A_1(g) & * & \cdots & * \\ 0 & A_2(g) & \ddots & \vdots \\ \vdots & \ddots & \ddots & * \\ 0 & \cdots & 0 & A_r(g) \end{bmatrix} \quad \text{for all } g \in G,$$

where $A_i(g)$ describes the (irreducible) representation of G on V_i/V_{i-1}.

3.1.3 Definition

A *character* of G is a homomorphism of algebraic groups from G into the multiplicative group $\mathbb{G}_m = \mathrm{GL}_1(k)$. Since $\mathbb{G}_m$ is a principal open subset of k, every character can be regarded as an element of $A[G]$. We will use frequently the fact that pairwise different characters are linearly independent in $A[G]$. This follows from *Dedekind's theorem* which says (in a completely general setting) that pairwise different homomorphisms of a group into the multiplicative group of a field are linearly independent (see Chapter VIII, §4, of Lang (1984)).

The set of all characters of G is a group (under pointwise defined multiplication), called the *character* group of G and denoted by $X(G)$. We write this group additively, that is, we have $(\chi_1 + \chi_2)(g) = \chi_1(g)\,\chi_2(g)$ for all $g \in G$. The identity element of $X(G)$ is given by the *unit character* $1_G \colon G \to \mathbb{G}_m,\ g \mapsto 1$.

3.1.4 Example

Consider the group $G = \mathbb{G}_m$ itself. For any $m \in \mathbb{Z}$, we have an element $\chi_m \in X(G)$ given by $\chi_m(x) = x^m$ for $x \in \mathbb{G}_m$. We claim that

$$X(G) = \{\chi_m \mid m \in \mathbb{Z}\} \cong \mathbb{Z}.$$

Indeed, first note that $A[G] = k[X^{\pm 1}]$ is the ring of Laurent polynomials in one variable X. Thus, we have $\chi_m = X^m$ under the identification $X(G) \subseteq A[G]$. Now let $\chi \in X(G)$. Then we can find some non-zero $f \in k[X^{\pm 1}]$ such that $\chi(x) = f(x)$ for all $x \in G$. Now write $f = \sum_{i=1}^r a_i X^{m_i}$, where $a_i \in k$ and $m_i \in \mathbb{Z}$. Then we obtain the relation $\chi = \sum_{i=1}^r a_i \chi_{m_i}$. So Dedekind's theorem shows that we must have $\chi = \chi_{m_i}$ for some i, and the above claim is proved.

More generally, let $n \geqslant 1$ and define $D_n(k) = \mathbb{G}_m \times \cdots \times \mathbb{G}_m$ (n factors). Then we have $X(D_n(k)) \cong \mathbb{Z}^n$, and $A[D_n(k)]$ is spanned by $X(D_n(k))$ over k. This follows from the above statement and the fact that, if we have a direct product $G_1 \times G_2$, then there is a natural isomorphism $X(G_1 \times G_2) \cong X(G_1) \times X(G_2)$. (We leave this as an exercise for the reader.) Thus, we have $X(D_n(k)) = \langle \chi_1, \ldots, \chi_n \rangle$, where $\chi_i \in X(D_n(k))$ is the projection onto the ith factor.

3.1.5 Lemma   *Let $\rho\colon G \to \mathrm{GL}(V)$ be an algebraic representation. We define*

$$V_\chi := \{v \in V \mid \rho(g)(v) = \chi(g)v \text{ for all } g \in G\} \subseteq V$$

for any $\chi \in X(G)$. There are only finitely many $\chi \in X(G)$ such that $V_\chi \neq \{0\}$, and the sum of these non-zero subspaces is direct. These subspaces are the weight spaces of G on V.

Proof   Let $\chi_1, \ldots, \chi_r \in X(G)$ be pairwise different characters such that $V_{\chi_i} \neq 0$ for all i. We claim that the sum of the subspaces V_{χ_i} is direct. We proceed by induction on r. If $r = 1$, there is nothing to prove. Now let $r \geqslant 2$ and $v_i \in V_{\chi_i}$ be such that $v_1 + \cdots + v_r = 0$. We must show that $v_i = 0$ for all i. Let $i \geqslant 2$ and $g_i \in G$ be such that $\chi_1(g_i) \neq \chi_i(g_i)$. Then we have two equations:

$$\chi_1(g_i)v_1 + \chi_2(g_i)v_2 + \cdots + \chi_r(g_i)v_r = \rho(g_i)(v_1 + \cdots + v_r) = 0,$$
$$\chi_1(g_i)v_1 + \chi_1(g_i)v_2 + \cdots + \chi_1(g_i)v_r = \chi_1(g_i)(v_1 + \cdots + v_r) = 0.$$

Subtracting these equations, we obtain a linear relation among $v_2, \ldots, v_r$, where the coefficient of v_i is non-zero. So, by induction, we conclude that $v_i = 0$. This holds for $2 \leqslant i \leqslant r$ and, consequently, also for $i = 1$. Thus, the above claim is proved. Since $\dim V < \infty$, it is now clear that $\{\chi \mid V_\chi \neq \{0\}\}$ is finite. $\square$

The next result is a technical tool which will be used in the proof of Theorem 3.1.7 and later again in the proof of Theorem 3.4.2.

3.1.6 Lemma   *Let G be a connected affine algebraic group and V be an algebraic G-module. Let $H \subseteq G$ be a closed normal subgroup, and assume that there exists some $\psi \in X(H)$ such that $V_\psi \neq \{0\}$. Then V_ψ is G-invariant. If, furthermore, $H = [G, G]$, then ψ is the trivial character.*

Proof We set $W := \sum_{\chi \in X(H)} V_\chi \subseteq V$. Since $V_\psi \neq \{0\}$, we have $W \neq \{0\}$. Now, by Lemma 3.1.5, there are only finitely many $\chi \in X(H)$ such that $V_\chi \neq \{0\}$, and their sum is direct. Let $X_1 \subseteq X(H)$ be the finite set of all $\chi \in X(H)$ such that $V_\chi \neq \{0\}$. For $g \in G$, the map $\gamma_g \colon H \to H$, $h \mapsto ghg^{-1}$, is an isomorphism of algebraic groups. Hence, for each $\chi \in X(H)$, we also have $\chi^g := \chi \circ \gamma_g \in X(H)$, and a straightforward computation shows that $\rho(g)(V_\chi) = V_{\chi^g}$. Thus, each $g \in G$ induces a permutation of the elements in X_1. Then, since X_1 is finite, we see that

$$\mathrm{Stab}_G(\psi) = \{g \in G \mid \psi(gxg^{-1}) = \psi(x) \text{ for all } x \in G\}$$

is a closed subgroup of G of finite index. Hence, since G is connected, the above stabilizer must be all of G and so V_ψ is a G-invariant subspace.

Now assume that $H = [G, G]$, and consider the induced action of G on V_ψ. Let $\rho \colon G \to \mathrm{GL}(V_\psi)$ be the corresponding algebraic representation. Then we have $\rho(h) = \psi(h)\mathrm{id}_{V_\psi}$ for all $h \in H$. Furthermore, for all $x, y \in G$, we have $\det(\rho(x^{-1}y^{-1}xy)) = 1$, and so $\psi(h)^r = \det(\rho(h)) = 1$ for all $h \in H$, where $r = \dim_k V_\psi$. Hence the image of $\psi \colon H \to \mathbb{G}_m$ is finite. Since H is connected by Example 2.4.7, the image of ψ is irreducible. Consequently, the image is a singleton set, and so $\psi(h) = 1$ for all $h \in H$. $\qquad\square$

3.1.7 Theorem (Lie–Kolchin) *Assume that G is a connected solvable affine algebraic group, and let V be an algebraic G-module. Then there exists a basis $\{e_1, \ldots, e_n\}$ of V such that the corresponding matrices $(a_{ij}(g)) \in \mathrm{GL}_n(k)$ all lie in $B_n(k)$. In particular, every closed connected solvable subgroup of $\mathrm{GL}_n(k)$ is conjugate to a subgroup of $B_n(k)$.*

Proof First we show that there exists some non-zero $e_1 \in V$ such that $\langle e_1 \rangle$ is a G-invariant subspace of V. We do this by induction on $\dim G$. If $\dim G = 0$, then $G = \{1\}$ and there is nothing to prove. Now assume that $\dim G > 0$. Let $G' = [G, G]$; by Example 2.4.7, this is a closed connected subgroup of G. Since G is solvable, we have $\dim G' < \dim G$. So, by induction, there exists some G'-invariant 1-dimensional subspace of V. This means that we have $V_\psi \neq \{0\}$ for some $\psi \in X(G')$. By Lemma 3.1.6, V_ψ is G-invariant and $\psi = 1_{G'}$. So G' acts trivially on V_ψ. For each $g \in G$, denote by $\rho_g \colon V_\psi \to V_\psi$ the induced linear map. Then the fact that $G' = [G, G]$ acts trivially on V_ψ means that the operators ρ_g commute with each other. Hence, by a standard result in linear algebra (note also that k is algebraically closed), there exists a non-zero common eigenvector, that is, a non-zero $e_1 \in V$ such that $\langle e_1 \rangle$ is a G-invariant subspace of V. Thus, the above claim is proved.

Now we consider the algebraic G-module $V/\langle e_1 \rangle$. If this is non-zero, then we can find some $e_2 \in V \setminus \langle e_1 \rangle$ such that e_2 defines a 1-dimensional G-invariant subspace in $V/\langle e_1 \rangle$. Going on in this way, we obtain a basis $\{e_1, \ldots, e_n\}$ of V such that the matrices $(a_{ij}(g))$ are all upper triangular, as desired. $\qquad\square$

3.1.8 Remark   The assumption that G is connected is essential in the above result. For example, let $G = \mathfrak{S}_3$ be the symmetric group on three letters. Then

$$(1,2) \mapsto \begin{bmatrix} 1 & 1 \\ 0 & -1 \end{bmatrix}, \qquad (2,3) \mapsto \begin{bmatrix} -1 & 0 \\ 1 & 1 \end{bmatrix},$$

defines an irreducible representation $\rho \colon G \to \mathrm{GL}_2(\mathbb{C})$.

We shall now study diagonalizable groups and tori. Recall that $T_n(k)$ is the closed subgroup of all invertible diagonal matrices in $\mathrm{GL}_n(k)$. This group clearly is isomorphic to $D_n(k) := \mathbb{G}_m \times \cdots \times \mathbb{G}_m$ (n factors).

3.1.9 Proposition   *Let G be a connected affine algebraic group over k. We set $p := \mathrm{char}(k)$ if $\mathrm{char}(k) > 0$, and $p := 1$ if $\mathrm{char}(k) = 0$. Let*

$$G_{\mathit{fin}} := \{g \in G \mid g^d = 1 \text{ for some } d \geqslant 1 \text{ such that } \gcd(p, d) = 1\}.$$

Then the following conditions are equivalent.

(a) G is abelian and $G = \overline{G}_{\mathit{fin}}$.

(b) For every homomorphism of algebraic groups $\rho \colon G \to \mathrm{GL}_n(k)$, there exists some $T \in \mathrm{GL}_n(k)$ such that $T\rho(G)T^{-1} \subseteq T_n(k)$.

(c) There exists a closed embedding $G \subseteq \mathrm{GL}_n(k)$ such that $G \subseteq T_n(k)$.

(d) G is isomorphic to $D_d(k)$ for some $d \geqslant 1$.

A group G satisfying the above conditions will be called a *torus*.

Proof   '(a) $\Rightarrow$ (b)' Let $\rho \colon G \to \mathrm{GL}_n(k)$ be a homomorphism. For $g \in G_{\mathrm{fin}}$, we have $g^d = 1$, where d is coprime to p. This implies that the minimal polynomial of $\rho(g) \in \mathrm{GL}_n(k)$ has distinct roots, and so $\rho(g)$ is diagonalizable. Since G is abelian, we can even find some $P \in \mathrm{GL}_n(k)$ such that $P\rho(g)P^{-1} \in T_n(k)$ for all $g \in G_{\mathrm{fin}}$. (This is a standard result in linear algebra.) Thus, we may assume without loss of generality that $\rho(G_{\mathrm{fin}}) \subseteq T_n(k)$. Since ρ is continuous and $T_n(k)$ is closed in $\mathrm{GL}_n(k)$, we conclude that $\rho(G) = \rho(\overline{G_{\mathrm{fin}}}) \subseteq \overline{\rho(G_{\mathrm{fin}})} \subseteq T_n(k)$.

'(b) $\Rightarrow$ (c)' By Corollary 2.4.4, there exists a closed embedding $\rho\colon G \to \mathrm{GL}_n(k)$. Thus, by hypothesis (b), there exists some $T \in \mathrm{GL}_n(k)$ such that $T\rho(G)T^{-1} \subseteq T_n(k)$. Hence we obtain a closed embedding $G \to T_n(k)$.

'(c) $\Rightarrow$ (d)' By assumption, there exists a closed embedding $\varphi\colon G \to D_n(k)$. Consider the corresponding comorphism $\varphi^*\colon A[D_n(k)] \to A[G]$. First note that φ^* induces a group homomorphism $\varphi^*\colon X(D_n(k)) \to X(G)$. We claim that this induced homomorphism is surjective. This is seen as follows. Let $\psi \in X(D_n(k)) \subseteq A[G]$. Since φ is a closed embedding, $\varphi^*\colon A[D_n(k)] \to A[G]$ is surjective; see Proposition 2.2.1. So there exists some $f \in A[D_n(k)]$ such that $\varphi^*(f) = \psi$. By the discussion in Example 3.1.4, f can be written as a linear combination of characters: $f = \sum_{i=1}^{r} a_i\psi_i$ say, where $a_i \in k$. Then we obtain $\psi = \varphi^*(f) = \sum_{i=1}^{r} a_i\varphi^*(\psi_i)$. By Dedekind's theorem, we must have $\psi = \varphi^*(\psi_i)$ for some i, as required. Thus, as claimed, $\varphi^*\colon X(D_n(k)) \to X(G)$ is surjective, and $A[G]$ is spanned by $X(G)$. Now $X(G)$ is a finitely generated abelian group, since $X(D_n(k)) \cong \mathbb{Z}^n$. Furthermore, we claim that $X(G)$ is free abelian. Indeed, if this were not the case, then—as is well-known—there would exist some element $1_G \neq \chi \in X(G)$ of finite order, $e \geqslant 2$ say. But then $\chi(G) \subseteq \{x \in \mathbb{G}_m \mid x^e = 1\}$ would be a finite set, contradicting the assumption that G is connected. Thus, $X(G)$ is a free abelian group of finite rank.

Let $\{\varepsilon_1, \ldots, \varepsilon_d\}$ be a basis of $X(G)$, and consider the homomorphism $\psi\colon G \to D_d(k)$, $g \mapsto (\varepsilon_1(g), \ldots, \varepsilon_d(g))$. Then ψ^* will be surjective, since $X(G)$ spans $A[G]$. Hence ψ is a closed embedding by Proposition 2.2.1. In particular, ψ is injective and so $\dim G \leqslant d$. On the other hand, the fact that $X(G)$ is free abelian means that all monomials in $\varepsilon_1, \ldots, \varepsilon_d$ are linearly independent in $A[G]$. Hence we have $\dim G = \partial_k(A[G]) \geqslant d$ and so, finally, ψ is an isomorphism.

(d) $\Rightarrow$ (a) By assumption, G certainly is abelian. Furthermore, every element of finite order in G has order prime to p. It remains to show that the set of these elements is dense in G. For this purpose, it is actually enough to consider the case $d = 1$. But it is easily checked that $\mathbb{G}_m$ contains infinitely many roots of unity of order prime to p.

The closure of that set will be an algebraic subset of $\mathbb{G}_m$ of dimension $\geqslant 1$ and, hence, must be equal to $\mathbb{G}_m$. $\qquad\square$

3.1.10 Proposition (Rigidity of tori)   *Let G be an affine algebraic group and $T \subseteq G$ be a closed subgroup which is a torus. Then $\mathrm{N}_G(T)^\circ = \mathrm{C}_G(T)^\circ$.*

Proof   It is clear that $\mathrm{C}_G(T) \subseteq \mathrm{N}_G(T)$ and so $\mathrm{C}_G(T)^\circ \subseteq \mathrm{N}_G(T)^\circ$. To prove the reverse inclusion, let $N := \mathrm{N}_G(T)^\circ$ and consider the closed set $X := \{t \in T \mid ntn^{-1} = t \text{ for all } n \in N\}$. We must show that $X = T$. To prove this, it will be enough to show that X contains a dense subset of T. Now, by Proposition 3.1.9(a), the set of elements of finite order in T is dense. So it will be enough to show that any $t \in T$ of finite order lies in X. Assume that t has order d. Then consider the morphism $\psi_t \colon N \to T$, $n \mapsto ntn^{-1}$. We have $\psi_t(n)^d = (ntn^{-1})^d = nt^d n^{-1} = 1$ for all $n \in N$. Thus, the subset $\psi_t(N)$ of T consists of elements whose order divides d. Now, by Proposition 3.1.9, T is isomorphic to $D_n(k)$ for some $n \geqslant 1$. It is easily checked that, in $D_n(k)$, there are only finitely many elements of a given order (since this is the case in $\mathbb{G}_m$). Hence, $\psi_t(N)$ is a finite set. Since N is irreducible, $\psi_t(N)$ actually is a singleton set. Thus, we have $ntn^{-1} = t$ for all $n \in N$, as required. $\qquad\square$

3.1.11 The unipotent radical   Let G be a connected solvable affine algebraic group. The *unipotent radical* of G is defined as the subgroup
$$R_{\mathrm{u}}(G) := \bigcap_{\chi \in X(G)} \ker(\chi).$$

(For example, if $G = B_n(k)$, then we have $R_{\mathrm{u}}(G) = U_n(k)$.) Clearly, $R_{\mathrm{u}}(G)$ is a closed normal subgroup of G. Furthermore, since G is a noetherian topological space, there exist finitely many characters $\chi_1, \ldots, \chi_n \in X(G)$ such that $R_{\mathrm{u}}(G) = \bigcap_{i=1}^{n} \ker(\chi_i)$. Then we obtain a homomorphism $\varphi \colon G \to D_n(k)$ by sending $g \in G$ to $(\chi_1(g), \ldots, \chi_n(g))$, such that $\ker(\varphi) = R_{\mathrm{u}}(G)$.

3.1.12 Lemma   *Let G be connected and solvable. Then the unipotent radical $R_{\mathrm{u}}(G)$ is nilpotent. Furthermore, if $\rho \colon G \to \mathrm{GL}_n(k)$ is any homomorphism of algebraic groups, then the eigenvalues of $\rho(g)$ $(g \in R_{\mathrm{u}}(G))$ are all 1.*

Proof Let $\rho\colon G \to \mathrm{GL}_n(k)$ be any homomorphism of algebraic groups. By Theorem 3.1.7, we may assume that $\rho(G) \subseteq B_n(k)$. Now recall that $B_n(k) = U_n(k)T_n(k)$. For $1 \leqslant i \leqslant n$, let $\varepsilon_i \in X(B_n(k))$ be the character which assigns to a matrix in $B_n(k)$ its ith diagonal coefficient. Then we have $\varepsilon_i \circ \rho \in X(G)$ for all i. Now, if $g \in R_{\mathrm{u}}(G)$, then $\varepsilon_i(\rho(g)) = 1$ for all i, and so the diagonal coefficients of $\rho(g)$ are 1, as claimed. To show that $R_{\mathrm{u}}(G)$ is nilpotent, choose ρ to be a closed embedding. Then $R_{\mathrm{u}}(G)$ is isomorphic (as an abstract group) to a subgroup of $U_n(k)$ and, hence, $R_{\mathrm{u}}(G)$ is nilpotent (since $U_n(k)$ is). $\qquad\square$

3.1.13 Semisimple and reductive groups   Let G be a connected affine algebraic group. Then the *radical* of G is defined as

$$R(G) := \left\langle S \subseteq G \;\middle|\; \begin{array}{l} S \text{ is a closed connected} \\ \text{solvable normal subgroup} \end{array} \right\rangle \subseteq G.$$

It is clear that $R(G)$ is a normal subgroup. By Theorem 2.4.6, $R(G)$ is a closed connected subgroup of G; furthermore, there are finitely many closed connected solvable normal subgroups $S_1, \ldots, S_n$ such that $R(G) = S_1 \cdots S_n$. Now it is a general fact about (abstract) groups that the product of two normal solvable subgroups of a given group is again a normal solvable subgroup. Thus, we conclude that $R(G)$ is solvable. In fact, $R(G)$ is the unique maximal closed connected solvable normal subgroup of G. We say that G is a *semisimple group* if $R(G) = \{1\}$. We say that G is a *reductive group* if $R(G)$ is a torus.

3.1.14 Lemma   *Let G be a connected reductive affine algebraic group. Then $R(G) = Z(G)^\circ$, where $Z(G)$ is the centre of G. Furthermore, $[G, G]$ is a connected semisimple group and we have $|R(G) \cap [G, G]| < \infty$.*

Proof Since $R(G)$ is a torus, we can apply Proposition 3.1.10 and conclude that $G = \mathrm{N}_G(R(G)) = \mathrm{C}_G(R(G))$. So we have $R(G) \subseteq Z(G)^\circ$, while the reverse inclusion is clear anyway (since $Z(G)^\circ$ is a closed connected abelian normal subgroup). Now, $[G, G]$ certainly is connected by Example 2.4.7. We also note that the radical of $[G, G]$ is contained in $R(G)$. (Indeed, the radical of $[G, G]$ is also a closed connected solvable normal subgroup in G.) Hence, it will be enough to show that $|R(G) \cap [G, G]| < \infty$, For this purpose, we choose a closed embedding $G \subseteq \mathrm{GL}_n(k)$. Then $V = k^n$ is an algebraic

G-module and, by Proposition 3.1.9, we can write $V = \bigoplus_{i=1}^{r} V_i$, where $V_i := V_{\chi_i}$ and $\chi_i \in X(R(G))$. Since $R(G)$ lies in the centre of G, we have $g.V_i \subseteq V_i$ for all $g \in G$ and all i; see Lemma 3.1.6. Thus, as in §3.1.2, we actually obtain an injective representation

$$\rho\colon G \to \mathrm{GL}(V_1) \times \cdots \times \mathrm{GL}(V_r), \quad g \mapsto (\rho_1(g), \ldots, \rho_r(g)),$$

where $\rho_i(z) = \chi_i(z)\,\mathrm{id}_{V_i}$ for all $z \in R(G)$ and all i. Now, if we take an element $g \in [G, G]$, then we can write g as a product of commutators, and so $\rho_i(g)$ is seen to have determinant 1. If g also lies in $R(G)$, then $\rho_i(g)$ will be a scalar multiple of the identity for each i. But there are only finitely many scalar matrices whose determinant is 1. Thus, $|R(G) \cap [G, G]| < \infty$, as required. $\qquad\square$

3.1.15 Example The group $\mathrm{GL}_n(k)$ is reductive and $\mathrm{SL}_n(k)$ is semisimple.

Indeed, by Theorem 3.1.7, the radical of $\mathrm{GL}_n(k)$ is contained in $B_n(k)$. But $B_n(k)$ is conjugate to the subgroup $B_n^-(k)$ of all lower triangular invertible matrices, and so the radical will be contained in $B_n(k) \cap B_n^-(k) = T_n(k)$. Thus, if $g \in R(\mathrm{GL}_n(k))$, then xgx^{-1} must be a diagonal matrix for all $x \in \mathrm{GL}_n(k)$. Taking for x suitable upper unitriangular matrices, we can conclude that g must be a scalar matrix. Hence, we have $R(\mathrm{GL}_n(k)) = Z(G) \cong \mathbb{G}_m$ and so this is a torus. Now consider $\mathrm{SL}_n(k)$. Since $\det\colon \mathrm{GL}_n(k) \to k^\times$ is a group homomorphism, we have $[\mathrm{GL}_n(k), \mathrm{GL}_n(k)] \subseteq \mathrm{SL}_n(k)$. On the other hand, by Example 2.4.10, we have $\mathrm{SL}_n(k) = [\mathrm{SL}_n(k), \mathrm{SL}_n(k)]$ and so we must have $\mathrm{SL}_n(k) = [\mathrm{GL}_n(k), \mathrm{GL}_n(k)]$. It remains to apply Lemma 3.1.14.

In Definition 3.4.5, we will introduce a general class of reductive groups.

3.2 The main theorem of elimination theory

In the previous sections, we have studied actions of algebraic groups on affine varieties. The linear action of an algebraic group G on an algebraic G-module V will now lead us to consider the induced action on the projective space $\mathbb{P}(V)$. We will see that this basically amounts to working with affine algebraic sets which are defined by *homogeneous* polynomials. A fundamental result about such sets is the *main theorem of elimination theory*,[1] which we establish in Theorem 3.2.7.

3.2.1 Example Let G be an affine algebraic group, and assume that $\rho\colon G \rightarrow \mathrm{GL}(V)$ is an algebraic representation, as in Definition 3.1.1. Then, regarding V as an affine variety as in Exercise 2.7.1, we see that V is a G-variety, where the action is given by $g.v = \rho(g)(v)$ for $g \in G$ and $v \in V$. We can go one step further. Consider the multiplicative group $\mathbb{G}_m$ of k. We have a natural action of $\mathbb{G}_m$ on V, simply given by scalar multiplication. We can combine this action of $\mathbb{G}_m$ with that of G to obtain an action

$$(\mathbb{G}_m \times G) \times V \rightarrow V, \qquad ((\lambda, g), v) \mapsto \lambda\rho(g)(v).$$

Since the above map is still a morphism of affine varieties, we see that V is a $\mathbb{G}_m \times G$-variety. The set of non-zero $\mathbb{G}_m$-orbits on V is the *projective space*

$$\mathbb{P}(V) := \{\langle v \rangle \mid 0 \neq v \in V\}.$$

We have an induced action of G on $\mathbb{P}(V)$ given by $g.\langle v \rangle = \langle g.v \rangle$, where $g \in G$ and $\langle v \rangle \in \mathbb{P}(V)$. Then we see that we have a bijection

$$\{\text{non-zero } (\mathbb{G}_m \times G)\text{-orbits on } V\} \quad \longleftrightarrow \quad \{G\text{-orbits on } \mathbb{P}(V)\}.$$

3.2.2 The Zariski topology on $\mathbb{P}(V)$ Let V be a finite-dimensional k-vector space, and consider the corresponding *projective space* $\mathbb{P}(V)$, as defined above. Using the natural map $\pi\colon V \backslash \{0\} \rightarrow \mathbb{P}(V)$, $v \mapsto \langle v \rangle$, we could define a topology on $\mathbb{P}(V)$ by declaring a subset $U \subseteq \mathbb{P}(V)$ to be open if $\pi^{-1}(U)$ is open in the induced topology on $V \setminus \{0\}$. However, it is possible to define this topology on $\mathbb{P}(V)$ in a more direct way, using zero sets of polynomials as in the case of affine algebraic sets.

For this purpose, we choose a basis $\{e_1, \ldots, e_n\}$ of V and identify $V = k^n$ using that basis. Let $T_1, \ldots, T_n$ be indeterminates. A non-zero polynomial $f \in k[T_1, \ldots, T_n]$ is called *homogeneous* of degree $d \geqslant 0$ if f is a linear combination of monomials which all have total degree d. For such a polynomial f, we have $f(av_1, \ldots, av_n) = a^d f(v_1, \ldots, v_n)$ for all $a \in k$ and $v_i \in k$. Thus,

---

[1]For the relation with 'elimination theory' see e.g. Chapter 8, §5, of Cox *et al.* (1992).

for $\langle v \rangle \in \mathbb{P}(V)$, the condition $f(v_1, \ldots, v_n) = 0$ is well-defined, where $v = \sum_i v_i e_i$. For any subset $H \subseteq k[T_1, \ldots, T_n]$ consisting of homogeneous polynomials, we define

$$\mathbf{V}^{\mathrm{p}}(H) := \{\langle v_1 e_1 + \cdots + v_n e_n \rangle \in \mathbb{P}(V) \mid$$
$$f(v_1, \ldots, v_n) = 0 \text{ for all } f \in H\}.$$

By exactly the same arguments as in §1.1.6, we see that the sets $\mathbf{V}^{\mathrm{p}}(H)$ form the closed sets of a topology on $\mathbb{P}(V)$, which we call the *Zariski* topology. Note that this topology does not depend on the choice of the basis $\{e_1, \ldots, e_n\}$. This follows from the fact that any two bases of V can be transformed into each other by a linear transformation of V, and such a linear transformation also transforms homogeneous polynomials into homogeneous polynomials.

We summarize the above discussion in the following result.

3.2.3 Lemma *We set $V^{\sharp} := V \setminus \{0\}$. Then the map*

$$\pi \colon V^{\sharp} \to \mathbb{P}(V), \qquad v \mapsto \langle v \rangle,$$

is continuous and open. We have $\pi(\overline{Z} \setminus \{0\}) = \overline{\pi(Z)}$ for any $\mathbb{G}_m$-invariant subset $Z \subseteq V^{\sharp}$. If Z is closed, then $\pi(Z)$ also is closed.

Proof To check that π is continuous, we must check that $\pi^{-1}(Z)$ is closed for any closed subset $Z \subseteq \mathbb{P}(V)$. Let $H \subseteq k[T_1, \ldots, T_n]$ be a set of homogeneous polynomials such that $Z = \mathbf{V}^{\mathrm{p}}(H)$. But then it is clear that $\pi^{-1}(Z) = \mathbf{V}(H) \setminus \{0\}$, where $\mathbf{V}(H)$ denotes the usual affine algebraic set in $V = k^n$ defined by H.

To show that π is open, we must check that $\pi(U)$ is open for any open subset $U \subseteq V^{\sharp}$. Since $\pi(U) = \pi(\pi^{-1}(\pi(U)))$, it is enough to do this for all open sets which are unions of complete fibres of π, that is, which are $\mathbb{G}_m$-invariant. But this is equivalent to checking that $\pi(C)$ is closed for all $\mathbb{G}_m$-invariant closed subsets $C \subseteq V^{\sharp}$. Now, by Exercise 3.6.4, such a subset C is of the form $\mathbf{V}(H) \setminus \{0\}$ for some set $H \subseteq k[T_1, \ldots, T_n]$ of homogeneous polynomials. But then we see that $\pi(C) = \mathbf{V}^{\mathrm{p}}(H)$ is closed.

Finally, let $Z \subseteq V^{\sharp}$ be any $\mathbb{G}_m$-invariant subset. By Exercise 3.6.4 and §1.1.7, we have $\overline{Z} = \mathbf{V}(H)$, where $H \subseteq k[T_1, \ldots, T_n]$ is a set of homogeneous polynomials. In particular, $\overline{Z}$ is $\mathbb{G}_m$-invariant, and so $\pi(\overline{Z} \setminus \{0\}) = \mathbf{V}^{\mathrm{p}}(H)$ by the previous discussion. We certainly have $\pi(Z) \subseteq \mathbf{V}^{\mathrm{p}}(H)$ and so $\overline{\pi(Z)} \subseteq \pi(\overline{Z} \setminus \{0\})$. On the other hand, π is continuous and so $\pi(\overline{Z} \setminus \{0\}) \subseteq \overline{\pi(Z)}$. $\qquad\square$

3.2.4 Remark In Chapter 1, we defined the dimension of an algebraic set $V \subseteq k^n$ using the Hilbert polynomial of an ideal. In a similar way, one can also define the dimension of a projective algebraic set in $\mathbb{P}(V)$, using the 'projective' *Hilbert polynomial* of a homogeneous ideal; see Exercise 3.6.6. However, we shall not need this here; see Chapter 9 of Cox *et al.* (1992) for more details.

3.2.5 Proposition *Let V be an algebraic G-module, and consider the (abstract) operation of G on $\mathbb{P}(V)$ given by $g.\langle v \rangle = \langle g.v \rangle$, where $g \in G$ and $\langle v \rangle \in \mathbb{P}(V)$. This operation has the following properties:*

(a) For any non-zero $v \in V$, the subgroup $\mathrm{Stab}_G(\langle v \rangle) \subseteq G$ is closed.

(b) Let $O \subseteq \mathbb{P}(V)$ be a G-orbit. Then O is open in $\overline{O}$, and $\overline{O}$ is a union of orbits.

(c) Every non-empty G-invariant closed subset of $\mathbb{P}(V)$ contains a closed orbit. In particular, closed orbits exist.

Proof (a) Let $\langle v \rangle \in \mathbb{P}(V)$. Then we have $\mathrm{Stab}_G(\langle v \rangle) = \mathrm{Tran}_G(\langle v \rangle, \langle v \rangle)$, and so this is closed by Lemma 2.5.1(a).

(b) Let O be the G-orbit of $\langle v \rangle \in \mathbb{P}(V)$. As remarked in §3.2.2, we have $\pi^{-1}(O) = C$, where C is the $(\mathbb{G}_m \times G)$-orbit of v in V. By Lemma 3.2.3, we have $\pi(\overline{C}^{\sharp}) = \overline{O}$, where the superscript $\sharp$ denotes intersection with $V^{\sharp}$. Now we can argue as follows. By Proposition 2.5.2(a), $\overline{C}$ is in fact a union of $(\mathbb{G}_m \times G)$-orbits, and $\overline{C} \setminus C$ is closed. Let us write $\overline{C} = C \cup R$, where R is a closed $(\mathbb{G}_m \times G)$-invariant subset and $C \cap R = \varnothing$. The $\mathbb{G}_m$-invariance implies that $O \cap \pi(R^{\sharp}) = \pi(C^{\sharp}) \cap \pi(R^{\sharp}) = \varnothing$. Hence, we have $\overline{O} = \pi(\overline{C}^{\sharp}) = O \cup \pi(R^{\sharp})$, where the union is disjoint and $\pi(R^{\sharp})$ is closed (again, by Lemma 3.2.3). Hence O is open in $\overline{O}$.

(c) Let C be a G-invariant closed subset of $\mathbb{P}(V)$. Then C is a union of orbits which contains the closure of each of these orbits. The existence of a closed orbit in C follows from the fact that $\mathbb{P}(V)$ is noetherian (see Exercise 3.6.3), by exactly the same argument as in the proof of Proposition 2.5.2(c). $\qquad\square$

We now turn to another important property of projective space, namely, what is usually called the 'main theorem of elimination theory' or the 'completeness of projective space'. To prove this, we first need the following 'homogeneous' version of *Hilbert's nullstellensatz*.

3.2.6 Lemma *Recall our standing assumption that k is algebraically closed. Let $H \subseteq k[T_1, \ldots, T_n]$ be a set of homogeneous polynomials. Then $\mathbf{V}^{\mathrm{p}}(H) = \varnothing$ if and only if there exists some $s \geqslant 1$ such that all monomials of degree s lie in the ideal generated by H.*

Proof First assume that there exists some $s \geqslant 1$ such that all monomials of degree s lie in (H). Then H contains T_i^s for $1 \leqslant i \leqslant n$, and so $\mathbf{V}^{\mathrm{p}}(H) \subseteq \mathbf{V}^{\mathrm{p}}(\{T_1^s, \ldots, T_n^s\}) = \varnothing$. Conversely, assume that $\mathbf{V}^{\mathrm{p}}(H) = \varnothing$. Then we consider the algebraic set in k^n defined by H, that is, $\mathbf{V}(H) \subseteq k^n$. Then the condition $\mathbf{V}^{\mathrm{p}}(H) = \varnothing$ means that $\mathbf{V}(H) \subseteq \{0\}$ and so $\mathbf{I}(\{0\}) \subseteq \mathbf{I}(\mathbf{V}(H)) = \sqrt{(H)}$, where the last equality holds by Hilbert's nullstellensatz; see Theorem 2.1.5. On the other hand, we certainly have $T_i \in \mathbf{I}(\{0\})$, and so $T_i \in \sqrt{(H)}$ for $1 \leqslant i \leqslant n$. Hence there exist some $m_i \geqslant 1$ such that $T_i^{m_i} \in (H)$ for all i. Now set $m := \max\{m_i\}$ and $s := nm \geqslant 1$. Then, if $T_1^{\alpha_1} \cdots T_n^{\alpha_n}$ is any monomial of degree s, there exists some i such that $\alpha_i \geqslant m \geqslant m_i$. Since $T_i^{m_i} \in (H)$, we conclude that $T_1^{\alpha_1} \cdots T_n^{\alpha_n} \in (H)$, as desired. $\square$

To state the following result, we introduce the following notation. Let (X, A) be an affine variety over k and V be a finite-dimensional k-vector space. Let $\{e_1, \ldots, e_n\}$ be a basis of V, and consider the polynomial ring $A[T_1, \ldots, T_n]$, where $T_1, \ldots, T_n$ are indeterminates. Then, for any $x \in X$, the evaluation homomorphism $\varepsilon_x \colon A \to k$ canonically extends to a k-algebra homomorphism

$$\tilde{\varepsilon}_x \colon A[T_1, \ldots, T_n] \to k[T_1, \ldots, T_n],$$

simply by applying ε_x to the coefficients of polynomials. Thus, for any subset $H \subseteq A[T_1, \ldots, T_n]$ consisting of homogeneous polynomials and any $x \in X$, we can define

$$\tilde{\mathbf{V}}(H) := \{(\langle v \rangle, x) \in \mathbb{P}(V) \times X \mid \tilde{\varepsilon}_x(f)(v) = 0 \text{ for all } f \in H\}.$$

As in §1.1.6, it is readily checked that the sets $\tilde{\mathbf{V}}(H)$ form the closed subsets of a topology on $\mathbb{P}(V) \times X$.

3.2.7 Theorem (Main theorem of elimination theory)
Recall that k is algebraically closed. In the above setting, the second projection map $\mathrm{pr}_2 \colon \mathbb{P}(V) \times X \to X$ is a closed map, that is, it sends closed subsets to closed subsets.

Proof Let $H \subseteq A[T_1, \ldots, T_n]$ be a collection of homogeneous polynomials. We must show that $\mathrm{pr}_2(\tilde{\mathbf{V}}(H)) \subseteq X$ is a closed set. First note that we may assume without loss of generality that H is a finite set and that all polynomials $f \in H$ are homogeneous of the same degree. Indeed, since A is a noetherian ring, the same is also true for $A[T_1, \dots, T_n]$ (by Hilbert's basis theorem) and so there exist finitely many homogeneous polynomials $f_1, \dots, f_r \in H$ such that the ideal $(H) \subseteq A[T_1, \dots, T_n]$ is generated by $f_1, \dots, f_r$. Then it is straightforward and easy to check that $\tilde{\mathbf{V}}(H) = \tilde{\mathbf{V}}(\{f_1, \dots, f_r\})$. Assume that each f_i is homogeneous of degree $d_i \geqslant 0$, say. Then set $d := \max \{d_i\}$, and define $g_{ij} := T_j^{d-d_i} f_i \in A[T_1, \dots, T_n]$ for $1 \leqslant i \leqslant r$ and $1 \leqslant j \leqslant n$. Again, it is straightforward and easy to check that $\tilde{\mathbf{V}}(H) = \tilde{\mathbf{V}}(\{g_{ij} \mid 1 \leqslant i \leqslant r, 1 \leqslant j \leqslant n\})$. Thus, we are reduced to the case where H consists of finitely many polynomials which are all homogeneous of the same degree, d say. If $d = 0$, then all $f \in H$ are constant and there is nothing to prove. So let us assume that $d \geqslant 1$.

Let us fix $x \in X$. Then we have $x \in \mathrm{pr}_2(\tilde{\mathbf{V}}(H))$ if and only if $V_x(H) := \mathbf{V}^{\mathrm{p}}(\{\tilde{\varepsilon}_x(f) \mid f \subset H\}) \neq \varnothing$. We begin by trying to find a more convenient characterization of the condition $V_x(H) = \varnothing$. By Lemma 3.2.6, we have $V_x(H) = \varnothing$ if and only if, for some $s \geqslant 1$, all monomials of degree s lie in the ideal $(\tilde{\varepsilon}_x(f) \mid f \in H) \subseteq k[T_1, \dots, T_n]$. As in Section 1.2, we write

$$M_s := \{\alpha = (\alpha_1, \dots, \alpha_n) \in \mathbb{Z}_{\geqslant 0}^n \mid \alpha_1 + \cdots + \alpha_n = s\},$$

and denote the monomial $T_1^{\alpha_1} \cdots T_n^{\alpha_n}$ simply by T^α. With this notation, we see that the condition $V_x(H) = \varnothing$ is equivalent to the condition that there exists some $s \geqslant 1$ such that, for any $\alpha \in M_s$, we have

$$T^\alpha = \sum_{f \in H} h_f^\alpha \, \tilde{\varepsilon}_x(f), \qquad \text{where } h_f^\alpha \in k[T_1, \dots, T_n].$$

Expressing all h_f^α as linear combinations of monomials, and noting that all $\tilde{\varepsilon}_x(f)$ are homogeneous of degree d (or 0) and T^α is just a monomial of degree s, we see that the above condition is equivalent to the condition that there exists some $s \geqslant d$ such that, for any $\alpha \in M_s$, we have

$$T^\alpha = \sum_{f \in H} \sum_{\beta \in M_{s-d}} c_{f,\beta}^\alpha \, T^\beta \tilde{\varepsilon}_x(f), \qquad \text{where } c_{f,\beta}^\alpha \in k.$$

To simplify notation, denote by $R_x^{(s)}$ the subspace of $k[T_1, \dots, T_n]$ spanned by the homogeneous polynomials $\{T^\beta \tilde{\varepsilon}_x(f) \mid \beta \in M_{s-d},$

$f \in H\}$. Then the above condition can be rephrased by saying that $R_x^{(s)}$ is the whole space of homogeneous polynomials of degree s, for some $s \geqslant d$. Thus, we have:

$$V_x(H) = \varnothing \quad \Leftrightarrow \quad \dim_k R_x^{(s)} = |M_s| \text{ for some } s \geqslant d.$$

Hence we have $\{x \in X \mid V_x(H) \neq \varnothing\} = \bigcap_{s \geqslant d}\{x \in X \mid \dim_k R_x^{(s)} < |M_s|\}$. So it will be enough to show that, for a fixed $s \geqslant d$, the set of all $x \in X$ with $\dim_k R_x^{(s)} < |M_s|$ is closed. This is seen as follows. For $f \in H$, we write $f = \sum_{\gamma \in M_d} f_\gamma T^\gamma$, where $f_\gamma \in A$. Then we have

$$T^\beta f = \sum_{\alpha \in M_s} f_{\alpha-\beta} T^\alpha \qquad \text{for any } \beta \in M_{s-d}$$

(where we set $f_{\alpha-\beta} = 0$ if some coefficient in $\alpha - \beta$ is negative) and so

$$T^\beta \tilde{\varepsilon}_x(f) = \sum_{\alpha \in M_s} f_{\alpha-\beta}(x) T^\alpha \qquad \text{for any } f \in H.$$

Now let us arrange the coefficients $f_{\alpha-\beta}$ (for various f, α, β) in a matrix, where the rows are labelled by the various pairs (f, β) ($f \in H$, $\beta \in M_{s-d}$) and the columns are labelled by the various $\alpha \in M_s$. Then the condition $\dim_k R_x^{(s)} < |M_s|$ is equivalent to the condition that the rank of that matrix is less than $|M_s|$. The latter condition can be expressed by the vanishing of certain minors of that matrix, that is, by determinantal expressions in the coefficients $f_{\alpha-\beta}$. These determinantal expressions yield a collection of regular functions in A. The above discussion shows that the closed set of X defined by that collection is precisely the set of all $x \in X$ with $\dim_k R_x^{(s)} < |M_s|$, as desired. $\qquad\qquad\square$

3.2.8 Example Consider the polynomials $F_1 := T_1 + T_2 Y$, $F_2 := T_1 + T_1 Y$ in $k[T_1, T_2, Y]$. These polynomials are homogeneous in T_1 and T_2, and so the set

$$\tilde{\mathbf{V}}(\{F_1, F_2\}) = \left\{ (\langle x_1, x_2 \rangle, y) \in \mathbb{P}(k^2) \times k \; \middle| \; \begin{array}{l} F_1(x_1, x_2, y) = 0 \\ F_2(x_1, x_2, y) = 0 \end{array} \right\}$$

is well-defined. It is easily checked that the above set just contains the two elements $(\langle 0, 1 \rangle, 0)$ and $(\langle 1, 1 \rangle, -1)$. Thus, we have

$$\mathrm{pr}_2(\tilde{\mathbf{V}}(\{F_1, F_2\})) = \{0, -1\}.$$

We conclude that $\mathrm{pr}_2(\tilde{\mathbf{V}}(\{F_1, F_2\})) = \mathbf{V}(\{Y(Y+1)\})$. It is a natural question to ask if there is a general procedure for finding defining polynomials for $\mathrm{pr}_2(\tilde{\mathbf{V}}(H))$. See Exercise 3.6.5 for an answer to this question.

3.2.9 Lemma *Let $V^\sharp = V \setminus \{0\}$ and X be an affine variety. Consider the map*

$$\pi \times \mathrm{id} \colon V^\sharp \times X \to \mathbb{P}(V) \times X, \qquad (v, x) \mapsto (\langle v \rangle, x).$$

Let $C \subseteq V^\sharp \times X$ be a subset satisfying the following two conditions.

(a) C is closed in the induced topology on $V^\sharp \times X$ as a subset of $V \times X$.

(b) For any $(v, x) \in C$ and any $0 \neq t \in k$, we have $(tv, x) \in C$. Then $(\pi \times \mathrm{id})(C) \subseteq \mathbb{P}(V) \times X$ is closed.

Proof Using a basis $\{e_1, \ldots, e_n\}$ of V, we can identify $A[V \times X]$ with the polynomial ring $A[T_1, \ldots, T_n]$. Let $\overline{C}$ be the closure of C in $V \times X$. By §1.1.7, we have $\overline{C} = \mathbf{V}(\mathbf{I}(C))$. Now, by (b) and Exercise 3.6.4, $\mathbf{I}(C)$ is generated by homogeneous polynomials. Thus, we have $\overline{C} = \mathbf{V}(H)$, where $H \subseteq A[T_1, \ldots, T_n]$ is a set of polynomials which are homogeneous in $T_1, \ldots, T_n$. We claim that

$$(\pi \times \mathrm{id})(C) = \tilde{\mathbf{V}}(H) \subseteq \mathbb{P}(V) \times X.$$

Indeed, the inclusion '$\subseteq$' is clear by definition. Conversely, let $(\langle v \rangle, x) \in \tilde{\mathbf{V}}(H)$. Then $F(v, x) = 0$ for all $F \in H$, and so $(v, x) \in \overline{C} = \mathbf{V}(H) \subseteq V \times X$. Since $v \neq 0$, we also have $(v, x) \subseteq V^\sharp \times X$, and so $(v, x) \in C$, using (a). $\square$

We now describe applications of the above results to algebraic groups.

3.2.10 Definition Let G be an affine algebraic group. A closed subgroup $P \subseteq G$ is called a *parabolic* subgroup if there exists an algebraic G-module V and a non-zero vector $v \in V$ such that

(a) P is the stabilizer of $\langle v \rangle \in \mathbb{P}(V)$ and
(b) the orbit $\{\langle g.v \rangle \mid g \in G\} \subseteq \mathbb{P}(V)$ is closed.

Here, we consider the induced action of G on $\mathbb{P}(V)$, as in Proposition 3.2.5. For the time being, this is a rather technical condition. In Theorem 3.4.3, we will obtain a purely group-theoretic characterization of parabolic subgroups.

For example, $P = G$ always is a parabolic subgroup. (To see this, we simply take the trivial G-module $V = k$, where each $g \in G$ acts as the identity.) A similar argument also shows that, if $P \subseteq G$ is a parabolic subgroup and H is any affine algebraic group, then $H \times P$ is a parabolic subgroup of $H \times G$.

Now we can prove a basic property of parabolic subgroups, which is essential in many applications; see, for example, the two results at the end of Section 3.4.

3.2.11 Theorem *Let X be an affine variety, G an affine algebraic group, and $P \subseteq G$ a parabolic subgroup. Assume that $Z \subseteq G \times X$ is closed and that*

$$(g, x) \in Z \quad \Rightarrow \quad (gp, x) \in Z \quad \text{for all } p \in P. \qquad (*)$$

Then $\mathrm{pr}_2(Z) \subseteq X$ is closed, where $\mathrm{pr}_2 \colon G \times X \to X$ is the second projection.

Proof The statement already bears some similarity to that in Theorem 3.2.7. To establish the link, we proceed as follows. By definition, there exists an algebraic G-module V and some non-zero $v_0 \in V$ such that the G-orbit $C := O_{\langle v_0 \rangle} \subseteq \mathbb{P}(V)$ is closed, and we have $P = \mathrm{Stab}_G(\langle v_0 \rangle)$. Then we also have

$$\mathrm{pr}_2(Z) = \mathrm{pr}_2((\varphi_{\langle v_0 \rangle} \times \mathrm{id})(Z)),$$

where $\varphi_{\langle v_0 \rangle} \colon G \to \mathbb{P}(V)$, $g \mapsto \langle g.v_0 \rangle$. Hence, by Theorem 3.2.7, it will be enough to show that the image of Z under $\varphi_{\langle v_0 \rangle} \times \mathrm{id} \colon G \times X \to \mathbb{P}(V) \times X$ is closed. For this purpose, we begin with the following preparations. Let us set $\tilde{G} = \mathbb{G}_m \times G$. As we have seen in §3.2.1, we may also regard V as an algebraic $\tilde{G}$-module, where $(t, g) \in \tilde{G}$ acts by $v \mapsto tg.v$ ($v \in V$). Let $\tilde{\varphi}_{v_0} \colon \tilde{G} \to V$, $(t, g) \mapsto tg.v_0$, be the corresponding orbit map. Then we have

$$(\varphi_{\langle v_0 \rangle} \times \mathrm{id})(Z) = (\pi \times \mathrm{id})((\tilde{\varphi}_{v_0} \times \mathrm{id})(\tilde{Z})),$$

where $\pi \times \mathrm{id}$ is as in Lemma 3.2.9 and

$$\tilde{Z} = \{((t, g), x) \in \tilde{G} \times X \mid t \in \mathbb{G}_m, (g, x) \in Z\}.$$

Thus, it will be sufficient to show that the set $(\tilde{\varphi}_{v_0} \times \mathrm{id})(\tilde{Z}) \subseteq V^{\sharp} \times X$ satisfies the two conditions in Lemma 3.2.9. By construction, it is clear that condition (b) is satisfied. Hence it remains to show that

$$(\tilde{\varphi}_{v_0} \times \mathrm{id})(\tilde{Z}) \text{ is closed in } V^{\sharp} \times X \subseteq V \times X. \qquad (\dagger)$$

To see this, consider the canonical map $\pi \colon V^{\sharp} \to \mathbb{P}(V)$, $v \mapsto \langle v \rangle$, and let $C_V = \pi^{-1}(C) \cup \{0\}$. As we have seen in §3.2.2, C_V is a closed subset in V. Hence, since $\tilde{\varphi}_{v_0}(\tilde{G}) \subseteq C_V$, it will be enough to show that

$$(\tilde{\varphi}_{v_0} \times \mathrm{id})(\tilde{Z}) \text{ is closed in } C_V^{\sharp} \times X \subseteq C_V \times X, \qquad (\dagger')$$

where $C_V^{\sharp} := C_V \setminus \{0\}$. To prove this, first note that $\tilde{Z}$ is a closed subset of $\tilde{G} \times X$. Furthermore, by Corollary 2.5.7, we already know that

$$\tilde{\varphi}_{v_0} \times \mathrm{id} \colon \tilde{G} \times X \to C_V \times X$$

is an open map. Hence, in order to prove $(\dagger')$, it is enough to show that $\tilde{Z} \subseteq \tilde{G} \times X$ consists of complete fibres of the map $\tilde{\varphi}_{v_0} \times \mathrm{id}$. To check the latter property, let $((t,g),x) \in \tilde{Z}$ and $((t',g'),x') \in \tilde{G} \times X$ be such that $(tg.v_0, x) = (t'g'.v_0, x')$. Then we certainly have $x = x'$ and $g^{-1}g'.v_0 \in \langle v_0 \rangle$. This implies $g^{-1}g' \in \mathrm{Stab}_G(\langle v_0 \rangle) = P$, and so $g' = gp$ for some $p \in P$. The fact that Z satisfies condition $(*)$ now implies that we also have $((t',g'),x') \in \tilde{Z}$, as required. Thus, $(\dagger')$ is proved. $\qquad \square$

3.2.12 Corollary *Let P be a parabolic subgroup of G.*

(a) Let X be a G-variety and $Y \subseteq X$ be a P-invariant closed subset. Then

$$G.Y = \{g.y \mid g \in G, y \in Y\} \subseteq X \text{ is closed.}$$

(b) Let V be an algebraic G-module and $C \subseteq \mathbb{P}(V)$ be a P-invariant closed subset. Then $G.C = \{\langle g.v \rangle \mid g \in G, \langle v \rangle \in C\} \subseteq \mathbb{P}(V)$ is closed.

Proof (a) Consider the direct product $G \times X$, and set

$$Z := \{(g,x) \in G \times X \mid g^{-1}.x \in Y\}.$$

Then Z is closed. Furthermore, let $(g,x) \in Z$ and $p \in P$. Then we have $(gp)^{-1}.x = p^{-1}(g^{-1}.x) \in Y$ since Y is P-invariant. Thus, all the assumptions of Theorem 3.2.11 are satisfied, and so $G.Y = \mathrm{pr}_2(Z)$ is closed.

(b) As in the proof of Theorem 3.2.11, let $\tilde{G} = \mathbb{G}_m \times G$ and consider V as a $\tilde{G}$-variety. Furthermore, we set $C_V = \pi^{-1}(C) \cup \{0\}$, where $\pi \colon V^{\sharp} \to \mathbb{P}(V)$ is the canonical map. By §3.2.2, C_V is a closed set. We claim that $\tilde{G}.C_V \subseteq V$ is also closed. By the remarks following Definition 3.2.10, $\mathbb{G}_m \times P$ is a parabolic subgroup in $\tilde{G}$. Since C_V is invariant under $\mathbb{G}_m \times P$, the claim follows from (a). Since $\tilde{G}.C_V$ is $\mathbb{G}_m$-invariant, $G.C = \pi(\tilde{G}.C_V)$ is closed, as required. $\qquad \square$

3.2.13 Example   Let P be a parabolic subgroup of G and $U \subseteq P$ be a closed normal subgroup. Then

$$\bigcup_{x \in G} xUx^{-1} \text{ is a closed subset of } G.$$

To see this, we consider G as a G-variety via conjugation. Then U is a P-invariant closed subset of G, and we have $G.U = \{xux^{-1} \mid x \in G, u \in U\} = \bigcup_{x \in G} xUx^{-1}$. By Corollary 3.2.12, this is a closed subset of G.

3.3   Grassmannian varieties and flag varieties

The purpose of this section is to describe various constructions of G-varieties arising from the linear action of an affine algebraic group G on a vector space. This will lead to the important examples of grassmannian varieties and flag varieties. We assume that the reader has some familiarity with the basic facts concerning tensor products and exterior powers of vector spaces; see, for example, Appendix B of Fulton and Harris (1991). As a first application, we prove a result due to Chevalley which asserts the existence of certain algebraic representations.

3.3.1 Partial flag varieties   Let V be an algebraic G-module. Then G acts naturally on various sets constructed from V.

Indeed, let $0 < n_1 < n_2 < \cdots < n_d \leqslant \dim V$ be a sequence of strictly increasing integers. Then the corresponding partial *flag variety* is defined by

$$\mathcal{F}_{n_1,\ldots,n_d}(V) := \left\{ (E_1,\ldots,E_d) \ \middle| \ \begin{array}{l} E_r \subseteq V \text{ subspace, } \dim_k E_r = n_r \\ \text{for all } r, \text{ and } E_1 \subseteq \cdots \subseteq E_d \end{array} \right\}.$$

The given action of G on V certainly induces an action on $\mathcal{F}_{n_1,\ldots,n_d}(V)$. Just note that, for $g \in G$ and a subspace $E \subseteq V$, the set $g.E = \{g.v \mid v \in E\} \subseteq V$ is a subspace with $\dim_k E = \dim_k g.E$ (since $\pi_g \colon V \to V$ is a bijective linear map).

We note the following special cases of the above construction. Let $n = \dim V$. Taking the sequence $0 < 1 < 2 < \cdots < n-1 < n$, the corresponding set

$$\mathcal{F}(V) := \mathcal{F}_{1,2,\ldots,n}(V) \qquad\qquad (a)$$

is called the *flag variety*. For $0 \leqslant r \leqslant n$, the corresponding set

$$\mathcal{G}_r(V) := \mathcal{F}_r(V) = \{E \subseteq V \mid E \text{ subspace with } \dim_k E = r\} \qquad \text{(b)}$$

is called a *grassmannian variety*. For $r = 1$, we just obtain the *projective space* $\mathbb{P}(V) := \{\langle v \rangle \mid 0 \neq v \in V\}$.

However, note that the induced actions on the above sets are only defined on a purely abstract level. The aim of this section is to show that the sets $\mathcal{F}_{n_1,\ldots,n_d}(V)$ can be identified with some closed subsets in a suitably chosen projective space.

We begin by showing that the notion of algebraic G-modules is compatible with the usual linear algebra constructions: tensor products and exterior powers.

3.3.2 Tensor products Let V and W be algebraic G-modules. Then the tensor product $V \otimes W$ also is an algebraic G-module, with G-action given by

$$g.(v \otimes w) = (g.v) \otimes (g.w), \qquad \text{where } g \in G, \, v \in V, \text{ and } w \in W.$$

Indeed, first note that the above formula defines an (abstract) action of G on $V \otimes W$ which is linear. Now, by assumption, V has a basis $\{e_1, \ldots, e_n\}$ such that the action of G on V is given by the formula in §3.1.1, with regular functions $a_{ij} \in A[G]$. Similarly, there is a basis $\{f_1, \ldots, f_m\}$ of W such that the action of G on W is given by regular functions $b_{rs} \in A[G]$. Now it is well-known that $\{e_i \otimes f_r \mid 1 \leqslant i \leqslant n, 1 \leqslant r \leqslant m\}$ is a basis of $V \otimes W$. Then we have the formula

$$g.(e_j \otimes f_s) = \sum_{i=1}^{n} \sum_{r=1}^{m} a_{ij}(g) b_{rs}(g) \, e_i \otimes f_r,$$

which shows that the action of G on $V \otimes W$ is given by the functions $a_{ij} b_{rs} \in A[G]$. More generally, if $V_1, \ldots, V_d$ are algebraic G-modules, then the tensor product $V_1 \otimes \cdots \otimes V_d$ is an algebraic G-module, with G-action given by

$$g.(v_1 \otimes \cdots \otimes v_d) = (g.v_1) \otimes \cdots \otimes (g.v_d),$$

where $g \in G$ and $v_i \in V_i$ for all i.

3.3.3 Lemma (the Segre embedding) *Let $V_1, \ldots, V_d$ be algebraic G-modules. Then the map*

$$\sigma : \begin{array}{ccc} \mathbb{P}(V_1) \times \cdots \times \mathbb{P}(V_d) & \to & \mathbb{P}(V_1 \otimes \cdots \otimes V_d) \\ (\langle v_1 \rangle, \ldots, \langle v_d \rangle) & \mapsto & \langle v_1 \otimes \cdots \otimes v_d \rangle \end{array}$$

is well-defined, injective, and G-equivariant, and its image is closed. Here, the G-action on the left-hand side is given by $g.(\langle v_1 \rangle, \ldots, \langle v_d \rangle) = (\langle g.v_1 \rangle, \ldots, \langle g.v_d \rangle)$.

See also Exercise 3.6.7 where we give an explicit description of the subsets of $\mathbb{P}(V_1) \times \cdots \times \mathbb{P}(V_d)$ which correspond to closed subsets under the Segre embedding.

Proof By induction on d, it is enough to prove this in the case where $d = 2$. Let $\{e_1, \ldots, e_n\}$ be a basis of $V = V_1$ and $\{f_1, \ldots, f_m\}$ be a basis of $W = V_2$. Then $\{e_i \otimes f_j \mid 1 \leqslant i \leqslant n, 1 \leqslant j \leqslant m\}$ is a basis of $V \otimes W$. If we write $v = \sum_i v_i e_i$ with $v_i \in k$, and $w = \sum_j w_j f_j$ with $w_j \in k$, then

$$v \otimes w = \sum_{i=1}^{n} \sum_{j=1}^{m} v_i w_j \, (e_i \otimes f_j).$$

Thus, if $v \neq 0$ and $w \neq 0$, then $v_i \neq 0$ for some i, and $w_j \neq 0$ for some j, and so $v \otimes w \neq 0$. Furthermore, if we replace v by av for $0 \neq a \in k$, and w by bw for some $0 \neq b \in k$, then $v \otimes w$ is replaced by $ab(v \otimes w)$. This shows that the map σ is well-defined. The above formula for $v \otimes w$ also shows how the coordinates of v and w can be recovered from the coordinates of $v \otimes w$ (up to scalar multiples). Thus, σ is injective. It is also clear that σ is G-equivariant.

Now let us prove that the image of σ is closed. With respect to the basis $\{e_i \otimes f_j\}$ of $V \otimes W$, the closed sets in $\mathbb{P}(V \otimes W)$ are given by the zero sets of homogeneous polynomials in the nm variables Y_{ij} $(1 \leqslant i \leqslant n, 1 \leqslant j \leqslant m)$. We claim that $\sigma(\mathbb{P}(V) \times \mathbb{P}(W)) = \mathbf{V}^\mathrm{P}(H) \subseteq \mathbb{P}(V \otimes W)$, where

$$H := \{Y_{ij} Y_{rs} - Y_{is} Y_{rj} \mid 1 \leqslant i, r \leqslant n, 1 \leqslant j, s \leqslant m\}.$$

To see this, let $0 \neq z = \sum_{i,j} c_{ij} e_i \otimes f_j \in V \otimes W$. If $z = v \otimes w$ for some $v = \sum_i v_i e_i$, and $w = \sum_j w_j f_j$, then $c_{ij} = v_i w_j$, and so $f(c_{ij}) = 0$ for all $f \in H$. Conversely, assume that $f(c_{ij}) = 0$ for all $f \in H$. Now, there exist some i_0 and j_0 such that $c_{i_0 j_0} \neq 0$; we may even assume that $c_{i_0 j_0} = 1$. Considering $Y_{i j_0} Y_{i_0 j} - Y_{ij} Y_{i_0 j_0} \in H$ shows that $c_{i j_0} c_{i_0 j} = c_{ij} c_{i_0 j_0} = c_{ij}$ for all i and j, and so $z = v \otimes w$, where $v := \sum_i c_{i j_0} e_i$ and $w := \sum_j c_{i_0 j} f_j$. $\qquad\square$

3.3.4 Exterior powers Let V be an algebraic G-module. We set

$$T^r(V) := \underbrace{V \otimes \cdots \otimes V}_{r \text{ times}} \qquad \text{for any } r \geqslant 0, \text{ where } T^0(V) = k.$$

By §3.3.2, $T^r(V)$ also is an algebraic G-module. Recall that if $\{e_1, \ldots, e_n\}$ is a basis of V, then $\{e_{j_1} \otimes \cdots \otimes e_{j_r} \mid 1 \leqslant j_1, \ldots, j_r \leqslant n\}$ is a basis of $T^r(V)$; in particular, we have $\dim T^r(V) = n^r$. Now consider the subspace

$$I^r(V) = \langle v_1 \otimes \cdots \otimes v_r \mid v_i = v_j \text{ for some } i \neq j \rangle_k \subseteq T^r(V).$$

Clearly, $I^r(V)$ is G-invariant, and so the *exterior power* $\bigwedge^r V := T^r(V)/I^r(V)$ is an algebraic G-module. As usual, the natural map $T^r(V) \to \bigwedge^r V$ is written as

$$v_1 \wedge \cdots \wedge v_r := v_1 \otimes \cdots \otimes v_r + I^r(V),$$

and then the G-action is given by $g.(v_1 \wedge \cdots \wedge v_r) = (g.v_1) \wedge \cdots \wedge (g.v_r)$. It is well known that if $\{e_1, \ldots, e_n\}$ is a basis of V, then

$$\{e_{j_1} \wedge \cdots \wedge e_{j_r} \mid 1 \leqslant j_1 < \cdots < j_r \leqslant n\}$$

is a basis of $\bigwedge^r V$. In particular, we have $\bigwedge^r V = \{0\}$ for $r \geqslant n+1$, while $\dim \bigwedge^r V = \binom{n}{r}$ for $0 \leqslant r \leqslant n$. Given $v_1, \ldots, v_r \in V$ and writing $v_j = \sum_{i=1}^n a_{ij} e_i$, where $A = (a_{ij}) \in k^{n \times r}$, we have

$$v_1 \wedge \cdots \wedge v_r = \sum_{1 \leqslant i_1 < \cdots < i_r \leqslant n} p_{i_1 i_2 \ldots i_r}(A)\, e_{i_1} \wedge \cdots \wedge e_{i_r}, \qquad \text{(a)}$$

where $p_{i_1 i_2 \ldots i_r}(A)$ is the determinant of the $r \times r$ submatrix of A given by the i_1th, i_2th,$\ldots$, i_rth rows of A. The numbers $p_{i_1 i_2 \ldots i_r}(A)$ are called the *Plücker coordinates* of $v_1 \wedge \cdots \wedge v_r$ with respect to the given basis of V. Since A has full rank if and only if some $p_{i_1 i_2 \ldots i_r}(A)$ is non-zero, we see that

$$v_1 \wedge \cdots \wedge v_r \neq 0 \text{ if and only if } v_1, \ldots, v_r \text{ are linearly independent.}$$
$$\text{(b)}$$

As a first application, we shall construct a certain bijective homomorphism of the *symplectic group* $\mathrm{Sp}_4(k)$ into itself, where k has characteristic 2.

3.3.5 The group $\mathrm{Sp}_4(k)$, char$(k) = 2$ Assume that k is an algebraically closed field of characteristic 2, and let $G = \mathrm{Sp}_4(k)$. Then one can show that $G = \langle U, U' \rangle$, where U and U' are the subgroups of upper and lower unitriangular matrices in G, respectively. (See Example 2.4.11 or Section 4.6.) By Lemma 1.5.9, U consists of all matrices of the form

$$
u(t_1, t_2, t_3, t_4) := u\left(\begin{bmatrix} 1 & t_1 \\ 0 & 1 \end{bmatrix}, \begin{bmatrix} t_2 & t_3 \\ t_3 & t_4 \end{bmatrix} \right) = \begin{bmatrix} 1 & t_1 & t_3 + t_1 t_2 & t_4 + t_1 t_3 \\ 0 & 1 & t_2 & t_3 \\ 0 & 0 & 1 & t_1 \\ 0 & 0 & 0 & 1 \end{bmatrix},
$$

where $t_1, t_2, t_3, t_4 \in k$. Similarly, U' consists of the matrices $u(t_1, t_2, t_3, t_4)^{\mathrm{tr}}$. Now consider $V = k^4$ as an algebraic G-module, via the embedding $G \subseteq \mathrm{GL}(V)$. Then, by §3.3.4, $\bigwedge^2 V$ also is an algebraic G-module. Let $\{e_1, e_2, e_3, e_4\}$ be the standard basis of V. Then the vectors $e_{12}, e_{13}, e_{14}, e_{23}, e_{24}, e_{34}$ form a basis of $\bigwedge^2 V$, where we set $e_{ij} := e_i \wedge e_j$ for $1 \leqslant i < j \leqslant 4$. Then the action of $u(t_1, t_2, t_3, t_4)$ on $\bigwedge^2 V$ is given by the matrix

$$
\tilde{u}(t_1, t_2, t_3, t_4) = \begin{bmatrix} 1 & t_2 & t_3 & t_3 & t_4 & t_3^2 + t_2 t_4 \\ 0 & 1 & t_1 & t_1 & t_1^2 & t_4 + t_1^2 t_2 \\ 0 & 0 & 1 & 0 & t_1 & t_3 + t_1 t_2 \\ 0 & 0 & 0 & 1 & t_1 & t_3 + t_1 t_2 \\ 0 & 0 & 0 & 0 & 1 & t_2 \\ 0 & 0 & 0 & 0 & 0 & 1 \end{bmatrix};
$$

similarly, the action of $u(t_1, t_2, t_3, t_4)^{\mathrm{tr}}$ is given by the transpose of the above matrix. Now we notice that the subspace $W := \langle e_{12}, e_{13}, e_{14} + e_{23}, e_{24}, e_{34} \rangle \subseteq \bigwedge^2 V$ is invariant under the action of G. The action of $u(t_1, t_2, t_3, t_4)$ and $u(t_1, t_2, t_3, t_4)^{\mathrm{tr}}$ on W (with respect to the specified basis) is given by the matrices

$$
\begin{bmatrix} 1 & t_2 & 0 & t_4 & t_3^2 + t_2 t_4 \\ 0 & 1 & 0 & t_1^2 & t_4 + t_1^2 t_2 \\ 0 & 0 & 1 & t_1 & t_3 + t_1 t_2 \\ 0 & 0 & 0 & 1 & t_2 \\ 0 & 0 & 0 & 0 & 1 \end{bmatrix} \quad \text{and} \quad \begin{bmatrix} 1 & 0 & 0 & 0 & 0 \\ t_2 & 1 & 0 & 0 & 0 \\ t_3 & t_1 & 1 & 0 & 0 \\ t_4 & t_1^2 & 0 & 1 & 0 \\ t_3^2 + t_2 t_4 & t_4 + t_1^2 t_2 & 0 & t_2 & 1 \end{bmatrix},
$$

respectively. Next we notice that the subspace generated by the third basis vector in the above representation is invariant under the action of G. The action of $u(t_1, t_2, t_3, t_4)$ on the quotient space is given by the matrix

$$\hat{u}(t_1, t_2, t_3, t_4) = \begin{bmatrix} 1 & t_2 & t_4 & t_3^2 + t_2 t_4 \\ 0 & 1 & t_1^2 & t_4 + t_1^2 t_2 \\ 0 & 0 & 1 & t_2 \\ 0 & 0 & 0 & 1 \end{bmatrix};$$

similarly, one checks that the action of $u(t_1, t_2, t_3, t_4)^{\mathrm{tr}}$ is given by the transpose of the above matrix. This leads us to the following result.

3.3.6 Theorem (Chevalley) *In the above setting, there exists a unique homomorphism of algebraic groups* $\theta\colon \mathrm{Sp}_4(k) \to \mathrm{Sp}_4(k)$ *such that*

$$u(t_1, t_2, t_3, t_4) \mapsto \begin{bmatrix} 1 & t_2 & t_4 & t_3^2 + t_2 t_4 \\ 0 & 1 & t_1^2 & t_4 + t_1^2 t_2 \\ 0 & 0 & 1 & t_2 \\ 0 & 0 & 0 & 1 \end{bmatrix}$$

and $u(t_1, t_2, t_3, t_4)^{\mathrm{tr}}$ *is sent to the transpose of the above matrix. The homomorphism* θ *is bijective, and we have*

$$\theta^2(u(t_1, t_2, t_3, t_4)) = u(t_1^2, t_2^2, t_3^2, t_4^2),$$

$$\theta^2(u(t_1, t_2, t_3, t_4)^{\mathrm{tr}}) = u(t_1^2, t_2^2, t_3^2, t_4^2)^{\mathrm{tr}}.$$

Proof Let $G = \mathrm{Sp}_4(k)$. By the construction in §3.3.5, the assignment

$$u(t_1, t_2, t_3, t_4) \mapsto \hat{u}(t_1, t_2, t_3, t_4), \quad u(t_1, t_2, t_3, t_4)^{\mathrm{tr}} \mapsto \hat{u}(t_1, t_2, t_3, t_4)^{\mathrm{tr}},$$

defines a homomorphism of algebraic groups $\theta\colon G \to \mathrm{GL}_4(k)$. Now we notice that $\hat{u}(t_1, t_2, t_3, t_4)$ and $\hat{u}(t_1, t_2, t_3, t_4)^{\mathrm{tr}}$ actually lie in G. Thus, $\theta(G) \subseteq G$. It is straightforward to check that θ^2 is given by the formulas specified above. This yields that θ^2 is the map which sends a matrix $(a_{ij}) \in G$ to the matrix $(a_{ij}^2) \in G$. This map certainly is bijective. Consequently, θ also is bijective. $\qquad\square$

We shall study θ in more detail in Section 4.6, where it will be used to define the finite *Suzuki groups*. Now we continue with some representation-theoretic constructions.

3.3.7 The Plücker embedding Let V be an algebraic G-module. Consider the grassmannian variety $\mathcal{G}_r(V)$, where $0 \leqslant r \leqslant \dim V$.

We wish to define a G-equivariant map (for the natural G-actions defined in §3.3.1 and §3.3.4)

$$\psi_r \colon \mathcal{G}_r(V) \to \mathbb{P}(\textstyle\bigwedge^r V).$$

This is done as follows. Let $E \in \mathcal{G}_r(V)$. Then choose a basis of E, say $\{v_1, \ldots, v_r\}$, and consider $v_1 \wedge \cdots \wedge v_r \in \bigwedge^r V$. First of all, this element in $\bigwedge^r V$ is non-zero by §3.3.4(b). We claim that $\langle v_1 \wedge \cdots \wedge v_r \rangle \in \mathbb{P}(\bigwedge^r V)$ depends only on E and not on the choice of the basis. To see this, fix a basis $\{e_1, \ldots, e_n\}$ of V and express $v_1 \wedge \cdots \wedge v_r$ as in §3.3.4(a). Choosing another basis of E amounts to multiplying the matrix A on the right by a matrix $P \in \mathrm{GL}_r(k)$. But then we have $p_{i_1 i_2 \ldots i_r}(AP) = p_{i_1 i_2 \ldots i_r}(A) \det(P)$, and so $v_1 \wedge \cdots \wedge v_r$ is determined up to multiplication by a non-zero scalar. Thus, $\langle v_1 \wedge \cdots \wedge v_r \rangle \in \mathbb{P}(\bigwedge^r V)$ is well-defined. The fact that ψ_r is G-equivariant is clear by the definition of the G-actions on $\mathcal{G}_r(V)$ and $\bigwedge^r V$.

In the following statement, we use the observation that we have a natural map $\bigwedge^r V \times V \to \bigwedge^{r+1} V$ given by sending $(v_1 \wedge \cdots \wedge v_r, v)$ to $v_1 \wedge \cdots \wedge v_r \wedge v$. Thus, any fixed $\omega \in \bigwedge^r V$ defines a linear map

$$\varphi(\omega) \colon V \to \textstyle\bigwedge^{r+1} V, \qquad v \mapsto \omega \wedge v.$$

3.3.8 Lemma *The map $\psi_r \colon \mathcal{G}_r(V) \to \mathbb{P}(\bigwedge^r V)$ is injective, and its image is a closed subset of $\mathbb{P}(\bigwedge^r V)$. More precisely, the following hold.*

(a) *If $E \in \mathcal{G}_r(V)$, then we have $E = \ker(\varphi(\omega))$, where $\langle \omega \rangle = \psi_r(E)$.*

(b) *Given any $0 \neq \omega \in \bigwedge^r V$, we have*

$$\langle \omega \rangle \in \psi_r(\mathcal{G}_r(V)) \quad \Leftrightarrow \quad \varphi(\omega) \text{ has rank at most } \dim V - r.$$

Proof Let $n = \dim V$. We first prove (a) and (b).

(a) Let $\{v_1, \ldots, v_r\}$ be a basis of E, and set $\omega = v_1 \wedge \cdots \wedge v_r$. Let $v \in V$. Then $v \in E$ if and only if $\{v_1, \ldots, v_r, v\}$ is linearly dependent, and this is equivalent to $\omega \wedge v = v_1 \wedge \cdots \wedge v_r \wedge v = 0$. Thus, we have $E = \ker(\varphi(\omega))$, as claimed.

(b) The implication '$\Rightarrow$' is clear by (a). Conversely, assume that $\varphi(\omega)$ has rank at most $n - r$. Then $\ker(\varphi(\omega))$ contains a subspace $E \subseteq V$ with $\dim E = r$. We take a basis $\{e_1, \ldots, e_n\}$ of V such that

$\{e_{n-r+1}, \ldots, e_n\}$ is a basis of E. Let

$$\omega = \sum_{1 \leqslant i_1 < \cdots < i_r \leqslant n} c_{i_1 i_2 \ldots i_r}\, e_{i_1} \wedge \cdots \wedge e_{i_r}, \quad \text{where } c_{i_1 i_2 \ldots i_r} \in k.$$

Now each $e_{n-r+1}, \ldots, e_n$ lies in the kernel of $\varphi(\omega)$. Let us write out what this means. To start with, the condition $\omega \wedge e_n = 0$ reads

$$0 = \omega \wedge e_n = \sum_{1 \leqslant i_1 < \cdots < i_r \leqslant n} c_{i_1 i_2 \ldots i_r}\, e_{i_1} \wedge \cdots \wedge e_{i_r} \wedge e_n.$$

This implies that $c_{i_1 i_2 \ldots i_r} = 0$ whenever $i_r < n$. Thus, we have

$$\omega = \sum_{1 \leqslant i_1 < \cdots < i_{r-1} < n} c_{i_1 i_2 \ldots i_{r-1} n}\, e_{i_1} \wedge \cdots \wedge e_{i_{r-1}} \wedge e_n.$$

Now the condition $\omega \wedge e_{n-1} = 0$ reads

$$0 = \omega \wedge e_{n-1} = \sum_{1 \leqslant i_1 < \cdots < i_{r-1} < n} c_{i_1 i_2 \ldots i_{r-1}, n}\, e_{i_1} \wedge \cdots \wedge e_{i_{r-1}} \wedge e_n \wedge e_{n-1}$$

$$= - \sum_{1 \leqslant i_1 < \cdots < i_{r-1} < n} c_{i_1 i_2 \ldots i_{r-1}, n}\, e_{i_1} \wedge \cdots \wedge e_{i_{r-1}} \wedge e_{n-1} \wedge e_n.$$

So we conclude that $c_{i_1 i_2 \ldots i_{r-1}, n} = 0$ whenever $i_{r-1} < n - 1$. Thus, we have

$$\omega = \sum_{1 \leqslant i_1 < \cdots < i_{r-2} < n-1} c_{i_1 i_2 \ldots i_{r-2}, n-1, n}\, e_{i_1} \wedge \cdots \wedge e_{i_{r-2}} \wedge e_{n-1} \wedge e_n.$$

Continuing in this way, we finally get $\omega = c\, e_{n-r+1} \wedge e_{n-r+2} \wedge \cdots \wedge e_n$ for some $c \in k$, and so $\langle \omega \rangle \in \psi_r(E)$, as required.

Thus, (a) and (b) are proved. Now note that the statement in (a) certainly implies that ψ_r is injective. Similarly, the statement in (b) implies that the image of ψ_r is a closed subset of $\mathbb{P}(\bigwedge^r V)$. Indeed, let us fix a basis $\{e_1, \ldots, e_n\}$ of V. Then we must express the condition that $\mathrm{rank}(\varphi(\omega)) \leqslant n - r$ in terms of homogeneous polynomials in the Plücker coordinates of ω. For this purpose, consider any $\omega \in \bigwedge^r V$ and write

$$\omega = \sum_{1 \leqslant i_1 < \cdots < i_r \leqslant n} c_{i_1 i_2 \ldots i_r}\, e_{i_1} \wedge \cdots \wedge e_{i_r}, \quad \text{where } c_{i_1 i_2 \ldots i_r} \in k.$$

Let $b^l_{i_1 i_2 \ldots i_r}$ be the Plücker coordinates of $\varphi(\omega)(e_l)$ for $1 \leqslant l \leqslant n$. We have

$$\varphi(\omega)(e_l) = \omega \wedge e_l = \sum_{1 \leqslant i_1 < \cdots < i_r \leqslant n} c_{i_1 i_2 \ldots i_r}\, e_{i_1} \wedge \cdots \wedge e_{i_r} \wedge e_l,$$

and so

$$b^l_{i_1 i_2 \ldots i_r} = \begin{cases} 0 & \text{if } l = i_j \text{ for some } j, \\ (-1)^{r-s} c_{i_1 i_2 \cdots i_s, l, i_{s+1} \cdots i_r} & \text{if } i_s < l < i_{s+1} < \cdots < i_r. \end{cases}$$

Thus, the Plücker coordinates of $\varphi(\omega)(e_l)$ can be expressed by homogeneous linear expressions in the coordinates of ω. So, if we denote by A_ω the matrix of $\varphi(\omega)$, then the coefficients of A_ω are given by homogeneous (even linear) polynomials in the Plücker coordinates of ω. The condition that the rank of $\varphi(\omega)$ be at most $n - r$ is equivalent to the vanishing of all $(n - r + 1) \times (n - r + 1)$ minors of A_ω, which are again given by homogeneous polynomials in the coefficients of A_ω. Thus, there exist homogeneous polynomials in the $\binom{n}{r}$ variables $X_{i_1 < i_2 < \cdots < i_r}$ such that $\psi_r(\mathcal{G}_r(V))$ is defined by the vanishing of these polynomials. $\qquad\square$

Here is the first application to the existence of certain representations.

3.3.9 Proposition (Chevalley) *Let G be an affine algebraic group and $H \subseteq G$ be a closed subgroup. There exists an algebraic G-module V such that $H = \mathrm{Stab}_G(\langle v \rangle)$ for some non-zero $v \in V$.*

Proof Let $I \subseteq A[G]$ be the vanishing ideal of H. Since $A[G]$ is noetherian, there exist $f_1, \ldots, f_m \in A[G]$ such that $I = (f_1, \ldots, f_m)$. Then, by Theorem 2.4.3, there exists a finite-dimensional subspace $E \subseteq A[G]$ with $f_1, \ldots, f_m \in E$ and such that $\rho_g^*(E) \subseteq E$ for all $g \in G$. We have a linear action of G on E given by $g.f = \rho_g^*(f)$ for $g \in G$ and $f \in E$. Furthermore, with respect to a basis $\{e_1, \ldots, e_n\}$ of E, this action is given by a formula as in §3.1.1, with regular functions $a_{ij} \in A[G]$. Thus, E is an algebraic G-module. We claim that

$$H = \{g \in G \mid g.F \subseteq F\}, \quad \text{where } F := E \cap I \subseteq E.$$

Indeed, if $h \in H$ and $f \in F$, then $h.f \in E$ and $h.f(x) = \rho_h^*(f)(x) = f(xh) = 0$ for all $x \in H$, since $f \in I$. Thus, we also have $h.f \in I$ and so $h.f \in F$. Conversely, assume that $g.F \subseteq F$. By construction, we have $f_i \in F$ and $\rho_g^*(f_i) = g.f_i \in F \subseteq I$ for $i = 1, \ldots, m$. Since ρ_g^* is an algebra homomorphism, we conclude that $\rho_g^*(I) \subseteq I$. Consequently, we have $f(g) = \rho_g^*(f)(1) = 0$ for all $f \in I$, and so $g \in \mathbf{V}_G(I) = H$. Thus, the above claim is established.

Now consider the algebraic G-module $V := \bigwedge^r E$, where $r = \dim_k F$. We have $\psi_r(F) = \langle v \rangle$, where $0 \neq v \in V$. By Lemma 3.3.8, $\psi_r \colon \mathcal{G}_r(E) \to \mathbb{P}(\bigwedge^r E)$ is injective and G-equivariant. Hence $H = \{g \in G \mid g.F \subseteq F\} = \mathrm{Stab}_G(\langle v \rangle)$. $\qquad\square$

3.3.10 Corollary *Let G be a connected affine algebraic group and $H \subseteq G$ be a closed normal subgroup. Then there exists an algebraic representation $\gamma \colon G \to \mathrm{GL}(W)$ such that $H = \ker(\gamma)$.*

Proof Let V and a non-zero $v \in V$ be as in Proposition 3.3.9. Then we have $\langle v \rangle \subseteq V_\chi$ for some character $\chi \in X(H)$. By Lemma 3.1.6, V_χ is a G-invariant subspace of V and, hence, we have an algebraic representation $\rho \colon G \to \mathrm{GL}(V_\chi)$. We now consider the k-vector space $W := \mathrm{End}_k(V_\chi)$ and define a map $\gamma \colon G \to \mathrm{GL}(W)$ by $\gamma(g)(\varphi) = \rho(g) \circ \varphi \circ \rho(g)^{-1}$, where $g \in G$ and $\varphi \in W$. First of all, it is clear that γ is a homomorphism of abstract groups. Furthermore, choosing a basis of V_χ yields a basis of $W = \mathrm{End}_k(V_\chi)$, and then we see that the coordinates of $\gamma(g)$ are given by polynomial functions in the coordinates of $\rho(g)$ and $\rho(g^{-1})$. Hence γ is an algebraic representation. It remains to check that $\ker(\gamma) = H$.

If $h \in H$, then $\rho(h)$ is a scalar multiple of the identity, and so $\rho(h)$ commutes with every $\varphi \in W$. Hence $\gamma(h)$ is the identity in $\mathrm{GL}(W)$, and so $h \in \ker(\gamma)$. Conversely, assume that $g \in G$ is such that $\gamma(g)$ is the identity in $\mathrm{GL}(W)$. This means that $\rho(g)$ commutes with any k-endomorphism of V_χ. But it is well known that this implies that $\rho'(g)$ must be a scalar multiple of the identity on V_χ. In particular, we have $g.\langle v \rangle = \langle v \rangle$ and so $g \in H$. $\qquad\square$

The above result remains true (with minor changes in the proof) without the assumption that G is connected; see §11.1 of Goodman and Wallach (1998).

As promised in §3.3.1, we can now identify the partial flag varieties with closed sets in some projective space.

3.3.11 Theorem *Let V be an algebraic G-module and $0 < n_1 < \cdots < n_d(\leqslant \dim V)$ be a sequence of strictly increasing integers. Then the map*

$$\pi \colon \begin{array}{ccc} \mathcal{F}_{n_1,\ldots,n_d}(V) & \to & \mathbb{P}(\bigwedge^{n_1} V \otimes \cdots \otimes \bigwedge^{n_d} V) \\ (E_1, \ldots, E_d) & \mapsto & \langle \psi_{n_1}(E_1) \otimes \cdots \otimes \psi_{n_d}(E_d) \rangle \end{array}$$

is G-equivariant and injective, and its image is closed. Here, the G-actions on both sides are the natural actions induced by the given G-action on V.

Proof The above map is obtained by composing the G-equivariant injective map

$$\pi' : \begin{array}{ccc} \mathcal{F}_{n_1,\ldots,n_d}(V) & \to & \mathbb{P}(\bigwedge{}^{n_1} V) \times \cdots \times \mathbb{P}(\bigwedge{}^{n_d} V) \\ (E_1, \ldots, E_d) & \mapsto & (\psi_{n_1}(E_1), \ldots, \psi_{n_d}(E_d)) \end{array}$$

given by Lemma 3.3.8 with the G-equivariant injective map

$$\sigma : \mathbb{P}(\bigwedge{}^{n_1} V) \times \cdots \times \mathbb{P}(\bigwedge{}^{n_d} V) \to \mathbb{P}(\bigwedge{}^{n_1} V \otimes \cdots \otimes \bigwedge{}^{n_d} V)$$

given by Lemma 3.3.3. Thus, we already have that π is G-equivariant and injective. It remains to show that the image is closed. By Exercise 3.6.7, we have a description of the subsets of $\mathbb{P}(\bigwedge{}^{n_1} V) \times \cdots \times \mathbb{P}(\bigwedge{}^{n_d} V)$ which correspond to closed subsets of $\mathbb{P}(\bigwedge{}^{n_1} V \otimes \cdots \otimes \bigwedge{}^{n_d} V)$ under the Segre embedding. Thus, we must show that the image of π' can be characterized as the zero set of a collection of certain multi-homogeneous polynomials.

Let $1 \leqslant i < j \leqslant d$, and fix $\omega_i \in \bigwedge{}^{n_i} V$ and $\omega_j \in \bigwedge{}^{n_j}(V)$. Then, using the notation of Lemma 3.3.8, we consider the linear map

$$\varphi(\omega_i, \omega_j) : V \to (\bigwedge{}^{n_i+1} V) \oplus (\bigwedge{}^{n_j+1} V), \quad v \mapsto \varphi(\omega_i)(v) \oplus \varphi(\omega_j)(v).$$

We have $\ker(\varphi(\omega_i, \omega_j)) = \ker(\varphi(\omega_i)) \cap \ker(\varphi(\omega_j))$. Now let $E_i := \ker(\varphi(\omega_i))$ and $E_j := \ker(\varphi(\omega_j))$. Then we have $E_i \subseteq E_j$ if and only if $E_i = E_i \cap E_j$, which is equivalent to the condition that the rank of $\varphi(\omega_i, \omega_j)$ is less than or equal to the rank of $\varphi(\omega_i)$. In combination with Lemma 3.3.8, this yields the following description of the image of π'. We have

$$(\langle \omega_1 \rangle, \ldots, \langle \omega_d \rangle) \in \pi'(\mathcal{F}_{n_1,\ldots,n_d}(V)) \ (\text{where } 0 \neq \omega_i \in \bigwedge{}^{n_i}(V) \text{ for all } i)$$

if and only if

$$\left\{ \begin{array}{rcll} \mathrm{rank}_k(\varphi(\omega_i)) & \leqslant & \dim V - n_i & \text{for } 1 \leqslant i \leqslant d \\ \mathrm{rank}_k(\varphi(\omega_i, \omega_j)) & \leqslant & \mathrm{rank}_k(\varphi(\omega_i)) & \text{for } 1 \leqslant i < j \leqslant d \end{array} \right\}.$$

Now, as in the proof of Lemma 3.3.8, the first of the above two rank conditions can be expressed by the vanishing of certain determinantal expressions in the Plücker coordinates of $\omega_1, \ldots, \omega_d$ with respect to a fixed basis $\{e_1, \ldots, e_n\}$ of V. It is easily checked that this also holds for the second condition. These determinantal expressions yield the required multi-homogeneous polynomials. $\square$

3.3.12 Example Consider the group $G = \mathrm{GL}_n(k)$. Then $V = k^n$ is naturally an algebraic G-module, where the action is given by multiplying a matrix in $\mathrm{GL}_n(k)$ by a column vector in k^n. Let $0 < n_1 < n_2 < \cdots < n_{d-1} < n_d (= n)$ be a strictly increasing sequence of integers, and consider the corresponding (partial) flag variety $\mathcal{F}_{n_1,\ldots,n_d}(V)$. First note that G certainly acts transitively on $\mathcal{F}_{n_1,\ldots,n_d}(V)$. (Indeed, let $E_1 \subset \cdots \subset E_d$ be a chain of subspaces with $\dim_k E_r = n_r$ for all r; then choose an adapted basis of k^n such that E_r is spanned by the first n_r vectors in that basis. If $E_1' \subset \cdots \subset E_d'$ is another chain of subspaces, we can find an element in G which sends the first n_r basis vectors into E_r' for all r.) Now consider the standard basis $\{e_1, \ldots, e_n\}$ of $V = k^n$. For $1 \leqslant r \leqslant d$, let E_r be the subspace spanned by $e_1, \ldots, e_{n_r}$. Then the stabilizer of $(E_1, \ldots, E_d) \in \mathcal{F}_{n_1,\ldots,n_d}(V)$ is the subgroup $P_{n_1,\ldots,n_d}(k) \subseteq G$ consisting of all block-triangular matrices of the form

$$
\begin{bmatrix}
A_1 & * & \cdots & * \\
0 & A_2 & \ddots & \vdots \\
\vdots & \ddots & \ddots & * \\
0 & \cdots & 0 & A_d
\end{bmatrix}
\qquad \text{with } A_i \in \mathrm{GL}_{n_i - n_{i-1}}(k)
$$

(where we set $n_0 = 0$). Hence we have a bijection

$$
G/P_{n_1,\ldots,n_d}(k) \xrightarrow{\sim} \mathcal{F}_{n_1,\ldots,n_d}(V), \quad gP_{n_1,\ldots,n_d}(k) \mapsto (g.E_1, \ldots, g.E_d).
$$

Note that all the groups $P_{n_1,\ldots,n_d}(k)$ are connected. (Check this!) In the special case where we take the full flag variety (corresponding to the sequence $0 < 1 < 2 < \cdots < n-1 < n$), the corresponding subgroup is the subgroup $B_n(k) \subseteq G$ consisting of all upper triangular matrices. It is not difficult to check that the various groups $P_{n_1,\ldots,n_d}(k)$ are precisely the subgroups of G which contain $B_n(k)$.

3.4 Parabolic subgroups and Borel subgroups

Throughout this section, let G be an affine algebraic group over an algebraically closed field k. Using the results on flag varieties in the previous section, we can now establish some basic structure theorems. One of the aims of this section is to obtain a purely group-theoretic characterization of parabolic subgroups. This will involve the following important class of subgroups of G.

3.4.1 Definition   A maximal closed connected solvable subgroup of G is called a *Borel* subgroup of G.

Note that Borel subgroups always exist, for dimensional reasons. In Proposition 3.4.6, we will show that, if G admits a BN-pair where B is a closed connected solvable subgroup, then B is a Borel subgroup. Thus, by Theorems 1.7.4 and 1.7.8, we have a description of a Borel subgroup in any group of classical type. The key to the study of Borel subgroups is provided by the following result.

3.4.2 Theorem (Steinberg)   *Let V be an algebraic G-module and $Z \subseteq \mathbb{P}(V)$ be a non-empty G-invariant closed subset. Assume that G is connected and solvable. Then there exists some $\langle v \rangle \in Z$ such that $g.\langle v \rangle = \langle v \rangle$ for all $g \in G$.*

Proof   Let $\{e_1, \ldots, e_n\}$ be a basis of V such that the action of G on V is given by the formula in §3.1.1, with regular functions $a_{ij} \in A[G]$. To simplify notation, we will identify $V = k^n$ using that basis. Let $\rho\colon G \to \mathrm{GL}_n(k)$ be the corresponding algebraic homomorphism. Now, the G-invariant closed subset $Z \subseteq \mathbb{P}(V)$ is defined by a set of homogeneous polynomials in $k[T_1, \ldots, T_n]$. Let $C \subseteq k^n$ be the corresponding affine algebraic set. Then, as in §3.2.1, C is a G-invariant union of 1-dimensional subspaces. We must show that at least one of these 1-dimensional subspaces is G-invariant. We will do this in two steps.

Step 1. First we assume that $\rho(G) \subseteq \mathrm{GL}_n(k)$ is abelian. Then we will proceed by induction on n. If $n = 1$, there is nothing to prove. Now assume that $n > 1$. By Theorem 3.1.7, there exists at least some G-invariant 1-dimensional subspace $V_1 \subseteq V$. We must show that this can be chosen inside C. Conjugating $\rho(G) \subseteq \mathrm{GL}_n(k)$ by a suitable matrix, we may assume without loss of generality that V_1 is spanned by the first standard basis vector $e_1 = (1, 0, \ldots, 0) \in k^n$.

Now let us first settle the case where $n = 2$. If we also have $\dim C = 2$, then $C = V$ by Proposition 1.2.20, and so $V_1 \subseteq C$. Thus, we are trivially done in this case. If $\dim C = 1$ then each 1-dimensional subspace in C is an irreducible component of C (since a 1-dimensional subspace is a closed irreducible subset of dimension 1). Consequently, C is just a union of finitely many 1-dimensional subspaces. Now G permutes these finitely many subspaces in C, and therefore the stabilizer in G of such a subspace is a closed subgroup of finite index. Since G is connected, that stabilizer must be all of G, and so each 1-dimensional subspace in C is G-invariant. So we are done in this case as well.

Now let $n > 2$. If $V_1 \subseteq C$, we are done. Now assume that $V_1 \not\subseteq C$. Let $V' := k^{n-1}$ and consider the projection $\pi\colon V \to V'$, $(x_1, \ldots, x_n) \mapsto (x_2, \ldots, x_n)$. We have $\pi(C) \neq \{0\}$ since $V_1 \not\subseteq C$. We claim that $\pi(C) \subseteq V'$ is a closed subset. By Lemma 2.2.5(b), it is enough to show that the restriction $\pi\colon C \to V'$ is a finite morphism. So let us consider the homomorphism $\pi^*\colon k[X_2, \ldots, X_n] \to A[C]$. We write $A[C] = k[X_1, \ldots, X_n]/\mathbf{I}(C)$ and denote the image of X_i by $\bar{X}_i$. Then $\pi^*(X_i) = \bar{X}_i$ for $2 \leqslant i \leqslant n$. So it will be enough to show that $\bar{X}_1$ is integral over $\pi^*(k[X_2, \ldots, X_n])$. Now, since $V_1 \not\subseteq C$, we have $\mathbf{I}(C) \not\subseteq \mathbf{I}(V_1)$, and so there exists a non-zero polynomial $F \in k[X_1, \ldots, X_n]$ such that F is identically zero on C but $F(1, 0, \ldots, 0) \neq 0$. By Exercise 3.6.4, $\mathbf{I}(V)$ is generated by homogeneous polynomials. So we may choose F to be a homogeneous polynomial. Assume that F has degree d and write $F = \sum_{i=0}^{d} f_i X_1^{d-i}$, where $f_i \in k[X_2, \ldots, X_n]$ is homogeneous of degree i. Then we have

$$0 = F(\bar{X}_1, \ldots, \bar{X}_n) = \sum_{i=0}^{d} f_i(\bar{X}_2, \ldots, \bar{X}_n)\, \bar{X}_1^{d-i} = \sum_{i=0}^{d} \pi^*(f_i)\, \bar{X}_1^{d-i}$$

$$= \pi^*(f_d) + \pi^*(f_{d-1})\, \bar{X}_1 + \cdots + \pi^*(f_1)\, \bar{X}_1^{d-1} + \pi^*(f_0)\, \bar{X}_1^{d}.$$

Now, we have $f_0 = F(1, 0, \ldots, 0) \neq 0$. Hence, dividing the above equation by the constant $\pi^*(f_0)$, we obtain an equation of integral dependence for $\bar{X}_1$ over $\pi^*(k[X_2, \ldots, X_n])$. It follows that $\pi\colon C \to V'$ is a finite morphism, as claimed.

Now $\pi(C) \subseteq V'$ is a closed subset and it is a cone in V'. Furthermore, V' also is an algebraic G-module. So, by induction on n, there exists a G-invariant 1-dimensional subspace $V_1' \subseteq \pi(C)$. Let $V_2 := \pi^{-1}(V_1') \subseteq V$. Then V_2 is a G-invariant 2-dimensional subspace, and $C \cap V_2 \neq \{0\}$ is a G-invariant non-zero closed cone. Thus, we have reduced our problem to a subspace of dimension 2. But here we have already solved our problem. This completes the proof of the theorem in the special case where $\rho(G) \subseteq \mathrm{GL}_n(k)$ is abelian.

Step 2. We show that the general case can be reduced to the above special case. We proceed by induction on $\dim G$. If $\dim G = 0$, then there is nothing to prove. Now assume that $\dim G > 0$ and consider the commutator subgroup $G' \subseteq G$. Since G is solvable, we have $G' \subsetneq G$. By Example 2.4.7, G' is closed and connected and so $\dim G' < \dim G$. By induction, there exists a G'-invariant

1-dimensional subspace $V_1 \subseteq C$. Let $\chi \in X(G')$ be such that $V_1 \subseteq V_\chi \cap C$. By Lemma 3.1.6, V_χ is a G-invariant subspace on which G' acts trivially. Since $V_\chi \cap C$ is a non-zero closed cone in V_χ, it will be enough to prove the existence of a G-invariant 1-dimensional subspace inside $V_\chi \cap C$. So we can just work with the restricted representation $\rho'\colon G \to \mathrm{GL}(V_\chi)$. But, since G' acts trivially on V_χ, we have that $\rho'(G) \subseteq \mathrm{GL}(V_\chi)$ is abelian, and we can apply Step 1.
$\qquad\square$

3.4.3 Theorem (Borel) *Let G be an affine algebraic group. Then all Borel subgroups of G are conjugate. Furthermore, a closed subgroup of G is parabolic if and only if it contains a Borel subgroup. In particular, every Borel subgroup of G is parabolic.*

Proof We proceed in several steps.

Step 1. There exists a parabolic subgroup $P \subseteq G$ such that $B := P^\circ$ is a Borel subgroup and such that all Borel subgroups of G are conjugate to B.

This is proved as follows. By Theorem 2.4.4, there exists some $n \geqslant 1$ and an injective homomorphism of algebraic groups $\varphi\colon G \to \mathrm{GL}_n(k)$. Then $V = k^n$ is an algebraic G-module, and G acts on the flag variety $\mathcal{F}(V)$. We identify $\mathcal{F}(V)$ with a closed subset in $\mathbb{P}(\bigwedge^1 V \otimes \bigwedge^2 V \otimes \cdots)$ as in Theorem 3.3.11, and we regard $\mathcal{F}(V)$ as a topological space using that identification. With this convention, by Proposition 3.2.5, there exists some $F \in \mathcal{F}(V)$ such that the G-orbit of F is closed. We set $P := \mathrm{Stab}_G(F)$ and $B := P^\circ$. Then P is parabolic by construction. Furthermore, since B stabilizes a flag and φ is injective, $\varphi(B)$ is conjugate to a subgroup of $B_n(k)$ and, hence, solvable. Consequently, B is a closed connected solvable subgroup of G.

Now let $B' \subseteq G$ be any Borel subgroup of G. Applying Theorem 3.4.2 to the G-orbit of F, we see that B' has a fixed point on that orbit, that is, there exists some $g \in G$ such that $B'g.F = g.F$. In other words, this means that $g^{-1}B'g \subseteq \mathrm{Stab}_G(F)$. Since $g^{-1}B'g$ is connected, we must have $g^{-1}B'g \subseteq B$. By the maximality of B', we conclude that $g^{-1}B'g = B$. Hence, every Borel subgroup is conjugate to B. In particular, B itself is a Borel subgroup.

Step 2. Next we show that any Borel subgroup of G is parabolic. By Step 1, there exists a parabolic subgroup $P \subseteq G$ such that $B := P^\circ$ is a Borel subgroup. We claim that B must be parabolic. To see this, let V be an algebraic G-module such that $B = \mathrm{Stab}_G(\langle v \rangle)$ for some non-zero $v \in V$; see Proposition 3.3.9. We want to show that the G-orbit of $\langle v \rangle$ in $\mathbb{P}(V)$ is closed. Now, since B has finite index in P, the P-orbit of $\langle v \rangle$ in $\mathbb{P}(V)$ is a finite set and, hence, closed. Denote that P-orbit by C. Then C is P-invariant, and $G.C$ is the G-orbit of $\langle v \rangle$. The assertion now follows from Corollary 3.2.12(b). Thus, B is parabolic. Since all Borel subgroups are conjugate, it easily follows that they are all parabolic.

Step 3. Now let P be any parabolic subgroup of G. We want to show that P contains a Borel subgroup. By definition, there exists an algebraic G-module V and a non-zero $v \in V$ such that $P = \mathrm{Stab}_G(\langle v \rangle)$ and the G-orbit of $\langle v \rangle$ in $\mathbb{P}(V)$ is closed. Now let $B \subseteq G$ be any Borel subgroup. Then, by Theorem 3.4.2, B has a fixed point on the G-orbit of $\langle v \rangle$, and so there exists some $g \in G$ such that $b.(g.\langle v \rangle) = g.\langle v \rangle$ for all $b \in B$. This implies $g^{-1}bg \in \mathrm{Stab}_G(\langle v \rangle) = P$ for all $b \in B$, and so $g^{-1}Bg \subseteq P$. Now, clearly, $g^{-1}Bg$ also is a Borel subgroup of G. Thus, we have shown that every parabolic subgroup contains a Borel subgroup.

Step 4. Finally, assume that P is a closed subgroup containing a Borel subgroup, B say. Let V be an algebraic G-module such that $P = \mathrm{Stab}_G(\langle v \rangle)$ for some $0 \neq v \in V$. We want to show that the G-orbit of $\langle v \rangle$ in $\mathbb{P}(V)$ is closed. Since $B \subseteq \mathrm{Stab}_G(\langle v \rangle)$, this immediately follows from Corollary 3.2.12(b), using the fact that B is parabolic. $\qquad\square$

3.4.4 Example Let $G = \mathrm{GL}_n(k)$. By Theorem 3.1.7, every connected closed solvable subgroup of G is conjugate to a subgroup of $B_n(k)$. Since $B_n(k)$ is connected and solvable, we conclude that $B_n(k)$ is a Borel subgroup in G.

Furthermore, every parabolic subgroup of G is conjugate to some subgroup $P_{n_1,\ldots,n_d}(k)$ as in §3.3.12. Indeed, we have already seen in §3.3.12 that all of the subgroups $P_{n_1,\ldots,n_d}(k)$ are parabolic. Conversely, let $P \subseteq G$ be a parabolic subgroup. By Theorem 3.4.3, P contains some Borel subgroup $B \subseteq G$. However, all Borel subgroups of G are conjugate and $B_n(k)$ is a Borel subgroup. Thus, conjugating P by some $g \in G$ if necessary, we may assume that $B_n(k) \subseteq P$. But then P must be one of the subgroups $P_{n_1,\ldots,n_d}(k)$, as remarked in §3.3.12.

Now recall from Section 1.6 the definition of a group with a BN-pair. Our next aim is to show that, under suitable hypotheses, the subgroup B will be a Borel subgroup. The required conditions are isolated in the following definition.

3.4.5 Definition   Let G be an affine algebraic group over k. Assume that G contains closed subgroups B, N which form a split BN-pair. Write $B = UH$, where $H = B \cap N$ and U is a normal subgroup of B; see Definition 1.6.13. We say that G is a group with a *reductive* BN-pair if the following conditions hold.

(a) The group H is a torus (see Proposition 3.1.9) such that $C_G(H) = H$.

(b) The group U is closed connected and nilpotent.

The above conditions imply, first of all, that $B = UH$ is a closed connected solvable subgroup. (Indeed, the natural map $U \times H \to B$ given by multiplication is surjective; so it remains to use Proposition 1.3.8(a) and Remark 1.3.2.) Consequently, G is connected; see Remark 1.6.7. Furthermore, by the rigidity of tori (Proposition 3.1.10), $W = N/H$ is a finite group.

The BN-pair for $\mathrm{GL}_n(k)$ in Example 1.6.10 and the BN-pairs for classical groups in Theorems 1.7.4 and 1.7.8 are all reductive. (It only remains to check that $C_G(H) = H$, which is easy.) We just remark here that every *group of Lie type* as in Example 1.5.15 has a reductive BN-pair; see p. 58, Theorem 6, of Steinberg (1967).

3.4.6 Proposition   *Let G be an affine algebraic group with a reductive BN-pair. Then the following hold.*

(a) We have $\mathrm{N}_G(U) = B$, $[B, B] = U$, and $R_{\mathrm{u}}(B) = U$.

(b) B is a Borel subgroup and G is a reductive group; see §3.1.13.

(c) Let $G' := \langle gUg^{-1} \mid g \in G \rangle \subseteq G$. Then $G' = [G, G]$ and $G = H.G'$; note that G' is a semisimple group by Lemma 3.1.14.

Proof   (a) First we show that $\mathrm{N}_G(U) = B$. Indeed, let $P := \mathrm{N}_G(U) \supseteq B$ and assume, if possible, that $B \neq P$. Then, as in the proof of Corollary 1.6.4, we see that $n_s \in P$ for some $s \in S$. This would imply $n_s U n_s^{-1} = U$. Since n_s also normalizes H, we conclude that $n_s B n_s^{-1} = n_s B n_s = B$, contradicting (BN3). Thus, we must have $B = \mathrm{N}_G(U)$. The fact that $[B, B] = U$ follows from Exercises 3.6.1 and 2.7.13. Finally, consider $R_{\mathrm{u}}(B)$. Since $B/R_{\mathrm{u}}(B)$ is abelian, we have $U = [B, B] \subseteq R_{\mathrm{u}}(B)$. Now, by Corollary 3.3.10, there exists a homomorphism $\rho \colon B \to \mathrm{GL}_n(k)$ of algebraic groups such that $\ker(\rho) = U$. Since $\rho(B) = \rho(H)$ is diagonalizable, we may assume that $\rho(B)$ consists of diagonal matrices. Now let $u \in R_{\mathrm{u}}(B)$.

Then $\rho(u)$ is diagonal and, by Lemma 3.1.12, all its eigenvalues are 1. Thus, $\rho(u) = 1$ and so $u \in \ker(\rho) = U$, as required.

(b) Since B is closed, connected, and solvable, we know at least that there exists some Borel subgroup $B' \subseteq G$ such that $B \subseteq B'$. Let $U' := R_{\mathrm{u}}(B')$. The (abstract) quotient $B/B \cap U'$ is isomorphic to a subgroup of B'/U' and, hence, is abelian. Using (a), we conclude that $U \subseteq U'$. If $U = U'$, then $B' \subseteq \mathrm{N}_G(U') = \mathrm{N}_G(U) = B$ and we are done. So let us assume, if possible, that $U \neq U'$. Then, since U' is nilpotent, we must have $V := \mathrm{N}_{U'}(U) \supsetneq U$ (see Theorem III.2.3 of Huppert 1983). Consequently, we have $U \subsetneq V \subseteq \mathrm{N}_G(U) = B$, where $V \subseteq U'$. Let $v \in V \setminus U$ and write $v = uh$, where $u \in U$ and $1 \neq h \in H$. Now let $\rho \colon B' \to \mathrm{GL}_n(k)$ be a closed embedding; since B' is connected and solvable, we may assume that $\rho(B') \subseteq B_n(k)$; see Theorem 3.1.7. Then $\rho(v)$ and $\rho(u)$ are triangular matrices with 1 on the diagonal, by Lemma 3.1.12. Hence the same also holds for $\rho(h)$, since $h = u^{-1}v$. Thus, all eigenvalues of h are 1. But h is diagonalizable by Proposition 3.1.9, and so we must have $h = 1$, a contradiction. Thus, our assumption $U \neq U'$ was wrong and so we have $B' = B$, as desired.

Now we can also show that G is reductive. Indeed, since B is a Borel subgroup, we must have $R(G) \subseteq B$. Furthermore, since $R(G)$ is a normal subgroup, we have $R(G) \subseteq \bigcap_{n \in N} nBn^{-1} = H$. Thus, $R(G)$ is a closed connected subgroup of a torus, and hence $R(G)$ must be a torus itself.

(c) The subgroup $G' \subseteq G$ certainly is normal. Hence $G'.H$ is a subgroup containing B and normalized by N. Thus, we also have $n_0 B n_0^{-1} \subseteq G'.H$, where $n_0 \in N$ is a representative of the longest element $w_0 \in W$. So we have $G = G'.H$ by Corollary 1.6.6. Since H is abelian, we have $[G, G] \subseteq G'$. The reverse inclusion follows from (a), which shows that $U \subseteq [G, G]$. $\qquad\qquad\square$

3.4.7 Lemma *Assume that G is an algebraic group with a reductive BN-pair. Write $B = UH$ as in Definition 3.4.5.*

(a) Let $t \in H$ and $g \in G$ be such that $gtg^{-1} \in H$. Then there exists some $n \in N$ such that $gtg^{-1} = ntn^{-1}$.

(b) We have $N = \mathrm{N}_G(H)$.

Proof (a) We set $t' := gtg^{-1}$. By the Bruhat decomposition, there exists some $n \in N$ such that $BgB = BnB$. So we can write $bg = nb_1$, where $b, b_1 \in B$. Now consider the identity $bgtg^{-1}b^{-1} = nb_1tb_1^{-1}n^{-1}$. The left-hand side can be evaluated as follows. Let us write $b = uh$, where $u \in U$ and $h \in H$. Since H is abelian, this yields $bgtg^{-1}b^{-1} = uht'h^{-1}u^{-1} = ut'u^{-1} = t'\tilde{u}$, where $\tilde{u} := (t')^{-1}ut'u^{-1} \in U$. Now consider the right-hand side. We write $b_1 = u_1h_1$, where $u_1 \in U$ and $h_1 \in H$. Then we obtain

$$nb_1tb_1^{-1}n^{-1} = nu_1h_1th_1^{-1}u_1^{-1}n^{-1} = nu_1tu_1^{-1}n^{-1}$$

$$= ntn^{-1}.nt^{-1}u_1tu_1^{-1}n^{-1} = ntn^{-1}.n\tilde{v}n^{-1},$$

where $\tilde{v} := t^{-1}u_1tu_1^{-1} \in U$. Thus, our original identity can be rewritten as

$$t'\tilde{u} = bgtg^{-1}t^{-1} = nb_1tb_1^{-1}n^{-1} = ntn^{-1}.n\tilde{v}n^{-1}. \tag{$*$}$$

Now the left-hand side of $(*)$ lies in B. Hence so does the right-hand side. Since $ntn^{-1} \in H \subseteq B$, we deduce that $n\tilde{v}n^{-1} \in nUn^{-1} \cap B \subseteq U$, using Definition 1.6.13(b). Since $B = UH = HU$ is a semidirect product, we can now conclude that $t' = ntn^{-1}$, as desired.

(b) Let $g \in \mathrm{N}_G(H)$ and $n \in N$ be such that $BgB = BnB$. Now take any $t \in H$. Then we also have $gtg^{-1} \in H$ and so we can apply the above argument. Thus, we have $ntn^{-1} = gtg^{-1}$ and so $n^{-1}g \in \mathrm{C}_G(t)$. This holds for all $t \in H$, and so $n^{-1}g \in \mathrm{C}_G(H) = H$. Hence we have $g \in nH \subseteq N$, as desired. $\qquad\square$

We can now establish two basic properties of algebraic groups with a reductive BN-pair. The proofs rely in an essential way on the fact that Borel subgroups are parabolic and on Theorem 3.2.11. The first result gives a combinatorial description of the Zariski closures of the Bruhat cells $C(w) = Bn_wB$. Note that $B \times B$ acts on G by $(b_1, b_2).g = b_1gb_2^{-1}$ ($b_i \in B$, $g \in G$). Hence $C(w)$ is locally closed in G, being the $(B \times B)$-orbit of n_w.

3.4.8 Theorem (the Bruhat–Chevalley order) *Let G be an affine algebraic group and assume that G has a BN-pair where B is a Borel subgroup. Let $w \in W$ and fix a reduced expression $w = s_1 \cdots s_p$ where $s_i \in S$ and $p = l(w)$. We set*

$$\mathcal{S}(w) := \left\{ s_{i_1}s_{i_2} \cdots s_{i_q} \ \middle| \ \begin{array}{l} 1 \leqslant i_1 < i_2 < \cdots < i_q \leqslant p \\ \textit{for some } q \textit{ with } 0 \leqslant q \leqslant p \end{array} \right\}.$$

Then $\mathcal{S}(w)$ does not depend on the chosen expression for w and the Zariski closure of $C(w)$ is given by

$$\overline{C(w)} = \bigcup_{y \in \mathcal{S}(w)} C(y).$$

Proof We remark that the assertion concerning $\mathcal{S}(w)$ can be deduced directly from the fact that W is a Coxeter group; see for example §1.2.4 of Geck and Pfeiffer (2000). Note, however, that once we have shown that $\overline{C(w)}$ is a union as above, it will automatically follow that $\mathcal{S}(w)$ does not depend on the chosen expression for w. We now proceed in four steps.

Step 1. For any $s \in S$, we consider the corresponding subgroup $P_s = P_{\{s\}} = B \cup C(s)$ as in Lemma 1.6.2. We claim that P_s is closed and connected. For the proof, we note that $P_s = \langle B, n_s B n_s \rangle$. Indeed, the subgroup on the right-hand side is certainly contained in P_s but it is also strictly larger than B, by (BN3). Hence it must be equal to P_s by the Bruhat decomposition. Now B is a closed connected subgroup, and so is $n_s B n_s = n_s B n_s^{-1}$. Hence the assertion follows from Theorem 2.4.6.

Step 2. We claim that $P_{s_1}.P_{s_2} \ldots P_{s_p} = \bigcup_{y \in \mathcal{S}(w)} C(y)$, where the left-hand side is interpreted as B if $p = 0$. This will be proved by induction on p. If $p = 0$, there is nothing to prove. Now assume that $p > 0$ and set $w' = s_2 \cdots s_p$. Then we have $l(w') = p - 1$ and so, by induction, $P_{s_2} \ldots P_{s_p} = \bigcup_{y \in \mathcal{S}(w')} C(y)$. (This set equals B if $p = 1$.) Now note that we have

$$P_s.C(y) = (B \cup C(s)).C(y) = C(y) \cup C(sy)$$

for any $s \in S$ and $y \in W$; see Proposition 1.6.3. This yields that

$$P_{s_1}.(P_{s_2} \ldots P_{s_p}) = P_{s_1}.\left(\bigcup_{y \in \mathcal{S}(w')} C(y) \right) = \bigcup_{y \in \mathcal{S}(w')} (C(y) \cup C(s_1 y)r).$$

Now note that $\mathcal{S}(w) = \mathcal{S}(w') \cup \{s_1 y \mid y \in \mathcal{S}(w')\}$. Hence the union of sets on the right-hand side of the above equation equals $\bigcup_{y \in \mathcal{S}(w)} C(y)$, as required. Thus, the above claim is proved.

Step 3. We claim that $P_{s_1}.P_{s_2} \ldots P_{s_p}$ is a closed irreducible subset of G. This too will be shown by induction on p. If $p = 0$, then the assertion is clear since B is assumed to be connected. Now assume that $p > 0$. Again, we set $w' = s_2 \cdots s_p$. Then, by induction, $Y := P_{s_2} \ldots P_{s_p}$ is a closed irreducible subset of G. (We have $Y = B$ if $p = 1$.) Now we consider G as a P_{s_1}-variety via left multiplication. Then Y is a B-invariant closed subset of G. Furthermore, since B is a Borel subgroup in G, it is also a Borel subgroup in P_{s_1} and, hence, a parabolic subgroup of P_{s_1} by Theorem 3.4.3. Applying Corollary 3.2.12(a) yields that the set $P_{s_1}.Y = P_{s_1}.P_{s_2} \ldots P_{s_p}$ is closed in G, as required. Finally, that set is also irreducible since it is the image of the morphism $P_{s_1} \times Y \to G$ given by multiplication and since P_{s_1} is connected by step 1.

Step 4. Now we show by induction on $l(w)$ that $\overline{C(w)} = \bigcup_{y \in \mathcal{S}(y)} C(y)$. If $w = 1$, there is nothing to prove. Now assume that $l(w) > 0$ and fix an expression $w = s_1 \cdots s_p$, where $s_i \in S$ and $p = l(w)$. Let us set $X := P_{s_1} \ldots P_{s_p}$. Using step 2 and the fact that X is closed (see step 3), we obtain

$$X = \overline{X} = \overline{\bigcup_{y \in \mathcal{S}(w)} C(y)} = \bigcup_{y \in \mathcal{S}(w)} \overline{C(y)}.$$

Since X is also irreducible, we must have $X = \overline{C(y)}$ for some $y \in \mathcal{S}(w)$. In particular, this implies that $C(w) \subseteq \overline{C(y)}$. If we had $y \neq w$, then $l(y) < l(w)$ and so, by induction, $\overline{C(y)}$ is a union of double cosets $C(z)$ where $l(z) \leqslant l(y) < l(w)$, a contradiction. Thus, we must have $y = w$, as required. $\qquad\qquad\square$

3.4.9 Proposition *Let G be a connected affine algebraic group. Let B be a Borel subgroup and $H \subseteq B$ be a closed subgroup which is a torus such that $\mathbf{C}_G(H) = H$. Then every element of G lies in a Borel subgroup; that is, we have*

$$G = \bigcup_{x \in G} xBx^{-1}.$$

In particular, this holds if G has a reductive BN-pair.

Proof Let $X := \bigcup_{x \in G} xBx^{-1}$. Since B is a parabolic subgroup, the set X is closed in G; see Example 3.2.13. Hence it will be enough to show that X contains a dense subset of G. This will follow from the condition that $\mathbf{C}_G(H) = H$. Indeed, consider the closed set $Y := \{(x, g) \in G \times G \mid x^{-1}gx \in H\} \subseteq G \times G$. First note that Y is irreducible, being the image of the morphism $G \times H \to G \times G$, $(x, h) \mapsto (x, xhx^{-1})$. Now we compute $\dim Y$ in two different ways. Let $\mathrm{pr}_1 \colon Y \to G$ be the first projection. Then $\mathrm{pr}_1^{-1}(x) = \{(x, g) \mid g \in xHx^{-1}\} \cong H$ for any $x \in G$. Thus, since pr_1 is surjective, Corollary 2.2.9 yields

$$\dim Y = \dim G + \dim H.$$

Now consider the second projection $\mathrm{pr}_2 \colon Y \to G$. Note that we certainly have

$$\mathrm{pr}_2(Y) \subseteq X \quad \text{and so} \quad \dim X \geqslant \dim \overline{\mathrm{pr}_2(Y)}.$$

Hence it is enough to show that $\dim \overline{\mathrm{pr}_2(Y)} \geqslant \dim G$. Now, by Corollary 2.2.9, there exists some non-empty open subset $U \subseteq G$ such that $U \subseteq \mathrm{pr}_2(Y)$ and

$$\dim \overline{\mathrm{pr}_2(Y)} = \dim G + \dim H - \dim(\mathrm{pr}_2^{-1}(g)) \quad \text{for all } g \in U,$$

where we also used the formula $\dim Y = \dim G + \dim H$. Furthermore, since H is a torus, the set $D := \{h \in H \mid \mathrm{C}_G(H) = \mathrm{C}_G(h)\}$ is dense in H; see Exercise 3.6.1. Let $h \in D$ and $x \in G$ be such that $h \in xHx^{-1}$; then $xHx^{-1} \subseteq \mathrm{C}_G(h) = H$ and so $x \in \mathrm{N}_G(H)$. This shows that $\mathrm{pr}_2^{-1}(h) \subseteq \{(x, h) \mid x \in \mathrm{N}_G(H)\}$ and so $\dim \mathrm{pr}_2^{-1}(h) \leqslant \dim \mathrm{N}_G(H)$. Using Proposition 3.1.10, we conclude that

$$\dim \mathrm{pr}_2^{-1}(h) \leqslant \dim H \quad \text{for all } h \in D.$$

Since D is dense in H, we deduce that $\bigcup_{x \in G} xDx^{-1}$ is dense in $\overline{\bigcup_{x \in G} xHx^{-1}} = \overline{\mathrm{pr}_2(Y)}$. But then $\bigcup_{x \in G} xDx^{-1}$ must have a non-empty intersection with the open set $U \subseteq \mathrm{pr}_2(Y)$. Consequently, there exists some $g \in U$ such that $\dim \mathrm{pr}_2^{-1}(g) \leqslant \dim H$ and so $\dim \overline{\mathrm{pr}_2(Y)} \geqslant \dim G$, as required. $\qquad\square$

3.5 On the structure of Borel subgroups

Throughout this section, let $k = \overline{\mathbb{F}}_p$ be an algebraic closure of the finite field with p elements, where p is a prime number. We shall establish some basic structural results on connected solvable groups over k. (The results that we will prove are also valid over an arbitrary algebraically closed ground field, but the proofs are considerably simpler in the case where $k = \overline{\mathbb{F}}_p$.)

Let G be an affine algebraic group over $\overline{\mathbb{F}}_p$. Then there exists a closed embedding of algebraic groups $\varphi \colon G \to \mathrm{GL}_n(\overline{\mathbb{F}}_p)$; see Corollary 2.4.4. Thus, we can always assume that G is isomorphic to a closed subgroup of some $\mathrm{GL}_n(\overline{\mathbb{F}}_p)$. Now, since every element of $\overline{\mathbb{F}}_p$ lies in a finite extension of $\mathbb{F}_p$, we have

$$\mathrm{GL}_n(\overline{\mathbb{F}}_p) = \bigcup_{d \geqslant 1} \mathrm{GL}_n(\mathbb{F}_{p^d}).$$

Thus, every element of $\mathrm{GL}_n(\overline{\mathbb{F}}_p)$ lies in a finite group and so has finite order. Consequenly, every element in G has finite order. This simple observation will make it possible to bring to bear techniques from the theory of finite groups in obtaining results concerning the structure of G.

First of all, we can write any $g \in G$ uniquely in the form $g = us = su$, where $u \in G$ is a *unipotent element* (that is, the order of u is a power of p) and $s \in G$ is a *semisimple element* (that is, the order of s is prime to p). We call this the *Jordan decomposition* of g; note that u and s can in fact be expressed as powers of g. If we embed G as a closed subgroup in some $\mathrm{GL}_n(\overline{\mathbb{F}}_p)$, we have

$$g \in G \text{ unipotent} \iff \text{ all eigenvalues of } g \text{ are } 1,$$

$$g \in G \text{ semisimple} \iff g \text{ is diagonalizable.}$$

3.5.1 Lemma *Let G be an affine algebraic group over $\overline{\mathbb{F}}_p$. Then there exists some $m \geqslant 1$ such that $u^{p^m} = 1$ for every unipotent element $u \in G$. (Note that, in general, there is no finite upper bound for the order of a semisimple element.)*

Proof By embedding G into some $\mathrm{GL}_n(\overline{\mathbb{F}}_p)$, we are reduced to proving the assertion for $G = \mathrm{GL}_n(\overline{\mathbb{F}}_p)$. Now, if $u \in \mathrm{GL}_n(\overline{\mathbb{F}}_p)$ is unipotent, then u has characteristic polynomial $(X - 1)^n$ (where X is an indeterminate) and so, by Cayley–Hamilton, $(u - I_n)^n = 0$ (where I_n is the identity matrix). Hence, if $m \geqslant 1$ is such that $p^m \geqslant n$, we have $u^{p^m} - I_n = (u - I_n)^{p^m} = 0$ and so $u^{p^m} = 1$. $\qquad\square$

As a first application of the use of finite-group techniques, we shall prove the following result. Recall from Example 2.4.9 that $U_n(\overline{\mathbb{F}}_p)$ is the subgroup of $\mathrm{GL}_n(\overline{\mathbb{F}}_p)$ consisting of all upper triangular matrices with 1 on the diagonal.

3.5.2 Proposition (Lie's theorem) *Let G be a subgroup of $\mathrm{GL}_n(\overline{\mathbb{F}}_p)$ which consists only of unipotent elements. Then G is conjugate to a subgroup of $U_n(\overline{\mathbb{F}}_p)$. In particular, G is solvable (even nilpotent; see Example 2.4.9).*

Proof Via the embedding $G \subseteq \mathrm{GL}_n(\overline{\mathbb{F}}_p)$, the group acts linearly on $\overline{\mathbb{F}}_p^n$. As in the proof of the Lie–Kolchin Theorem 3.1.7, it will be enough to show that there exists some non-zero $v \in \overline{\mathbb{F}}_p^n$ which is fixed by all elements of G. (Then the proof is completed using an induction on n.) Consider the $\overline{\mathbb{F}}_p$-subspace of $M_n(\overline{\mathbb{F}}_p)$ generated by G. Since $M_n(\overline{\mathbb{F}}_p)$ has finite dimension, we have

$$\langle G \rangle_{\overline{\mathbb{F}}_p} = \langle g_1, \ldots, g_d \rangle_{\overline{\mathbb{F}}_p} \subseteq M_n(\overline{\mathbb{F}}_p) \qquad \text{for some } g_1, \ldots, g_d \in G. \quad (*)$$

Let $P \subseteq \mathrm{GL}_n(\overline{\mathbb{F}}_p)$ be the subgroup generated by $g_1, \ldots, g_d$. Since $\overline{\mathbb{F}}_p$ is the union of finite fields, there exists some $m \geqslant 1$ such that $g_1, \ldots, g_d$ are invertible matrices with coefficients in $\mathbb{F}_q \subseteq \overline{\mathbb{F}}_p$, where $q = p^m$. Hence, $P = \langle g_1, \ldots, g_d \rangle \subseteq \mathrm{GL}_n(\mathbb{F}_q)$ and so P is finite. Furthermore, P is contained in G, and so the order of every element of P is a power of p. So, by Sylow's theorem, the order of P is a power of p. Then the required assertion for P follows from a standard argument about finite p-groups acting linearly on a vector space over a field of characteristic p. Indeed, consider the natural action of $P \subseteq \mathrm{GL}_n(\overline{\mathbb{F}}_q)$ on $\mathbb{F}_q^n$, and decompose $\mathbb{F}_q^n$ into orbits under that action. The cardinality of any orbit is a power of p; the sum of all these cardinalities equals $|\mathbb{F}_q^n| = q^n$; and $\{0\}$ is an orbit of cardinality 1. So there must be another orbit of cardinality 1. This means that there exists some non-zero $v \in \mathbb{F}_q^n$ such that $g_i.v = v$ for $1 \leqslant i \leqslant d$. Using $(*)$, we obtain $g.v = \lambda_g v$ for any $g \in G$, where $\lambda_g \in \overline{\mathbb{F}}_p$. Since g is unipotent, we have $\lambda_g = 1$, as required. Finally, $G \subseteq U_n(\overline{\mathbb{F}}_q)$ is nilpotent by Example 2.4.9. $\qquad\square$

3.5.3 Lemma *Let G be an abelian affine algebraic group over $\overline{\mathbb{F}}_p$. Let G_{s} be the set of semisimple elements in G, and G_{u} the set of unipotent elements in G. Then G_{s} and G_{u} are closed subgroups, and we have $G = G_{\mathrm{s}} \times G_{\mathrm{u}}$. Furthermore, we have $G_{\mathrm{s}}^{\circ} \times G_{\mathrm{u}}^{\circ} = G^{\circ}$, where the superscript $\circ$ denotes the identity component.*

Proof Since G is abelian, it is clear that G_{s} and G_{u} are subgroups. By the Jordan decomposition of elements, we have $G = G_{\mathrm{s}} G_{\mathrm{u}}$ and $G_{\mathrm{s}} \cap G_{\mathrm{u}} = \{1\}$, and so $G = G_{\mathrm{s}} \times G_{\mathrm{u}}$. We now prove that G_{s} and G_{u} are closed subsets. By Lemma 3.5.1, there exists some $m \geqslant 1$ such that $G_{\mathrm{u}} = \{g \in G \mid g^{p^m} = 1\}$; so G_{u} is closed. Now consider the map $\varphi \colon G \to G$, $g \mapsto g^{p^m}$. Since G is abelian, φ is a group homomorphism. It is easily checked that $\varphi(G) = G_{\mathrm{s}}$. Thus, G_{s} is closed by Proposition 2.2.14. Finally, consider the connected components of G_{s} and G_{u}. The group $G_{\mathrm{s}}^{\circ} \times G_{\mathrm{u}}^{\circ}$ is connected (see Proposition 1.3.8) and of finite index in G. Hence we must have $G_{\mathrm{s}}^{\circ} \times G_{\mathrm{u}}^{\circ} = G^{\circ}$ by Proposition 1.3.13. $\qquad\square$

3.5.4 Proposition *Let B be a connected solvable affine group over $\overline{\mathbb{F}}_p$. Let U be the set of unipotent elements in B. Then U is a closed* *connected normal subgroup, and B/U is an abelian group all whose elements have order prime to p. (Here, the quotient B/U is just regarded as an abstract group.) Furthermore, we have $U = R_{\mathrm{u}}(B)$ where $R_{\mathrm{u}}(B)$ is the unipotent radical introduced in §3.1.11.*

Proof First we show that $U = R_{\mathrm{u}}(B)$. Indeed, by Lemma 3.1.12, all elements in $R_{\mathrm{u}}(B)$ are unipotent, and so $R_{\mathrm{u}}(B) \subseteq U$. On the other hand, by §3.1.11, there exists a homomorphism of algebraic groups $\varphi \colon B \to T$, where T is a torus and $\ker(\varphi) = R_{\mathrm{u}}(B)$. Thus, all elements in $B/R_{\mathrm{u}}(B)$ have order prime to p, and so $U \subseteq \ker(\varphi) = R_{\mathrm{u}}(B)$.

It remains to show that U is connected. By Corollary 3.3.10, there exists an algebraic representation $\gamma \colon B \to \mathrm{GL}(W)$ such that $U^\circ = \ker(\gamma)$. Let us set $G := \gamma(B) \subseteq \mathrm{GL}(W)$. Then G is a closed subgroup of $\mathrm{GL}(W)$ (see Proposition 2.2.14) and γ restricts to a surjective homomorphism of affine algebraic groups $\gamma \colon B \to G$. Since B is connected, so is G. Furthermore, G is abelian. Indeed, since B/U is abelian, we have $[B, B] \subseteq U$. On the other hand, $[B, B]$ is connected (see Example 2.4.7) and so $[B, B] \subseteq U^\circ$. It follows that $G \cong B/U^\circ$ is abelian, as claimed. Applying Lemma 3.5.3, we obtain that $G = G_{\mathrm{s}} \times G_{\mathrm{u}}$, where G_{s} and G_{u} are connected. On the other hand, it is clear that $\gamma(U) = G_{\mathrm{u}}$. Since U/U° is a finite group and $U^\circ = \ker(\gamma)$, we have $|G_{\mathrm{u}}| = [U : U^\circ] < \infty$. Since G_{u} is connected, this implies that $U = U^\circ$, as required. $\square$

The next step is to show that the normal subgroup U has a complement in B. The proof that we will give for this result is in fact a variation of the proof of the Schur–Zassenhaus theorem in finite-group theory.

First recall that if G is any group and $N \subseteq G$ is a normal subgroup, then a *complement* of N is a subgroup $T \subseteq G$ such that $N \cap T = \{1\}$ and $N \cdot T = G$. In this case, G is a *semidirect product* with normal subgroup N and complement T.

3.5.5 Proposition (a variant of the Schur–Zassenhaus theorem) *Let G be a solvable group and $N \subseteq G$ be a normal subgroup such that G/N is finite. Assume that the following two conditions hold, where p is a fixed prime number.*

(a) There exists some $m \geqslant 1$ such that $u^{p^m} = 1$ for all $u \in N$.
(b) The quotient G/N is an abelian group of order prime to p.

Then there exists a complement H to N in G. Furthermore, if $S \subseteq G$ is any subgroup such that $N \cap S = \{1\}$, then there exists some $u \in N$ such that $uSu^{-1} \subseteq H$. In particular, any two complements to N are conjugate in G.

Proof First we prove the existence and conjugacy of complements to N in G. If G itself is a finite group, then this is just a special case of the usual *Schur–Zassenhaus theorem*; see §I.18 of Huppert (1983). It is readily checked that all the arguments go through under the weaker finiteness assumptions (a) and (b). Indeed, the crucial step of the proof is the case where N is abelian. We just sketch the relevant argument, following Chapter 1, Exercise 70 of Huppert (1983). Assume that N is abelian. Denote $\bar{G} = G/N$ and let $\tau \colon \bar{G} \to G$ be a map such that $G = \coprod_{x \in \bar{G}} \tau(x) N$. For $x_1, x_2 \in \bar{G}$, we have $\tau(x_1)\tau(x_2) = \tau(x_1 x_2) c(x_1, x_2)$ for a unique $c(x_1, x_2) \in N$. The above assumptions imply that there exists some $z \in \mathbb{Z}$ such that $z\,|\bar{G}| \equiv 1 \bmod p^m$. Now set

$$c(x) := \prod_{y \in \bar{G}} c(y, x)^{-z} \in N.$$

One checks that $H := \{\tau(x)c(x) \mid x \in \bar{G}\}$ is a subgroup of G such that $G = NH$ and $N \cap H = \{1\}$. Now assume that $H' \subseteq G$ is another subgroup such that $N \cap H' = \{1\}$ and $N H' = G$. Let $\tau \colon \bar{G} \to H$ and $\tau' \colon \bar{G} \to H'$ be isomorphisms. Then we can write $\tau'(x) = \tau(x)\,a(x)$ for a unique $a(x) \in N$. Now set

$$u := \prod_{x \in \bar{G}} a(x)^{-z} \in N.$$

One checks that $\tau'(x) = u\tau(x)u^{-1}$ and so $H' = uHu^{-1}$. This completes the proof in the case where N is abelian. In order to deal with the general case, we use an induction. Contrary to the case of finite groups, we clearly cannot use induction on the order of G. Instead, we proceed as follows. Since G is solvable and G/N is abelian, we have $[G, G] \subseteq N$. So there exists a chain of subgroups

$$\{1\} = N_d \subset N_{d-1} \subset \cdots \subset N_0 = N$$

such that N_i is normal in G and N_i/N_{i+1} is abelian for all i. We now proceed by induction on d. If $d = 0$, there is nothing to prove. If $d = 1$, then N is abelian and we have already settled this case. So let us now assume that $d > 1$ and consider the group G/N_1. Then N/N_1 is an abelian normal subgroup in G/N_1, and the quotient is a finite abelian group of order prime to p. Hence we already know that there exists a complement to N/N_1 in G/N_1. Writing that complement as G'/N_1, where G' is a subgroup in G, we see that $G' \cap N = N_1$ and $G = NG'$. We have $G'/N_1 = G'/(G' \cap N) \cong N G'/N = G/N$, and this is a finite abelian group of order prime to p. Thus, all the assumptions are satisfied for G' and the normal subgroup N_1. Since the chain of normal subgroups inside N_1 is strictly shorter than that in N, we can apply induction. So there exists a subgroup $H \subseteq G'$ such that $N_1 \cap H = \{1\}$ and $G' = N_1 H$. The group H consists of elements of order prime to p, and we have $G/N \cong G'/N_1 \cong H$. So we conclude that $N \cap H = \{1\}$ and $G = NH$. Thus, we have found a complement to N in G.

Now let $H' \subseteq G$ be another complement to N in G. We write $\bar{G} = G/N_1$, $\bar{N} = N/N_1$, $\bar{H}_1 = HN_1/N_1$, and $\bar{H}'_1 = H'N_1/N_1$. We claim that $\bar{H}_1$ and $\bar{H}'_1$ are complements to $\bar{N}$ in $\bar{G}$. Indeed, $\bar{H}_1$ and $\bar{H}_1$ certainly are finite abelian groups of order prime to p, and so $\bar{H} \cap \bar{N} = \bar{H}_1 \cap \bar{N} = \{1\}$. Furthermore, the fact that $G = NH = NH'$ implies that we also have $\bar{G} = \bar{N}\bar{H} = \bar{N}\bar{H}'$, as required. Thus, we already know that $\bar{H}$ and $\bar{H}'$ are conjugate in $\bar{G}$ (since $\bar{N}$ is abelian). So there exists some $g \in G$ such that $N_1 H' = g(N_1 H)g^{-1} = N_1 gHg^{-1}$. Now note that H' and gHg^{-1} are complements to N_1 in $N_1 H'$. Since the chain of normal subgroups inside N_1 is shorter than that in N, we can apply induction. This yields that there exists some $x \in N_1 H'$ such that $xH'x^{-1} = gHg^{-1}$. Thus, H and H' are conjugate in G. Since $G = NH$, it is clear that we may choose the conjugating element inside N. This completes the proof of the existence and the conjugacy of complements in G.

Finally, let $S \subseteq G$ be any subgroup such that $N \cap S = \{1\}$. Then consider the subgroup $G' = N S$. Since G/N is abelian, G' is a normal subgroup in G. Let $H \subseteq G$ be a complement to N in G and set $H' := G' \cap H$. We have $G = G'H$ and so $G/G' = G'H/G' \cong H/H'$. Since $G/N \cong H$ is finite, this implies that $[G' : N] = [NH' : N]$ and so $G' = NH'$. Thus, S and H' are two complements of N in G'. We have already shown that there exists some $u \in N'$ such that $uSu^{-1} = H' \subseteq H$. $\qquad\square$

3.5.6 Theorem *Let B be a connected solvable affine algebraic group over $\bar{\mathbb{F}}_p$. Let $U = R_u(B)$; see Proposition 3.5.4. Then the following hold.*

(a) *There exists a closed subgroup $H \subseteq B$ which is torus such that $U \cap H = \{1\}$ and $B = UH$. (For the definition of tori see Proposition 3.1.9.)*

(b) *Let $S \subseteq B$ be any set of pairwise commuting semisimple elements in B. Then there exists some $u \in U$ such that $uSu^{-1} \subseteq H$.*

In particular, every semisimple element in B is conjugate to an element in H and every complement of U in B is conjugate to H.

Proof First note that B is a countable set. This follows by embedding B into some $\mathrm{GL}_n(\overline{\mathbb{F}}_p)$ and noting that the latter group is countable. So let us write $B = \{b_i \mid i \in \mathbb{N}\}$, where $b_1 = 1$. Then we define subgroups

$$B_d := \langle U, b_1, \dots, b_d \rangle \subseteq B \qquad \text{for } d = 1, 2, \dots$$

We have $U = B_1 \subseteq B_2 \subset B_3 \subset \cdots$ and $B = \bigcup_{d \geqslant 1} B_d$. All the assumptions of Theorem 3.5.5 are satisfied for each B_d. Indeed, by Lemma 3.5.1, there exists some $m \geqslant 1$ such that $u^{p^m} = 1$ for all $u \in U$. Furthermore, since every element of B has finite order and B/U is abelian, it is easily seen that B_d/U is a finite group of order prime to p. Hence, there exist complements to U in B_d, and all these complements are conjugate in B_d.

(a) We construct a chain of subgroups $H_1 \subseteq H_2 \subseteq \cdots$ such that H_d is a complement of U in B_d. This is done as follows. Let H_1 be a complement of U in B_1. Then $H_1 \subseteq B_2$ and $U \cap H_1 = \{1\}$. Now let H_2 be a complement to U in B_2. Then, by Theorem 3.5.5, we have $xH_1x^{-1} \subseteq H_2$ for some $x \in B_2$. Replacing H_2 by $x^{-1}H_2x$, we can assume that $H_1 \subseteq H_2$. Now we have $H_2 \subseteq B_3$ and $U \cap H_2 = \{1\}$. Repeating the above argument, we obtain a complement H_3 to U in B_3 which contains H_2, and so on. Now we set

$$H := \bigcup_{d \geqslant 1} H_d \subseteq B.$$

Then H is an abelian subgroup all of whose elements are semisimple. This shows that $U \cap H = \{1\}$. Since $B = \bigcup_{d \geqslant 1} B_d$, we also have that $B = UH$.

To see that H is closed, we argue as follows. By Exercise 3.6.2, the Zariski closure $\bar{H} \subseteq B$ is also abelian. So, by Lemma 3.5.3, we can write $\bar{H} = \bar{H}_s \times \bar{H}_u$, where $\bar{H}_s$ and $\bar{H}_u$ are closed subgroups. Since H consists of semisimple elements, we have $H \subseteq \bar{H}_s$ and so

$\bar{H} = \bar{H}_s$. Let $t_1 \in \bar{H}$. Since $B = UH$, there exists some $u \in U$ and $t \in H$ such that $t_1 = ut$. This implies that $u = t_1^{-1} t \in \bar{H}_s$ is unipotent and semisimple. Thus, we must have $t_1 = t \in H$ and so $\bar{H} = H$, as required. Furthermore, H° is a closed subgroup of finite index in H. Hence, by Theorem 2.4.6, UH° is a closed connected subgroup in $UH = B$, and we have $[B : U H^\circ] = [H : H^\circ]$. Since B is connected, we must have $UH^\circ = B$ and so $H = H^\circ$. Thus, H is connected. Finally, let $\varphi \colon H \to \mathrm{GL}_n(\overline{\mathbb{F}}_p)$ be a closed embedding. Since H consists of semisimple elements and H is abelian, we can simultaneously diagonalize the matrices $\varphi(t)$ $(t \in H)$. Thus, H is isomorphic to a closed subgroup of the group of invertible diagonal matrices in $\mathrm{GL}_n(\overline{\mathbb{F}}_p)$. By definition, this means that H is diagonalizable.

(b) Since S is countable, we can find a chain of *finite* sets $S_1 \subseteq S_2 \subseteq \cdots$ such that $S = \bigcup_{d \geqslant 1} S_d$. We set

$$U_d := \{u \in U \mid uS_d u^{-1} \subseteq H\} = \mathrm{Tran}_B(S_d, H) \cap U \quad \text{for } d = 1, 2, \ldots$$

These sets are closed by Lemma 2.5.1(a). Furthermore, we have $U_1 \supseteq U_2 \supseteq \cdots$. Hence, since U is noetherian, there exists some $d_0 \geqslant 1$ such that $U_{d_0} \subseteq U_d$ for all $d \geqslant 1$. We claim that U_{d_0} is non-empty. This is seen as follows. Replacing S_{d_0} by the subgroup generated by S_{d_0}, we can assume without loss of generality that S_{d_0} is a finite abelian subgroup all of whose elements are semisimple. In particular, this means that $U \cap S_{d_0} = \{1\}$. Since S_{d_0} is finite, there exists some $d \geqslant 1$ such that $S_{d_0} \subseteq B_d$. Hence Theorem 3.5.5, applied to $B_d = UH_d$ as in (a), shows that there exists some $u \in U$ such that $uS_{d_0}u^{-1} \subseteq H_d \subseteq H$, as desired. Thus, we have $u \in U_{d_0}$. But this implies that $uSu^{-1} \subseteq H$. Indeed, let $s \in S$. Then there exists some $d \geqslant 1$ such that $s \in S_d$. We have $U_{d_0} \subseteq U_d$ and so $usu^{-1} \in H$, as desired. $\qquad\square$

The above result gives a good picture of the structure of a connected solvable group. We now discuss two applications.

Let G be any connected affine algebraic group over $\overline{\mathbb{F}}_p$. We define a *maximal torus* in G to be a maximal closed connected subgroup which is a torus (see Proposition 3.1.9). By Theorem 3.5.6, the maximal tori in a connected solvable affine algebraic group B are precisely the complements of U in B.

3.5.7 Corollary *All the maximal tori in a connected affine algebraic group G over $\overline{\mathbb{F}}_p$ are conjugate.*

Proof Let $T, T' \subseteq G$ be maximal tori. Since T, T' are connected and abelian, there exist Borel subgroups B, B' containing T, T', respectively. Now, by Theorem 3.4.3, B and B' are conjugate in G. Hence, replacing T' by a suitable conjugate, we may assume that $T, T' \subseteq B$. Now write $B = UH$ as in Theorem 3.5.6. Then T is conjugate to a subgroup of H and, hence, must be conjugate to H. The same argument also applies to T'. $\qquad\square$

We now adress another point which is the key to many connectedness statements. This will again be based on a variant of a result on finite groups.

3.5.8 Lemma (a variant of 'coprime action') *Keep the assumptions and the notation of Proposition 3.5.5. Let $S \subseteq G$ be an abelian finite subgroup consisting of elements of order prime to p. Assume that $N_1 \subseteq N$ is a subgroup normalized by S and that $u \in N$ is such that $s(uN_1)s^{-1} \subseteq uN_1$ for all $s \in S$. Then there exists some $u_1 \in uN_1$ such that $su_1s^{-1} = u_1$ for all $s \in S$.*

Proof If G itself is finite, this is a special case of a result due to Glauberman; see I.18.6 of Huppert (1983). We sketch the argument to see that it also works in our slightly different setting. We define a map

$$\varphi \colon S \to N_1, \qquad s \mapsto u^{-1}sus^{-1}.$$

(This is well-defined since N_1 is normalized by s.) We have

$$\varphi(s_1s_2)s_1s_2 = u^{-1}s_1s_2u = \varphi(s_1)s_1\varphi(s_2)s_2 \quad \text{for all } s_1, s_2 \in S.$$

Thus, $H := \{\varphi(s)s \mid s \in S\}$ is a subgroup of G. In fact, H is a complement to N_1 in the semidirect product N_1S. Since all the assumptions of Theorem 3.5.5 are satisfied for N_1S, there exists some $v \in N_1$ such that $vSv^{-1} = H$. Now consider the intersection $N_1s \cap H$ for a fixed $s \in S$. Since N_1H is a semidirect product, that intersection is a singleton set. Now $\varphi(s)s \in N_1s \cap H$ and so $N_1s \cap H = \{\varphi(s)s\}$. On the other hand, we also have $vsv^{-1} = vsv^{-1}s^{-1}s \in N_1s \cap H$ and so $vsv^{-1} = \varphi(s)s = u^{-1}su$. Consequently, $uvs = suv$ for all $s \in S$, and so $u_1 = uv \in uN_1$ is the required element. $\qquad\square$

3.5.9 Theorem *Let B be a connected solvable affine algebraic group over $\overline{\mathbb{F}}_p$. Write $B = UH$ as in Theorem 3.5.6. Let $S \subseteq H$ be any subset. Then $\mathrm{C}_B(S)$ is a connected subgroup of B. In particular, $\mathrm{C}_B(s)$ is connected for any $s \in H$.*

Proof First we reduce the proof to the special case where S is a singleton set. This can be done by induction on $\dim B$, as follows. Let $S \subseteq H$. If $\dim B = 0$ or if every element of S lies in the centre of B, there is nothing to prove. Now assume that $\dim B > 0$ and there exists some $s \in S$ which is not in the centre of B. Then we have $S \subseteq \mathrm{C}_B(S) \subseteq \mathrm{C}_B(s)$ and so $\mathrm{C}_B(S) = \mathrm{C}_{B'}(S)$, where $B' := \mathrm{C}_B(s) \subsetneq B$. If B' is known to be connected, we can apply induction and conclude that $\mathrm{C}_B(S)$ is connected, as required. Thus, it is enough to consider the case where $S = \{s\}$.

So let us fix some $s \in H$. We will prove that $\mathrm{C}_B(s)$ is connected by induction on $\dim B$. If $\dim B = 0$ or $U = \{1\}$, there is nothing to prove. Now assume that $\dim B > 0$ and $U \neq \{1\}$. Since B is solvable and $[B, B] \subseteq U$, there exists a non-trivial closed connected abelian subgroup $U_1 \subseteq U$ such that U_1 is normal in B. (Just take the last non-trivial term in the derived series of B; this is closed and connected by the discussion in Example 2.4.7.) By Corollary 3.3.10, there exists an algebraic representation $\gamma \colon B \to \mathrm{GL}(W)$ such that $U_1 = \ker(\gamma)$. Then $B' := \gamma(B) \subseteq \mathrm{GL}(W)$ is a closed connected subgroup and γ restricts to a surjective homomorphism of algebraic groups $\gamma_1 \colon B \to B'$. Being a homomorphic image of B, the group B' is also solvable. Let U' be the closed connected subgroup consisting of unipotent elements in B'. Then we have $B' = U'H'$ where $H' := \gamma(H)$ and $U' = \gamma(U)$. Let $s' := \gamma(s) \in H'$. Since $\dim B' < \dim B$, we can apply induction. This yields that $\mathrm{C}_{B'}(s')$ is connected. We certainly have $\gamma(\mathrm{C}_B(s)) \subseteq \mathrm{C}_{B'}(s')$ and so $\mathrm{C}_B(s)$ is contained in $G := \gamma^{-1}(\mathrm{C}_{B'}(s')) \subseteq B$. Since $\ker(\gamma) = U_1$ is connected, we have $\ker(\gamma) \subseteq G^\circ$. Hence we must have $G = G^\circ$, since $\mathrm{C}_{B'}(s')$ is connected (see Proposition 2.2.14(b)). Now, if $G < B$, then $\mathrm{C}_B(s) = \mathrm{C}_G(s)$ is connected by induction. Hence, we can assume that $G = B$. This means that $\mathrm{C}_{B'}(s') = B'$. We shall now prove that $B = \mathrm{C}_B(s)\,U_1$. Since $H \subseteq \mathrm{C}_B(s)$, it is enough to show that $U = \mathrm{C}_U(s)\,U_1$. So let $u \in U$. Since $\gamma(u) \in B'$ centralizes s', we have $\gamma(u^{-1}sus^{-1}) = 1$ and so $s(uU_1)s^{-1} \subseteq uU_1$. Applying Lemma 3.5.8 (with $N = U$, $N_1 = U_1$, and $S = \langle s \rangle$) we deduce that there exists some $u_1 \in uU_1$ such that $su_1s^{-1} = u_1$. Thus, we have $u \in \mathrm{C}_U(s)\,U_1$, as desired. We have now reached the conclusion that

$$U = \mathrm{C}_U(s)\,U_1 \quad \text{and} \quad B = \mathrm{C}_B(s)\,U_1, \tag{1}$$

as claimed. The next step is to refine this factorization. For this purpose, we consider the map $\varphi_s \colon U_1 \to U_1$, $u_1 \mapsto [u_1, s] = s^{-1}u_1^{-1}su_1$. Since U_1 is abelian, it is easily checked that φ_s is a group homomorphism; denote its image by M_s. Being the image of a homomorphism of algebraic groups, M_s is a closed subgroup of U_1; see Proposition 2.2.14. Furthermore, since U_1 is connected, M_s is also connected. We have $\ker(\varphi_s) = \mathrm{C}_{U_1}(s)$, and so Proposition 2.2.14 yields

$$\dim M_s = \dim U_1 - \dim \mathrm{C}_{U_1}(s). \tag{2}$$

Now we consider the map

$$\mu\colon \mathrm{C}_U(s) \times M_s \to U, \qquad (c, u) \mapsto cu.$$

We claim that this map is injective. To see this, let $c_1, c_2 \in \mathrm{C}_U(s)$ and $u_1, u_2 \in M_s$ be such that $c_1 u_1 = c_2 u_2$. Then we have $c_2^{-1} c_1 = u_2 u_1^{-1} \in \mathrm{C}_U(s) \cap M_s$. Hence it is enough to show that $\mathrm{C}_U(s) \cap M_s = \{1\}$. So let $u \in \mathrm{C}_U(s) \cap M_s$ and write $u = \varphi_s(u_1)$, where $u_1 \in U_1$. This yields $u = s^{-1} u_1^{-1} s u_1$ and so $su = u_1^{-1} s u_1$. Since u commutes with s, the left-hand side is the Jordan decomposition of an element in B. On the other hand, the right-hand side is semsimple. Hence we obtain $u = 1$, as required. So the above map is injective.

Consequently, by restriction, we also have an injective map $\mathrm{C}_{U_1}(s) \times M_s \to U_1$ and so $\dim \mathrm{C}_{U_1}(s) + \dim M_s \leqslant \dim U_1$; see Corollary 2.2.9. But now (2) implies that we must have equality. On the other hand, the restricted map is a group homomorphism, since U_1 is abelian. So the image $\mathrm{C}_{U_1}(s)\, M_s \subseteq U_1$ is closed in U_1 and has the same dimension as U_1. Since U_1 is connected, this implies $U_1 = \mathrm{C}_{U_1}(s)\, M_s$. Inserting this into (1), we obtain $U = \mathrm{C}_U(s)\, M_s$. Hence we see that μ also is surjective.

Thus, we have shown that μ is bijective. We claim that this implies that $\mathrm{C}_U(s)$ must be connected. To see this, recall that M_s is a closed subgroup in U. We now show that, in fact, M_s is normal in U. Indeed, since $U = \mathrm{C}_U(s)\, M_s$, it is enough to show that each element in $\mathrm{C}_U(s)$ normalizes M_s. So let $c \in \mathrm{C}_U(s)$ and $[u_1, s] \in M_s$. Since c commutes with s, we obtain $c[u_1, s]c^{-1} = [cu_1 c^{-1}, s] \in M_s$ since $cu_1 c^{-1} \in U_1$ (recall that U_1 is normal in B). Now we can argue as follows. Since M_s is normal, the product $\mathrm{C}_U(s)^\circ\, M_s$ is a subgroup of U. That product is closed by Theorem 2.4.6. Now, since μ is bijective, $\mathrm{C}_U(s)^\circ\, M_s$ has finite index in U, equal to $[\mathrm{C}_U(s) : \mathrm{C}_U(s)^\circ]$. But U is connected and so $\mathrm{C}_U(s) = \mathrm{C}_U(s)^\circ$.

Finally, since $H \subseteq \mathrm{C}_B(s)$, we have $\mathrm{C}_B(s) = \mathrm{C}_U(s)\, H$. Using once more Theorem 2.4.6, we see that $\mathrm{C}_B(s)$ is connected, as desired. $\qquad\square$

The complexity of the above proof shows that connectedness questions are, in general, quite subtle. See Exercise 3.6.11 for some applications. Note that, in general, centralizers of elements in G are not connected; see Exercise 2.7.12.

Here is another application of Lemma 3.5.8.

3.5.10 Proposition *Let G be an affine algebraic group over $\mathbb{F}_p$ and assume that G has a reductive BN-pair; see Definition 3.4.5. Write $B = UH$, where $U = R_{\mathrm{u}}(B)$ and H is a torus such that $C_G(H) = H$. Now let $V \subseteq U$ be a closed subgroup such that $tVt^{-1} \subseteq V$ for all $t \in H$. Then V is connected.*

Proof By Exercise 3.6.1, there exists some $s \in H$ such that $C_G(s) = C_G(H) = H$. Let $S := \langle s \rangle$, $N := V$ and $N_1 := V^\circ$. Then we are in a situation where Lemma 3.5.8 can be applied. (The group S is finite of order prime to p, and N, N_1 are normalized by s.) Now let $u \in N$. We claim that $s(uN_1)s^{-1} \subseteq uN_1$.

This is seen as follows. The group H acts on N by conjugation, and we have $tN_1t^{-1} \subseteq N_1$ for all $t \in H$. Thus, we have an induced (abstract) action of H on N/N_1. Since N/N_1 is a finite set, the orbit of the coset uN_1 is finite, and so $\mathrm{Stab}_H(uN_1)$ is a subgroup of finite index in N. On the other hand, uN_1 is a closed subset of N and so $\mathrm{Stab}_H(uN_1) = \mathrm{Tran}_H(uN_1)$ is a closed subgroup of H; see Lemma 2.5.1. Since H is connected, we conclude that $H = \mathrm{Stab}_H(uN_1)$ and so $t(uN_1)t^{-1} \subseteq uN_1$ for all $t \in H$. Taking $t = s$ yields the required assertion.

Now, by Lemma 3.5.8, there exists some $u_1 \in uN_1$ such that $su_1s^{-1} = u_1$ and so $u_1 \in C_G(s) = H$. Thus, we must have $u_1 = 1$ and so $u \in N_1$, as required. $\square$

3.6 Bibliographic remarks and exercises

The results in Section 3.1 are standard and can be found in any of the textbooks on algebraic groups. However, note that we have avoided the general definition of unipotent and semisimple elements.

The discussion of grassmannian and flag varieties in Section 3.3 follows §11.2.2 of Goodman and Wallach (1998). For a different treatment, see §10.3 of Borel (1991). Our proof of Theorem 3.2.7 is taken from §I.5.2 of Shafarevich (1994). The advantage of this proof is that it is completely elementary. The argument can in fact be expressed in purely algebraic terms, without any reference to closed sets in some projective space; see §I.9 of Mumford (1988). Example 3.2.8 and Exercise 3.6.5 are taken from §8.5 of Cox *et al.* (1992).

The homomorphism θ of $\mathrm{Sp}_4(k)$ into itself was first constructed in Exposé 22 of Chevalley (1956–8). The idea of using the second exterior power of the natural representation of $\mathrm{Sp}_4(k)$ is taken from §16.2 of Fulton and Harris (1991).

The results on Borel subgroups and parabolic subgroups are all proved by A. Borel in the 1950s; see §IV.11 of Borel (1991) for a modern exposition of this work. Here, we followed a slightly different approach, based on Steinberg (1977). First of all, our definition of parabolic subgroups in §3.2.10 is not the usual one! The usual definition involves the general notion of quotients and projective (or complete) varieties; see §11.2 of Borel (1991). However, our definition is seen to be equivalent to the usual one via the characterization in Theorem 3.4.3. Our approach has the advantage that it does not require the general construction of quotients. For many applications in the theory of algebraic groups, it is sufficient to use Theorem 3.2.11 (whose deduction from Theorem 2.5.5 seems to be new).

The argument in the proof of Lemma 3.4.7 is taken from p. 72 of Steinberg (1974). A different proof of Lemma 3.4.7(a), using the sharp form of the Bruhat decomposition, can be found in §3.7.1 of Carter (1985). The proof of Proposition 3.4.6 is a variant of that in §2.5.1 of Carter (1985). For the characterization of the Bruhat–Chevalley order in Theorem 3.4.8, we followed §8.5 of Springer (1998). The proof of Proposition 3.4.9 is an adaptation of the argument in Theorem 22.2 of Humphreys (1991). (The statement can actually be shown to hold in any connected algebraic group where B is a Borel subgroup.)

By Proposition 3.4.6, every algebraic group with a reductive BN-pair is reductive in the sense of §3.1.13; furthermore, $N = \mathrm{N}_G(H)$ by Lemma 3.4.7(b). Conversely, given a connected reductive algebraic group G, let $T \subseteq G$ be a maximal torus and $B \subseteq G$ be a Borel subgroup containing T. Then B and $\mathrm{N}_G(T)$ form a reductive BN-pair, but this is much harder to prove; see §14.15 of Borel (1991), §29.1 of Humphreys (1991), or Chapter 8 of Springer (1998).

The structure theorems on connected solvable groups in Section 3.5 also hold over any algebraically closed ground field k, but the proofs are more complicated. See, for example, Chapter 7 of Humphreys (1991). Note that already the definition of semisimple and unipotent elements is more complicated in general; see §15 of Humphreys (1991). The idea of using techniques from the theory of finite groups in our setting is taken from Exposé 6 of Chevalley (1956–8).

3.6.1 Exercise Let G be an affine algebraic group and $H \subseteq G$ be a closed connected subgroup which is a torus. The aim of this exercise is to show that

$$C_G(H) = C_G(h) \qquad \text{for some } h \in H.$$

For this purpose, choose a closed embedding $G \subseteq \mathrm{GL}_n(k)$ such that $H \subseteq T_n(k)$ (where $T_n(k)$ is the group of all diagonal invertible matrices). This is possible by Proposition 3.1.9. Thus, $V = k^n$ is an algebraic G-module, and we have $V = \bigoplus_{i=1}^m V_{\chi_i}$, where $\chi_1, \dots, \chi_m \in X(H)$.

 (a) Show that there exists some $h \in H$ such that $\chi_i(h) \neq \chi_j(h)$ for $i \neq j$.

 (b) Let $h \in H$ be as in (b). Show that $C_G(H) = C_G(h)$.

 (c) Show that $\{h \in H \mid C_G(H) = C_G(h)\}$ is dense in H.

[*Hint.* (a) For $i \neq j$, let $H_{ij} := \ker(\chi_i \chi_j^{-1})$. This is a proper closed subgroup of H. Since H is connected, there is some $h \in H$ which does not lie in the union of all these subgroups. (b) Check that $C_G(H) = \{g \in G \mid g(V_{\chi_i}) = V_{\chi_i} \text{ for all } i\}$ and then use (a) to show that $C_G(h) \subseteq C_G(H)$. (c) By (a) and (b), the set of all $h \in H$ such that $C_G(H) \subsetneq C_G(h)$ is a proper closed subset of H.]

3.6.2 Exercise Let G be an affine algebraic group. Assume that A is an abstract subgroup of G and let $\bar{A}$ be its closure in G. By Lemma 2.2.13, $\bar{A}$ is a subgroup of G. Show that, if A is abelian, then $\bar{A}$ is also abelian.

[*Hint.* For $s \in A$, consider the map $\varphi_s \colon \bar{A} \to \bar{A}$, $a \mapsto [a, s]$. Show that $\varphi_s(\bar{A}) = \{1\}$. Then take $t \in \bar{A}$ and consider the morphism $\varphi_t \colon \bar{A} \to \bar{A}$, $a \mapsto [a, t]$.]

3.6.3 Exercise Show that $\mathbb{P}(V)$ is a noetherian topological space and that any closed set in $\mathbb{P}(V)$ can be defined by finitely many homogeneous polynomials.

[*Hint.* Choose a basis $\{e_1, \dots, e_n\}$ of V and let $Z = \mathbf{V}^{\mathrm{P}}(H)$ be a closed subset, where $H \subseteq k[T_1, \dots, T_n]$ consists of homogeneous polynomials. We can find $f_1, \dots, f_r \in H$ such that $(H) = (f_1, \dots, f_r)$. Then check that $\mathbf{V}^{\mathrm{P}}(\{f_1, \dots, f_r\}) = \mathbf{V}^{\mathrm{P}}(H)$. To show that $\mathbb{P}(V)$ is a noetherian topological space, consider a chain of closed sets $Z_1 \supseteq Z_2 \supseteq \cdots$ in $\mathbb{P}(V)$. Write $Z_i = \mathbf{V}^\mathrm{p}(H_i)$, where $H_i \subseteq k[T_1, \ldots, T_n]$ consists of homogeneous polynomials. Since $Z_i \supseteq Z_{i+1}$, we can assume that $H_i \subseteq H_{i+1}$ for all i. Then $\bigcap_{i \geqslant 1} \mathbf{V}^\mathrm{p}(H_i) = \mathbf{V}^\mathrm{p}(\bigcup_{i \geqslant i} H_i)$ and the latter set can be defined by using only finitely many elements from $\bigcup_{i \geqslant 1} H_i$.]

3.6.4 Exercise Let $C \subseteq k^n$ be an affine algebraic set and assume that C is a cone, that is, a union of 1-dimensional subspaces. Show that there exists a set of homogeneous polynomials $H \subseteq k[T_1, \ldots, T_n]$ such that $C = \mathbf{V}(H)$.

More generally, let (X, A) be an affine variety and consider the direct product $k^n \times X$. Then we may identity $A[k^n \times X]$ with the polynomial ring $A[T_1, \ldots, T_n]$. Now let $C \subseteq k^n \times X$ be any subset such that

$$(v, x) \in C \quad \Rightarrow \quad (tv, x) \in C \quad \text{for all non-zero } t \in k.$$

Show that $\mathbf{I}(C)$ is generated by polynomials which are homogeneous in the T_i.

[*Hint.* Let the non-zero polynomial $F \in A[T_1, \ldots, T_n]$ be such that $F(v, x) = 0$ for all $(v, x) \in C$. Write $F = F_0 + \cdots + F_d$, where $F_j \in A[T_1, \ldots, T_n]$ is homogeneous of degree j. Now choose $d + 1$ different non-zero elements $t_0, \ldots, t_d \in k$, and let $(v, x) \in C$. Then $(t_i v, x) \in C$ for all i, and so $F(t_i v, x) = 0$. This yields equations $0 = F(t_i v, x) = \sum_{j=0}^d t_i^j F_j(v, x)$ for $0 \leqslant i \leqslant d$. The matrix (t_i^j) is a Vandermonde matrix and, hence, invertible. So we must have $F_j(v, x) = 0$ for all j.]

3.6.5 Exercise The purpose of this exercise is to provide a 'concrete' version of Theorem 3.2.7. Let $H \subseteq k[T_1, \ldots, T_n, Y_1, \ldots, Y_m]$ be a set of polynomials which are homogeneous in $T_1, \ldots, T_n$. Then the subset

$$\tilde{\mathbf{V}}(H) := \{(\langle t_1, \ldots, t_n \rangle, (y_1, \ldots, y_m)) \in \mathbb{P}(k^n) \times k^m \mid$$
$$h(t_1, \ldots, t_n, y_1, \ldots, y_m) = 0 \text{ for all } h \in H\}$$

is well-defined. Thus, by Theorem 3.2.7, the set $Y := \mathrm{pr}_2(\tilde{\mathbf{V}}(H)) \subseteq k^m$ is closed. Show that $Y = \mathbf{V}(\hat{H})$, where

$$\hat{H} := \left\{ f \in k[Y_1, \ldots, Y_m] \;\middle|\; \begin{array}{l} \text{for each } i \in \{1, \ldots, n\}, \text{ there exists} \\ \text{some } e_i \geqslant 0 \text{ such that } T_i^{e_i} f \in (H) \end{array} \right\}.$$

[*Hint.* Arguing as in Example 1.3.6, show that $\mathrm{pr}_2(\tilde{\mathbf{V}}(H)) =$

3.6.6 Exercise Consider the polynomial ring $R = k[T_1, \ldots, T_n]$, where k is any field. Assume that $I \subseteq R$ is a *homogeneous ideal*, that is, I is generated by homogeneous polynomials. For any $s \geqslant 0$, denote by R_s the subspace of R generated by all non-zero $f \in R$ which are homogeneous of degree s. Similarly, define I_s. We set

$$^{\mathrm{P}}\mathrm{HF}_I(s) := \dim_k(k[T_1, \ldots, T_n]_s / I_s).$$

This is called the projective *Hilbert function* of I.

(a) For $s \geqslant 1$, show that $I_{\leqslant s} = I_s \oplus I_{\leqslant (s-1)}$, where $I_{\leqslant s}$ is defined as in Definition 1.2.11. Deduce from this that $^{\mathrm{P}}\mathrm{HF}_I(s) = {}^{\mathrm{a}}\mathrm{HF}_I(s) - {}^{\mathrm{a}}\mathrm{HF}_I(s-1)$, where $^{\mathrm{a}}\mathrm{HF}_I(s)$ is the affine Hilbert function of I.

(b) Show that there exists a polynomial $^{\mathrm{P}}\mathrm{HP}_I(t) \in \mathbb{Q}[t]$ and some $s_0 \geqslant 0$ such that $^{\mathrm{P}}\mathrm{HF}_I(s) = {}^{\mathrm{P}}\mathrm{HP}_I(s)$ for all $s \geqslant s_0$. Show that $^{\mathrm{P}}\mathrm{HP}_I(t) = {}^{\mathrm{a}}\mathrm{HP}_I(t) - {}^{\mathrm{a}}\mathrm{HP}_I(t-1)$ where $^{\mathrm{a}}\mathrm{HP}_I(t)$ is the affine Hilbert polynomial.

(c) Determine the degree and the leading coefficient of $^{\mathrm{P}}\mathrm{HP}_I(t)$ in terms of the degree and the leading coefficient of $^{\mathrm{a}}\mathrm{HP}_I(t)$.

For more on $^{\mathrm{P}}\mathrm{HP}_I(t)$, see Chapter 9 of Cox *et al.* (1992).

[*Hint.* You will need the following characterization of homogeneous ideals. First note that any $f \in R$ can be written uniquely in the form $f = f_0 + \cdots + f_d$, where $f_i \in R$ is either zero or homogeneous of degree i. The non-zero terms f_i are called the homogeneous components of f. Show that an ideal $I \subseteq R$ is homogeneous if and only if I contains all homogeneous components of any polynomial in I.]

3.6.7 Exercise Let $m_1, \ldots, m_d$ be a sequence of positive integers, and consider the polynomial ring $R = k[X_i^{(r)} \mid 1 \leqslant r \leqslant d, 1 \leqslant i \leqslant m_r]$. We say that a non-zero $f \in R$ is *multi-homogeneous* if there exist $n_1, \ldots, n_d \geqslant 0$ such that f is a linear combination of monomials of the form

$$\prod_{r=1}^{d} \prod_{i=1}^{m_r} (X_i^{(r)})^{\alpha_{ri}} \quad \text{with} \quad \sum_{i=1}^{m_r} \alpha_{ri} = n_r \quad \text{for } r = 1, \ldots, d.$$

Thus, f is homogeneous of degree $n_1 + \cdots + n_r$ in the variables $X_i^{(r)}$ with the additional condition that every monomial occurring in f is homogeneous of degree n_r (for each fixed r) when regarded as a polynomial in the variables $X_1^{(r)}, \ldots, X_{m_r}^{(r)}$. We also say that f is $(n_1, \ldots, n_r)$-homogeneous.

Now let $V_1, \ldots, V_d$ be k-vector spaces with $\dim V_r = m_r$. Let $\{e_i^{(r)} \mid 1 \leqslant i \leqslant m_r\}$ be a basis of V_r. Let $v_r \in V_r$ and write $v_r = \sum_{i=1}^{m_r} v_i^{(r)} e_i^{(r)}$. Then, if $v_r \neq 0$ for all r, the condition $f(v_1^{(1)}, v_2^{(1)}, \ldots, v_{m_d}^{(d)}) = 0$ is well-defined for any multi-homogeneous polynomial $f \in R$. With this notation show that the image of a subset $C \subseteq \mathbb{P}(V_1) \times \cdots \times \mathbb{P}(V_d)$ under the Segre embedding in Lemma 3.3.3 is closed if and only if there exists a finite set $H \subseteq R$ of multi-homogeneous polynomials such that

$$C = \big\{ (\langle v_1 \rangle, \ldots, \langle v_d \rangle) \in \mathbb{P}(V_1) \times \cdots \times \mathbb{P}(V_d) \mid$$
$$f(v_1^{(1)}, v_2^{(1)}, \ldots, v_{m_d}^{(d)}) = 0 \quad \text{for all } f \in H \big\},$$

where we write $v_r = \sum_i v_i^{(r)} e_i^{(r)}$ as above.

[*Hint.* Consider the polynomial ring $S = k[Y_{i_1 i_2 \cdots i_d} \mid 1 \leqslant i_r \leqslant m_r$ for all $r]$. Then, with respect to the basis $\{e_{i_1} \otimes \cdots \otimes e_{i_d} \mid 1 \leqslant i_r \leqslant m_r$ for all $r\}$ of $V_1 \otimes \cdots \otimes V_d$, the closed subsets of $\mathbb{P}(V_1 \otimes \cdots \otimes V_d)$ are the sets $\mathbf{V}^{\mathrm{p}}(H)$, where $H \subseteq S$ consists of homogeneous polynomials. The Segre embedding is given by

$$\sigma(\langle v_1 \rangle, \ldots, \langle v_d \rangle) = \left\langle \sum_{r=1}^{d} \sum_{i_r=1}^{m_r} v_{i_1}^{(1)} \cdots v_{i_d}^{(d)} \; e_{i_1}^{(1)} \otimes \cdots \otimes e_{i_d}^{(d)} \right\rangle,$$

where we write $v_r = \sum_i v_i^{(r)} e_i^{(r)}$ for all r.

Now let $C \subseteq \mathbb{P}(V_1) \times \cdots \times \mathbb{P}(V_d)$ be a subset. First assume that $\sigma(C) = \mathbf{V}^{\mathrm{p}}(H)$ for a finite set of homogeneous polynomials $H \subseteq S$. Using the above formula for the coordinates of points in the image of σ, we see that C is the zero set of the multi-homogeneous polynomials obtained by setting $Y_{i_1 i_2 \cdots i_d} = X_{i_1}^{(1)} X_{i_2}^{(2)} \cdots X_{i_d}^{(d)}$ in the polynomials in H. Conversely, assume that C is defined by a finite set of multi-homogeneous polynomials $M \subseteq R$. For $f \in M$, define the numbers $n_r(f)$ by the condition that f is $(n_1(f), \ldots, n_r(f))$-homogeneous. We set $N := \max\{n_r(f) \mid f \in M, 1 \leqslant r \leqslant d\}$ and define

$$\tilde{H} := \{(X_i^{(r)})^{N - n_r(f)} f \mid f \in M, 1 \leqslant r \leqslant d, 1 \leqslant i \leqslant m_r.\}$$

First note that C is also defined as the zero set of the polynomials in $\tilde{H}$. Furthermore, each polynomial in $\tilde{H}$ is $(N, N, \ldots, N)$-homogeneous. Then show that, for each $f \in \tilde{H}$, there exists some $\hat{f} \in S$ such that f is obtained from $\hat{f}$ by setting $Y_{i_1 i_2 \cdots i_d} = X_{i_1}^{(1)} X_{i_2}^{(2)} \cdots X_{i_d}^{(d)}$. Then check that $\sigma(C) = \mathbf{V}^{\mathrm{p}}(\hat{f} \mid f \in \tilde{H}).$]

3.6.8 Exercise Let G be an affine algebraic group and $P \subseteq G$ be a parabolic subgroup. Show that $P \cap G^\circ$ is a parabolic subgroup of G°.

3.6.9 Exercise Let G_i be an affine algebraic group and $P_i \subseteq G_i$ be a parabolic subgroup ($i = 1, 2$). Show that $P_1 \times P_2 \subseteq G_1 \times G_2$ is a parabolic subgroup.

[*Hint.* There exists an algebraic G_i-module V_i and a non-zero $v_i \in V$ such that $P_i = \mathrm{Stab}_G(\langle v_i \rangle)$ and the G_i-orbit of $\langle v_i \rangle$ in $\mathbb{P}(V_i)$ is closed. Now consider $V_1 \otimes V_2$ as a $(G_1 \times G_2)$-module in the natural way and use the fact that $\mathbb{P}(V_1) \times \mathbb{P}(V_2)$ is a closed subset of $\mathbb{P}(V_1 \otimes V_2)$ via the Segre embedding; see Lemma 3.3.3.]

3.6.10 Exercise Let B be a connected solvable affine algebraic group over $\overline{\mathbb{F}}_p$ and write $B = UH$ as in Theorem 3.5.6. Let $S \subseteq H$ be a subgroup (not necessarily closed). Show that $\mathrm{C}_B(S) = \mathrm{N}_B(S)$.

[*Hint.* Each $b \in B$ is uniquely expressible as $b = ut$ with $u \in U$ and $t \in H$. Furthermore, the map $\pi \colon B \to H$, $ut \mapsto t$, is a group homomorphism whose restriction to H is the identity. (Here, we just consider B as an abstract group). Now let $b = ut \in \mathrm{N}_B(S)$. Then $bhb^{-1} \in S$ for all $h \in S$, and so $bhb^{-1} = \pi(bhb^{-1}) = tht^{-1}$. So we have $t^{-1}b \in \mathrm{C}_G(S)$ and, hence, $b \in \mathrm{C}_G(S)$.]

3.6.11 Exercise This exercise contains two applications of the connectedness statement in Theorem 3.5.9. Let G be an affine algebraic group over $\overline{\mathbb{F}}_p$ and $B \subseteq G$ be a Borel subgroup; write $B = UH$ as in Theorem 3.5.6 and assume that H is a torus such that $\mathrm{C}_G(H) = H$.

(a) Let $S \subseteq G$ be a closed connected diagonalizable subgroup. Show that $\mathrm{C}_G(S)$ is closed and connected.

(b) Let $g \in G$ and write $g = su = us$, where s is semisimple and u is unipotent. Show that $g \in \mathrm{C}_G(s)^\circ$.

[*Hint.* (a) It is clear that $\mathrm{C}_G(S)$ is closed. Let $x \in \mathrm{C}_G(S)$. By Proposition 3.4.9, there exists a Borel subgroup $B' \subseteq G$ such that $x \in B'$. Since B' is parabolic, there exist some algebraic G-module V and some non-zero $v_0 \in V$ such that $B' = \mathrm{Stab}_G(\langle v_0 \rangle)$ and the G-orbit C of $\langle v_0 \rangle$ in $\mathbb{P}(V)$ is closed. Let $X := \{\langle v \rangle \in C \mid \langle x.v \rangle = \langle v \rangle\}$. Note that $\langle v_0 \rangle \in X$ since $x \in B'$. Now use Theorem 3.4.2 to show that there exists some fixed point of S on X. Deduce from this that there exists a Borel subgroup $B_1 \subseteq G$ such that $S \subseteq B_1$ and $x \in B_1$. Thus, we have $x \in \mathrm{C}_{B_1}(S)$ and it remains to use Theorem 3.5.9.

(b) Choose a Borel subgroup $B' \subseteq G$ such that $su \in B'$. Then we also have $s \in B'$ and $u \in B'$ and so $g = us \in C_{B'}(s)$. Now apply Theorem 3.5.9.]

3.6.12 Exercise Let G be an affine algebraic group over $\mathbb{F}_p$. Assume that G has a reductive BN-pair. Write $B = UH$, where $H = B \cap N$ is a torus in B; let $W = N/H$ be the Weyl group of G. Then the conjugation action of N on H induces an action of W on H. Show that the following hold.

(a) Every semisimple element of G is conjugate to an element in H.

(b) The map $C \mapsto C \cap H$ defines a bijection between conjugacy classes of semisimple elements in G and W-orbits on H.

[*Hint.* (a) Proposition 3.4.9 and Theorem 3.5.6. (b) See Lemma 3.4.7(a).]

4
Frobenius maps and finite groups
of Lie type

In all of this chapter, k will be an algebraic closure of the finite field with p elements, where $p > 0$ is a prime number. We shall consider affine varieties and algebraic groups over k which are defined by polynomials with coefficients in the finite field $\mathbb{F}_q \subseteq k$ (where q is a power of p). For example, this is the case for all the classical groups that we considered in Chapter 1.

Section 4.1 gives a rigorous introduction to this notion, in the general framework of affine varieties and algebraic groups. If an algebraic group G is defined over $\mathbb{F}_q$, then we may consider the finite subgroup $G(\mathbb{F}_q)$ consisting of all elements with coordinates in $\mathbb{F}_q$. These groups are of central importance in finite group theory, especially in connection with the *classification of the finite simple groups*; see Gorenstein *et al.* (1994). The main tool for deriving properties of $G(\mathbb{F}_q)$ from properties of G is the Lang–Steinberg theorem, which we will prove in Section 4.1. A variety of applications will be discussed in Sections 4.2 and 4.3.

In Section 4.4, we consider the problem of studying the sequence of numbers $|X(\mathbb{F}_{q^n})|$ (for $n = 1, 2, 3, \ldots$), where X is an affine variety defined over $\mathbb{F}_q$. The most important result here is Grothendieck's trace formula, which gives a cohomological interpretation of these numbers. (This will be stated without proof.) In Section 4.5, we apply this to construct representations of finite groups of Lie type, following the fundamental work of Deligne and Lusztig (1976).

In the final Section 4.6, we illustrate many of these ideas by a detailed study of the finite *Suzuki groups*, which are defined as the fixed-point set of the symplectic group $\mathrm{Sp}_4(\overline{\mathbb{F}}_2)$ under a certain endomorphism with a finite fixed-point set.

4.1 Frobenius maps and rational structures

Throughout this section, we assume that k is an algebraic closure of the finite field with p elements, where $p > 0$ is a prime. Then, for any $a \geqslant 1$, there is a unique subfield of k with exactly $q = p^a$ elements; we denote that subfield by $\mathbb{F}_q \subseteq k$. In this situation, we have the corresponding *standard Frobenius map*

$$F_q \colon k^n \to k^n, \qquad (x_1, \ldots, x_n) \mapsto (x_1^q, \ldots, x_n^q).$$

Note that F_q is at the same time a dominant bijective regular map and an $\mathbb{F}_q$-linear map. (The latter follows from the fact that the map $x \mapsto x^q$ is a field automorphism of k with fixed-point set $\mathbb{F}_q$.) Now let $V \subseteq k^n$ be a closed subset, and assume that $V = \mathbf{V}(S)$, where $S \subseteq \mathbb{F}_q[X_1, \ldots, X_n]$. Then $F_q(V) \subseteq V$, and so F_q restricts to a morphism $F_q \colon V \to V$. Then the fixed-point set

$$V^{F_q} := \{ v \in V \mid F_q(v) = v \} = V \cap \mathbb{F}_q^n$$

is a finite subset of V. The same idea can also be applied to a linear algebraic group, giving rise to a finite subgroup.

Now, for theoretical purposes, we would certainly like to have an abstract definition for affine varieties and affine algebraic groups as in Definition 2.1.6 and §2.2.12. The aim of this section is to introduce Frobenius maps in this general setting. The main result will be the Lang–Steinberg Theorem 4.1.12.

To see what this general definition might look like, let us first consider the algebra homomorphism $F_q^* \colon R \to R$ induced by F_q, where $R = k[X_1, \ldots, X_n]$ is the algebra of regular functions on k^n. Since F_q is bijective, F_q^* is injective. Furthermore, $F_q^*(X_i) = X_i^q$ for all i, and so $F_q^*(R) = R^q$. Finally, note that any element of k is algebraic over the prime field of k and, hence, lies in some finite subfield of k. Consequently, for any $f \in R$, we can find some $m \geqslant 1$ such that all coefficients of f lie in $\mathbb{F}_{q^m} \subseteq k$. Then we have $(F_q^*)^m(f) = f^{q^m}$. Thus, F_q is a Frobenius map in the sense of the following definition.

4.1.1 Definition Let (X, A) be an affine variety over $k = \overline{\mathbb{F}}_p$ and $F \colon X \to X$ be a morphism. Assume that there is a power q of p such

that the following conditions hold for the algebra homomorphism
$F^* \colon A \to A$.

(a) F^* is injective and $F^*(A) = A^q$.

(b) For each $f \in A$ there exists some $m \geqslant 1$ such that
$(F^*)^m(f) = f^{q^m}$.

If these conditions hold, we say that (X, A) is defined over $\mathbb{F}_q$
(or that (X, A) admits an $\mathbb{F}_q$-*rational structure*) and that F is the
corresponding *Frobenius map*. Furthermore, we set

$$X^F := \{ x \in X \mid F(x) = x \}$$

for the fixed-point set of F. We also write this set as $X(\mathbb{F}_q)$ and call
it the set of $\mathbb{F}_q$-*rational points* in X.

Note that condition (a) shows that Frobenius maps are finite
morphisms.

We will see in Proposition 4.1.4 that an affine variety X which
is defined over $\mathbb{F}_q$ can always be embedded as an F_q-invariant closed
subset in some k^n such that the Frobenius map $F \colon X \to X$ is just
given by F_q under that embedding.

4.1.2 Remark   Let (X, A) be an affine variety defined over $\mathbb{F}_q$,
with corresponding Frobenius map F. Since $F^*(A) = A^q$ and F^*
is injective, we obtain a well-defined map $\sigma \colon A \to A$ such that
$\sigma(f) = (F^*)^{-1}(f^q)$ for all $f \in A$. This is called the *arithmetic
Frobenius map*. It has the following properties.

(a) The map $\sigma \colon A \to A$ is a bijective ring homomorphism
such that

$$\sigma(\xi f) = \xi^q \sigma(f) \qquad \text{for all } \xi \in k \text{ and } f \in A.$$

(b) For each $f \in A$, there exists some $m \geqslant 1$ such that
$\sigma^m(f) = f$.

Indeed, first note that $F^*(\sigma(f)) = f^q$ and also $\sigma(F^*(f)) = (F^*)^{-1}(F^*(f)^q) = f^q$, for all $f \in A$. So σ and F^* commute with
each other. Furthermore, since $f \mapsto f^q$ is a ring homomorphism and
F^* is a k-algebra homomorphism, we have that σ is a ring homomorphism such that $\sigma(\xi f) = \xi^q \sigma(f)$. The bijectivity of σ follows
from the facts that F^* is injective and that $F^*(A) = A^q$. Thus,
(a) holds. To prove (b), let $f \in A$. Then there exists some $m \geqslant 1$

such that $(F^*)^m(f) = f^{q^m}$. Since F^* and σ commute, this yields $(F^*)^m(\sigma^m(f)) = f^{q^m} = (F^*)^m(f)$. The injectivity of F^* then implies $\sigma^m(f) = f$, as required.

4.1.3 Lemma *Let (X, A) be an affine variety defined over $\mathbb{F}_q$, with corresponding Frobenius map F. Then, for any $f \in A$, the subspace*

$$\langle \sigma^j(f) \mid j = 0, 1, 2, \ldots \rangle_k \subseteq A$$

is finite-dimensional and spanned by elements which are fixed by σ, where σ is defined as in Remark 4.1.2. Furthermore,

$$A_0 := \{f \in A \mid \sigma(f) = f\} = \{f \in A \mid F^*(f) = f^q\}$$

is a finitely generated $\mathbb{F}_q$-subalgebra of A such that the natural map $k \otimes_{\mathbb{F}_q} A_0 \to A$ given by multiplication is an isomorphism.

Proof Let $f \in A$ and set $S = \langle \sigma^j(f) \mid j \geqslant 0 \rangle_k \subseteq A$. By Definition 4.1.1(b), there exists some $m \geqslant 1$ such that $(F^*)^m(f) = f^{q^m}$. Then we have $\sigma^m(f) = f$ by Remark 4.1.2(b), and so $\dim_k S < \infty$. To prove that S is spanned by elements which are fixed by σ, let $\{\xi_0, \ldots, \xi_{m-1}\}$ be an $\mathbb{F}_q$-basis of the subfield $\mathbb{F}_{q^m} \subseteq k$. Then we define elements

$$\tilde{f}_i := \sum_{j=0}^{m-1} \sigma^j(\xi_i f) = \sum_{j=0}^{m-1} \xi_i^{q^j} \sigma^j(f) \in S \qquad \text{for } 0 \leqslant i \leqslant m - 1.$$

We claim that each $\tilde{f}_i$ is fixed by σ. Indeed, using that $\sigma^m(f) = f$, we obtain

$$\sigma(\tilde{f}_i) = \sum_{j=0}^{m-1} \sigma^{j+1}(\xi_i f) = \sigma^m(\xi_i f) + \sum_{j=1}^{m-1} \sigma^j(\xi_i f) = \sum_{j=0}^{m-1} \sigma^j(\xi_i f) = \tilde{f}_i.$$

Now, the matrix $(\xi_i^{q^j})$ (where $0 \leqslant i, j \leqslant m - 1$) is a Vandermonde-type matrix. Since $\{\xi_i\}$ is an $\mathbb{F}_q$-basis of $\mathbb{F}_{q^m}$, it is easily seen that this matrix is invertible. (We leave this as an exercise to this reader.) Consequently, each $\sigma^j(f)$ is a k-linear combination of $\tilde{f}_0, \ldots, \tilde{f}_{m-1}$; in particular, this holds for $f = \sigma^0(f)$. Hence we have $S = \langle \tilde{f}_i \mid 0 \leqslant i \leqslant m - 1 \rangle_k$, as required.

Applying the above construction to a finite set of algebra generators for A, we obtain a new finite set $S \subseteq A$ of algebra generators which are all fixed by σ. This certainly implies that the natural map

$k \otimes_{\mathbb{F}_q} A_0 \to A$ is an isomorphism. Furthermore, we can find a collection of monomials in the elements of S which form a k-basis of A. Then these monomials also form an $\mathbb{F}_q$-basis of A_0, and so S is a finite generating set for A_0 as an $\mathbb{F}_q$-algebra. $\qquad\square$

4.1.4 Proposition *Assume that (X, A) is an affine variety defined over $\mathbb{F}_q$, with corresponding Frobenius map $F\colon X \to X$. Then there exist some $n \geqslant 1$ and a closed embedding $\iota\colon X \to k^n$ such that the following diagram is commutative:*

$$
\begin{array}{ccc}
X & \xrightarrow{\ \iota\ } & k^n \\
{\scriptstyle F}\big\downarrow & & \big\downarrow{\scriptstyle F_q} \\
X & \xrightarrow{\ \iota\ } & k^n \ .
\end{array}
$$

Furthermore, the vanishing ideal of $\iota(X)$ is generated by a set of polynomials in $\mathbb{F}_q[X_1, \ldots, X_n]$, the fixed-point set X^F is finite, and F is a bijective map.

Proof Let $A_0 \subseteq A$ be defined as in Lemma 4.1.3. Then A_0 is a finitely generated $\mathbb{F}_q$-algebra, and so we can write $A_0 = \mathbb{F}_q[X_1, \ldots, X_n]/I_0$ for some ideal I_0. Then we also have $A = k[X_1, \ldots, X_n]/I$, where I is the ideal in $k[X_1, \ldots, X_n]$ generated by I_0. Let $\pi\colon k[X_1, \ldots, X_n] \to A$ be the canonical projection. Then the restriction of π to $\mathbb{F}_q[X_1, \ldots, X_n]$ is the canonical projection onto A_0. (Note that $I \cap \mathbb{F}_q[X_1, \ldots, X_n] = I_0$.) Now we have $\pi = \iota^*$, where $\iota\colon X \to k^n$ is a morphism which is a closed embedding; see §2.2.1. Furthermore, we have $\iota(X) = \mathbf{V}(I)$ and so $\mathbf{I}(\iota(X)) = \sqrt{I} = I = (I_0)$; see Theorem 2.1.9. Finally, for any $f \in \mathbb{F}_q[X_1, \ldots, X_n]$, we have $\pi(F_q^*(f)) = \pi(f^q) = \pi(f)^q = F^*(\pi(f))$ and so $(F_q \circ \iota)^*(f) = (\iota \circ F)^*(f)$. Hence the above diagram is commutative.

The remaining assertions about F now follow from this commutativity and the analogous properties for F_q. Indeed, first note that any element in $\iota(X^F)$ is fixed by F_q and, hence, lies in the finite set $\mathbb{F}_q^n$. Since ι is injective, this yields $|X^F| < \infty$. This also implies that F is injective. Finally, let $x \in X$. Since F_q is bijective, there exists some $v \in k^n$ such that $F_q(v) = \iota(x)$. Then, for any $f \in I_0$, we have $f(v)^q = f(F_q(v)) = f(\iota(x)) = 0$ and so $v \in \iota(X)$. Writing $v = \iota(y)$ for some $y \in X$, we obtain $\iota(F(y)) = F_q(\iota(y)) = F_q(v) = \iota(x)$ and, hence, $F(y) = x$. Thus, F is also surjective. $\qquad\square$

4.1.5 Corollary *Let (X, A) be an affine variety defined over $\mathbb{F}_q$, with corresponding Frobenius map $F\colon X \to X$. Let $A_0 \subseteq A$ be defined as in Lemma 4.1.3. Then, for any closed subset $Y \subseteq X$, the following conditions are equivalent.*

(a) *We have $F(Y) \subseteq Y$.*
(b) *We have $F(Y) = Y$.*
(c) *The ideal $\mathbf{I}_A(Y)$ is generated by elements in A_0.*
(d) *We have $Y = \mathbf{V}_X(S)$ for some subset $S \subseteq A_0$.*

If these conditions hold, then $F|_Y\colon Y \to Y$ is a Frobenius map with respect to $\mathbb{F}_q$.

Proof First assume that the above conditions hold and set $I = \mathbf{I}_A(Y)$. For any $f \in I$ and $y \in Y$, we have $F^*(f)(y) = f(F(y)) = 0$ and so $F^*(f) \in I$. Thus, F^* induces a k-algebra homomorphism $\gamma\colon A/I \to A/I$. Identifying the algebra of regular functions on Y with A/I as in §2.1.12, we see that $\gamma = (F|_Y)^*$. So it remains to check the conditions in Definition 4.1.1. Since $F|_Y\colon Y \to Y$ is surjective, γ is injective. Furthermore, $\gamma(A/I) = \{f^q + I \mid f \in A\} = (A/I)^q$. Finally, for $f \in A$, we have $F^*(f) = f^{q^m}$ for some $m \geqslant 1$, and so $\gamma^m(f + I) = (f + I)^{q^m}$, as required.

We now prove the equivalence of the above four conditions.

'(a) $\Rightarrow$ (b)': Let $x \in Y$. By Exercise 4.7.3(b), there exists some $m \geqslant 1$ such that $F^m(x) = x$. Let $y := F^{m-1}(x)$. By (a), we have $y \in Y$. So $x = F(y)$, as required.

'(b) $\Rightarrow$ (c)': First we check that $I = \mathbf{I}_A(Y)$ is invariant under σ, where σ is defined as in Lemma 4.1.3. Let $f \in I$ and $g \in A$ be such that $F^*(g) = f^q$, that is, we have $\sigma(f) = g$. Then we have $g(F(y)) = F^*(g)(y) = f^q(y) = 0$ for any $y \in Y$, and so $g \in \mathbf{I}_A(F(Y)) = \mathbf{I}_A(Y)$, as required. Now consider a finite set S of ideal generators for I. Using Lemma 4.1.3 and the fact that I is invariant under σ, we may assume that S consists of elements fixed by σ.

'(c) $\Rightarrow$ (d)': Since $Y = \mathbf{V}_X(\mathbf{I}_A(Y))$, this is trivial.

'(d) $\Rightarrow$ (a)': For all $y \in Y$ and $f \in S$, we have $f(F(y)) = F^*(f)(y) = f^q(y) = f(y)^q = 0$ (since $S \subseteq A_0$) and so $F(y) \in Y$. $\quad\square$

4.1.6 Example Consider the standard Frobenius map $F_q\colon k^n \to k^n$. Then the subalgebra $A_0 \subseteq A = k[X_1, \ldots, X_n]$ defined in Lemma 4.1.3 is given by $A_0 = \mathbb{F}_q[X_1, \ldots, X_n]$. (Check this!) Now let

$V \subseteq k^n$ be an algebraic set. Then, by Corollary 4.1.5, the following conditions are equivalent.

(a) We have $F_q(V) \subseteq V$.
(b) We have $F_q(V) = V$.
(c) $\mathbf{I}(V)$ is generated by polynomials with coefficients in $\mathbb{F}_q$.
(d) We have $V = \mathbf{V}(\{f_1, \ldots, f_r\})$ for some $f_1, \ldots, f_r \in \mathbb{F}_q[X_1, \ldots, X_n]$.

In particular, F_q restricts to a Frobenius map on V. Hence, V is defined over $\mathbb{F}_q$ in the sense of Definition 4.1.1. We now see that this corresponds exactly to the intuitive meaning that V is defined by polynomials with coefficients with $\mathbb{F}_q$. (See the discussion at the beginning of this section; see also Exercise 4.7.2 for a 'concrete' version of this statement.) Furthermore, let us set

$$A_0[V] := \mathbb{F}_q[X_1, \ldots, X_n]/I_0, \quad \text{where } I_0 = \mathbf{I}(V) \cap \mathbb{F}_q[X_1, \ldots, X_n].$$

Using (c), we may identify $A_0[V]$ with an $\mathbb{F}_q$-subalgebra of $A[V]$, and this is precisely the subalgebra defined in Lemma 4.1.3.

To complete the picture, we will also want to know when a morphism of affine varieties is 'defined over $\mathbb{F}_q$'. The correct conditions are given as follows.

4.1.7 Lemma *Assume that (X, A) and (X', A') are affine varieties defined over $\mathbb{F}_q$, with corresponding Frobenius maps $F\colon X \to X$ and $F'\colon X' \to X'$. Let $\varphi\colon X \to X'$ be a morphism. Then the following conditions are equivalent.*

(a) φ commutes with F and F'; that is, we have $F' \circ \varphi = \varphi \circ F$.
(b) $\varphi^*(A_0') \subseteq A_0$, where $A_0 \subseteq A$ and $A_0' \subseteq A'$ are as in Lemma 4.1.3.

If these conditions hold, we say that φ is defined over $\mathbb{F}_q$.

Proof First assume that (a) holds. Then we also have $\varphi^* \circ (F')^* = F^* \circ \varphi^*$. Now let $f \in A_0'$; that is, we have $(F')^*(f) = f^q$. Then $F^*(\varphi^*(f)) = \varphi^*((F')^*(f)) = \varphi^*(f^q) = (\varphi^*(f))^q$, and so $\varphi^*(f) \in A_0$, as required. Conversely, assume that (b) holds. Then, for any $f \in A_0'$, we have $\varphi^*((F')^*(f)) = \varphi^*(f^q) = \varphi^*(f)^q$ which equals $F^*(\varphi^*(f))$ since $\varphi^*(f) \in A_0$. Since A_0' generates A' and A_0 generates A (as k-algebras), φ^* commutes with F^* and $(F')^*$. Consequently, (a) holds. $\qquad\square$

4.1.8 Example Consider a regular map $\varphi\colon V \to W$ between two algebraic sets $V \subseteq k^n$ and $W \subseteq k^m$. Denote by F_q the standard Frobenius maps on k^n and on k^m. Assume that V and W are defined over $\mathbb{F}_q$, that is, we have $F_q(V) = V$ and $F_q(W) = W$; see Example 4.1.6.

We claim that the morphism φ is defined over $\mathbb{F}_q$ if and only if there exist polynomials $f_1, \ldots, f_m \in \mathbb{F}_q[X_1, \ldots, X_n]$ such that $\varphi(v) = (f_1(v), \ldots, f_m(v))$ for all $v \in V$. Indeed, if such polynomials exist, then clearly φ commutes with F_q and so φ is defined over $\mathbb{F}_q$ by Lemma 4.1.7. Conversely, assume that the conditions in Lemma 4.1.7 hold, and write

$$A_0[V] = \mathbb{F}_q[X_1, \ldots, X_n]/I_0, \quad \text{where } I_0 = \mathbf{I}(V) \cap \mathbb{F}_q[X_1, \ldots, X_n],$$

$$A_0[W] = \mathbb{F}_q[Y_1, \ldots, Y_m]/J_0, \quad \text{where } J_0 = \mathbf{I}(W) \cap \mathbb{F}_q[Y_1, \ldots, Y_m];$$

see Example 4.1.6. Regarding $A_0[V]$ and $A_0[W]$ as $\mathbb{F}_q$-subalgebras of $A[V]$ and $A[W]$, respectively, we have $\varphi^*(A_0[W]) \subseteq A_0[V]$. For $1 \leqslant j \leqslant m$, choose $f_j \in \mathbb{F}_q[X_1, \ldots, X_n]$ such that $\varphi^*(\bar{Y}_j) = \bar{f}_j$. Arguing as in the proof of Proposition 1.3.4, we see that $\varphi(v) = (f_1(v), \ldots, f_m(v))$ for all $v \in V$, as required.

We now turn to algebraic groups defined over $\mathbb{F}_q$.

4.1.9 Definition Let G be an affine algebraic group over k. We say that G is defined over $\mathbb{F}_q$ if the affine variety G, the multiplication map $\mu\colon G \times G \to G$, and the inversion map $\iota\colon G \to G$ are defined over $\mathbb{F}_q$. Equivalently, there is a Frobenius map $F\colon G \to G$ with respect to $\mathbb{F}_q$ which commutes with μ and with ι, that is, F is a group homomorphism. Consequently, the fixed-point set

$$G^F = \{x \in G \mid F(x) = x\}$$

is a finite group which is called a *finite algebraic group*.

4.1.10 Example (a) Consider the algebraic group $\mathrm{GL}_n(k)$ with algebra of regular functions given by $\mathcal{A}_{\det} = k[X_{ij} \mid 1 \leqslant i, j \leqslant n]_{\det}$; see Example 2.4.1. It is easily checked that we have a well-defined k-algebra homomorphism $\lambda\colon \mathcal{A}_{\det} \to \mathcal{A}_{\det}$ such that $\lambda(X_{ij}) = X_{ij}^q$ and $\lambda(\det) = \det^q$. Furthermore, the two conditions in Definition 4.1.1 are satisfied, and we have

$$(\mathcal{A}_{\det})_0 = \{f \in \mathcal{A}_{\det} \mid \lambda(f) = f^q\} = \mathbb{F}_q[X_{ij} \mid 1 \leqslant i, j \leqslant n]_{\det}.$$

Thus, $\lambda = F_q^*$, where $F_q\colon \mathrm{GL}_n(k) \to \mathrm{GL}_n(k)$ is a Frobenius map which we call the *standard Frobenius map* on $\mathrm{GL}_n(k)$. Explicitly,

F_q is given by

$$F_q \colon \mathrm{GL}_n(k) \to \mathrm{GL}_n(k), \qquad (a_{ij}) \mapsto (a_{ij}^q).$$

The corresponding finite group is the finite general linear group $\mathrm{GL}_n(\mathbb{F}_q)$ of non-singular matrices over $\mathbb{F}_q$. It is well known that

$$|\mathrm{GL}_n(\mathbb{F}_q)| = q^{n(n-1)/2} \prod_{i=1}^{n} (q^i - 1).$$

(Recall that this is simply obtained by counting all vector space bases $(v_1, \ldots, v_n)$ of $\mathbb{F}_q^n$: for v_1, we can take any vector in $\mathbb{F}_q \setminus \{0\}$ and this gives $q^n - 1$ possibilities; for v_2, we can take any vector in $\mathbb{F}_q^n \setminus \langle v_1 \rangle$ and this gives $q^n - q$ possibilities; and so on.) See Exercise 1.8.21 for a structurally different formula.

(b) Let $G \subseteq \mathrm{GL}_n(k)$ be a closed subgroup such that $F_q(G) = G$. Then the restriction of F_q to G is a Frobenius map on G by Corollary 4.1.5. Hence G is an affine algebraic group defined over $\mathbb{F}_q$. This applies, for example, to the subgroup $\mathrm{SL}_n(k)$, and this leads to the finite special linear groups $\mathrm{SL}_n(k)^{F_q} = \mathrm{SL}_n(\mathbb{F}_q)$. Similarly, the subgroups $U_n(k)$ and $T_n(k)$ of all upper unitriangular matrices and all invertible diagonal matrices, respectively, are F_q-stable and we have

$$|U_n(\mathbb{F}_q)| = q^{n(n-1)/2} \quad \text{and} \quad |T_n(\mathbb{F}_q)| = (q-1)^n.$$

The defining equations of the symplectic and orthogonal groups in §1.3.15 and §1.3.16 are also seen to be given by polynomials with coefficients in $\mathbb{F}_q$. Thus, we obtain the finite symplectic and orthogonal groups

$$\begin{aligned}
\mathrm{Sp}_{2m}(k)^{F_q} &= \mathrm{Sp}_{2m}(\mathbb{F}_q), & &\text{any } m \geqslant 1, \text{ any } q; \\
\mathrm{O}_{2m+1}(k)^{F_q} &= \mathrm{O}_{2m+1}(\mathbb{F}_q), & &\text{any } m \geqslant 1,\, q \text{ odd}; \\
\mathrm{O}_{2m}^{+}(k)^{F_q} &= \mathrm{O}_{2m}^{+}(\mathbb{F}_q), & &\text{any } m \geqslant 1, \text{ any } q.
\end{aligned}$$

(c) Consider the map $\gamma \colon \mathrm{GL}_n(k) \to \mathrm{GL}_n(k)$, $g \mapsto n_0^{-1}(g^{\mathrm{tr}})^{-1}n_0$, where $n_0 = Q_n$; see Remark 1.5.12. Then γ is a homomorphism of algebraic groups which commutes with F_q and such that γ^2 is the identity. By Exercise 4.7.4, the map $F' := F_q \circ \gamma$ also is a Frobenius morphism on $\mathrm{GL}_n(k)$. To see which is the corresponding finite algebraic group, note that $(F')^2$ is the standard Frobenius map with respect to $\mathbb{F}_{q^2}$. Thus, we have $\mathrm{GL}_n(k)^{F''} \subseteq \mathrm{GL}_n(\mathbb{F}_{q^2})$. Now the restriction of F_q to $\mathrm{GL}_n(\mathbb{F}_{q^2})$ is an automorphism of order 2, which we denote by $g \mapsto \bar{g}$. Then we obtain

$$\mathrm{GL}_n(k)^{F'} = \mathrm{GU}_n(\mathbb{F}_q) := \{g \in \mathrm{GL}_n(\mathbb{F}_{q^2}) \mid \bar{g}^{\,\mathrm{tr}} n_0 g = n_0\},$$

the general unitary group with respect to the hermitian form defined by n_0.

(d) A similar construction can be applied to the even-dimensional orthogonal groups $\mathrm{SO}_{2m}^+(k)$. Consider the matrix

$$t_m := \left[\begin{array}{c|cc|c} I_{m-1} & \multicolumn{2}{c|}{0} & 0 \\ \hline 0 & 0 & 1 & 0 \\ & 1 & 0 & \\ \hline 0 & \multicolumn{2}{c|}{0} & I_{m-1} \end{array}\right] \in \mathrm{O}_{2m}^+(k).$$

This matrix normalizes $\mathrm{SO}_{2m}^+(k)$; let τ be the automorphism of $\mathrm{SO}_{2m}^+(k)$ defined by conjugation with t_m. It is readily checked that $\tau \circ F_q = F_q \circ \tau$. Thus, by Exercise 4.7.4, we have a Frobenius map for some $\mathbb{F}_q$-rational structure given by

$$F' : \mathrm{SO}_{2m}^+(k) \to \mathrm{SO}_{2m}^+(k), \qquad g \mapsto t_m^{-1} F_q(g) t_m.$$

Then $\mathrm{SO}_{2m}^-(\mathbb{F}_q) := \mathrm{SO}_{2m}^+(k)^{F'}$ is called the finite *non-split orthogonal group*.

We shall continue the discussion of the above examples in Section 4.2.

4.1.11 Proposition *Let* (G, A) *be an affine algebraic group defined over* $\mathbb{F}_q$, *with corresponding Frobenius map* F. *Then there exists a closed embedding of algebraic groups* $\varphi \colon G \to \mathrm{GL}_n(k)$ *such that the following diagram is commutative:*

$$\begin{array}{ccc} G & \xrightarrow{\ \varphi\ } & \mathrm{GL}_n(k) \\ {\scriptstyle F}\big\downarrow & & \big\downarrow{\scriptstyle F_q} \\ G & \xrightarrow{\ \varphi\ } & \mathrm{GL}_n(k)\,. \end{array}$$

Here, $F_q \colon \mathrm{GL}_n(k) \to \mathrm{GL}_n(k)$ *is defined as in Example 4.1.10.*

Proof By Corollary 2.4.4, there exists a linearly independent set $\{e_1, \ldots, e_n\}$ in A such that $A = k[e_1, \ldots, e_n]$ and

$$\mu^*(e_j) = \sum_{i=1}^{n} e_i \otimes a_{ij} \qquad \text{for all } j \in \{1, \ldots, n\}, \text{ where } a_{ij} \in A.$$

Furthermore, the map $\varphi \colon G \to \mathrm{GL}_n(k)$, $x \mapsto (a_{ij}(x))$, is a closed embedding. We now check that the e_i can be chosen such that $a_{ij} \in A_0$, with A_0 defined as in Lemma 4.1.3. Then the above relation implies that $\varphi^*((\mathcal{A}_{\det})_0) \subseteq A_0$ and so Proposition 4.1.7 yields the commutativity of the above diagram.

So let us recall the construction of the $\{e_i\}$ from the proof of Theorem 2.4.3. We begin with a set of algebra generators $f_1, \ldots, f_m$ for A. Since G is defined over $\mathbb{F}_q$, we may assume that $f_i \in A_0$ for all i. Then we obtain a finite-dimensional subspace $E \subseteq A$ as before. For any f_i, we write $\mu^*(f_i) = \sum_{j=1}^{m_i} g_{ij} \otimes h_{ij}$ with $m_i \geqslant 0$ minimal and $g_{ij}, h_{ij} \in A$. Since μ is defined over $\mathbb{F}_q$, we may in fact assume that $g_{ij}, h_{ij} \in A_0$. Repeating this for each algebra generator f_i, we obtain a k-basis $\{e_1, \ldots, e_n\}$ of E consisting of elements in A_0 and satisfying relations $\mu^*(e_j) = \sum_{i=1}^{n} e_i \otimes a_{ij}$ for $1 \leqslant j \leqslant n$, where all a_{ij} lie in A_0, as required. $\qquad \square$

Now we come to a result of fundamental importance for the study of groups G^F as in Definition 4.1.9. Since it actually holds in a slightly more general setting, it will be convenient to introduce the following notion. A homomorphism $F \colon G \to G$ of algebraic groups is called a *generalized Frobenius map* if some power of F is the Frobenius map for an $\mathbb{F}_q$-rational structure on G. We write

$$G^F := \{g \in G \mid F(g) = g\}$$

as before. First note that G^F certainly is a subgroup of G (since F is a group homomorphism). Secondly, note that if $d \geqslant 1$ is such that F^d is a Frobenius map, then we have $G^F \subseteq G^{F^d}$, and so G^F is a finite group (see Proposition 4.1.4).

4.1.12 Theorem (Lang–Steinberg) *Assume that G is a connected affine algebraic group over k and $F \colon G \to G$ is a generalized Frobenius map. Then*

$$L \colon G \to G, \qquad g \mapsto g^{-1}F(g),$$

is a dominant finite morphism. In particular, L is surjective.

Proof Let $d \geqslant 1$ be such that F^d is the Frobenius map for an $\mathbb{F}_q$-rational structure on G. We have already remarked above that G^F is a finite set. Furthermore, if $x, y \in G$ are such that $L(x) = L(y)$, then $xy^{-1} = F(xy^{-1})$ and so $xy^{-1} \in G^F$. Thus, all fibres of L are cosets of G^F and, hence, are finite. By Corollary 2.2.9, this implies that $\dim \overline{L(G)} = \dim G$. Since G is connected, the set $\overline{L(G)}$ is irreducible and we conclude that $G = \overline{L(G)}$. Thus, L is dominant. (Alternatively, this can be proved by computing the differential of L, using Lemma 1.5.7, and then applying the differential criterion for dominance in Theorem 1.4.15.) In order to show that L is a finite morphism, note that we have the following factorization of L. Define $\psi \colon G \to G$ by

$$\psi(g) := g \, F(g) \, F^2(g) \, \cdots \, F^{d-1}(g).$$

Then a straightforward computation shows that $L_d = \psi \circ L$, where $L_d(g) = g^{-1} F^d(g)$. Thus, if we can show that L_d is a finite morphism, then so are ψ and L, and we are done. So we can now assume without loss of generality that $d = 1$, that is, F is the Frobenius map for an $\mathbb{F}_q$-rational structure on G.

We must show that $A = A[G]$ is integral over the subring $L^*(A)$ or, in other words, that A is a finitely generated $L^*(A)$-module. Let $\{e_1, \ldots, e_n\} \subseteq A_0$ be as in the proof of Proposition 4.1.11. Now, by Lemma 4.1.3, we have $F^*(f) = f^q$ for all $f \in A_0$. Hence, for any $x \in G$, we have the following relation.

$$e_j^q(x) = F^*(e_j)(x) = e_j(F(x)) = e_j(xx^{-1}F(x)) = e_j(\mu(x, L(x)))$$

$$= \mu^*(e_j)(x, L(x)) = \sum_{j=1}^n e_i(x)a_{ij}(L(x)) = \Big(\sum_{j=1}^n e_i L^*(a_{ij})\Big)(x),$$

and so $e_j^q = \sum_{j=1}^n L^*(a_{ij})e_i$ for all j. It follows that $A = k[e_1, \ldots, e_n]$ is generated as an $L^*(A)$-module by the finite set $\{e_1^{r_1} \cdots e_n^{r_n} \mid 0 \leqslant r_i < q\}$, as required. $\qquad\square$

4.2 Frobenius maps and *BN*-pairs

In this section, we consider an algebraic group G over $k = \overline{\mathbb{F}}_p$ which has a split *BN*-pair. We shall see that, if B and N are invariant under a Frobenius map F on G, then B^F and N^F will form a split *BN*-pair in the finite group G^F. Furthermore, the sharp form of the Bruhat decomposition leads to a rather explicit order formula for G^F. We discuss this in detail for the finite classical groups. The Lang–Steinberg theorem enters in the discussion at several points.

4.2.1 F-stable BN-pairs Let G be a connected affine algebraic group over $\overline{\mathbb{F}}_p$ and $F\colon G \to G$ be a generalized Frobenius map. Assume that $B, N \subseteq G$ are closed F-stable subgroups which form a *reductive BN-pair*, with Weyl group W (see Definition 3.4.5). Let us write $B = UH$, where $U = R_{\mathrm{u}}(B)$ and $H = B \cap N$ is a torus. We wish to show that $F\colon G \to G$ satisfies the requirements in §1.7.1.

First note that F is a bijective morphism (see Proposition 4.1.4). Since B and N are assumed to be F-stable, the same also holds for H. Furthermore, U is F-stable, since U is characterized as the set of all unipotent elements in B (see Proposition 3.5.4). Thus, (BN$^\varphi$1) holds (where we also use the implication '(a) $\Rightarrow$ (b)' in Corollary 4.1.5). Next, consider (BN$^\varphi$2). Let $n \in N$ and assume that the coset Hn is invariant under F. Then we can argue as follows. The F-invariance implies that there exists some $t \in H$ such that $F(n) = t^{-1}n$. Now the restriction of F to H is a generalized Frobenius map on H by Corollary 4.1.5. Since H is connected, we can apply the Lang–Steinberg Theorem. This yields the existence of some $h \in H$ such that $t = h^{-1}F(h)$. Now we have $F(hn) = F(h)F(n) = (ht)(t^{-1}n) = hn$ and so hn is an element in the coset Hn which is fixed by F. Thus, (BN$^\varphi$2) holds, as required. So we can apply Proposition 1.7.2 which yields that we have $G^F = \langle B^F, N^F \rangle$ and

$$G^F = \coprod_{w \in W^F} U^F H^F n_w U_w^F \qquad \text{with uniqueness of expressions,}$$

where $n_w \in N^F$ for any $w \in W^F$. (Here, we denote the induced automorphism of W again by F.) This yields the order formula

$$|G^F| = |U^F| \cdot |H^F| \sum_{w \in W^F} |U_w^F|.$$

In Exercise 1.8.21, the terms in the above formula are evaluated in the case where $G^F = \mathrm{GL}_n(\mathbb{F}_q)$. In general, we don't even know yet if $U_w^F \neq \{1\}$. We will now establish some results which reduce the evaluation of the above sum to a purely combinatorial problem; see Corollary 4.2.5 below.

4.2.2 Lemma *Let G be an affine algebraic group over k with a reductive BN-pair. Then U_w is connected and $\dim U_w \geqslant l(w)$ for all $w \in W$. We have $\dim U_w = l(w)$ for all $w \in W$, if $\dim U = l(w_0)$, where $w_0 \in W$ is the longest element.*

Proof Write $B = UH$ as usual and recall that $U_w = U \cap n_{w_0 w}^{-1} U n_{w_0 w}$. Now, since $N \subseteq \mathrm{N}_G(H)$ and $H \subseteq \mathrm{N}_G(U)$, we conclude that U_w is normalized by all $t \in H$. Thus, Proposition 3.5.10 yields that U_w is connected.

We prove the assertion on $\dim U_w$ by induction on $l(w)$. If $l(w) = 0$, then $w = 1$ and so $U_1 = U \cap n_{w_0}^{-1} U n_{w_0} \subseteq H \cap U = \{1\}$ (see Lemma 1.6.16). If $l(w) = 1$, then $w \neq 1$ and so $U_w \neq \{1\}$ by Theorem 1.6.19. Since U_w is connected, this implies $\dim U_w \geqslant 1$, as required. Now assume that $l(w) > 1$ and write $w = ys$, where $s \in S$ and $y \in W$ is such that $l(ys) = l(y) + 1$. Using the decomposition $B_w = U_w H$ (see the proof of Theorem 1.6.19) and Lemma 1.6.17(c), we have $U_w = U_s.U_y^s$. Since $U_y^s \subseteq U_w$, we have $\dim U_w \geqslant \dim U_y^s = \dim U_y \geqslant l(w) - 1$ by induction. Assume, if possible, that $\dim U_w = l(w) - 1$. Then we would have $U_w = U_y^s$ and so $U_s \subseteq U_y^s$. But this would imply $n_s U_s n_s^{-1} \subseteq U_y$. On the other hand, using the definition of U_s, one checks that $n_s U_s n_s^{-1} \subseteq U^{w_0}$ and so $n_s U_s n_s^{-1} \subseteq U \cap U^{w_0} = \{1\}$ (see Lemma 1.6.16). Thus, we would conclude $U_s = \{1\}$, contradicting Theorem 1.6.19.

Finally, assume that $\dim U = l(w_0)$. Now, by Remark 1.6.20, the natural map $U_w \times U_{w_0 w} \to U$ given by multiplication is bijective. Hence, by Corollary 2.2.9, we have $\dim U_w + \dim U_{w_0 w} = \dim(U_w \times U_{w_0 w}) = \dim U = l(w_0)$. On the other hand, we know that $\dim U_w \geqslant l(w)$ and $\dim U_{w_0 w} \geqslant l(w_0 w) = l(w_0) - l(w)$, where the last equality holds by relation $(*)$ in the proof of Proposition 1.6.14. Consequently, we must have $\dim U_w = l(w)$, as claimed. $\qquad\square$

Next we need a result which is concerned with a certain class of homomorphisms. A surjective homomorphism $\varphi\colon G \to G_1$ of *connected* affine algebraic groups is called an *isogeny* if $\ker(\varphi)$ is finite. By Proposition 2.2.14, we have $\dim G = \dim G_1$. Furthermore, for any $x \in \ker(\varphi)$, the image of the map $G \to \ker(\varphi)$, $g \mapsto gxg^{-1}$, is finite and irreducible, and hence is a singleton set. Thus, $\ker(\varphi)$ lies in the centre of G. Now we can state the following result.

4.2.3 Proposition (Lusztig) *Let $\varphi\colon G \to G_1$ be an isogeny with kernel Z. Assume that we have generalized Frobenius maps*

$F\colon G \to G$, $F_1\colon G_1 \to G_1$ *such that* $F_1 \circ \varphi = \varphi \circ F$. *Let* $L\colon G \to G$, $g \mapsto g^{-1}F(g)$. *Then the following hold.*

(a) $L(Z)$ *is a normal subgroup of* Z.

(b) $\varphi(G^F) \subseteq G_1^{F_1}$ *is a normal subgroup and* $G_1^{F_1}/\varphi(G^F) \cong Z/L(Z)$.

(c) *We have* $|G^F| = |G_1^{F_1}|$.

Proof (a) Since φ commutes with F, F_1, we have $F(Z) \subseteq Z$. Furthermore, Z lies in the centre of G and so is abelian. Hence $L(Z) \subseteq Z$, and the restriction of L to Z is a group homomorphism.

(b) We define a map $\alpha\colon G_1^{F_1} \to Z/L(Z)$ as follows. Let $g_1 \in G_1^{F_1}$ and choose $g \in G$ such that $\varphi(g) = g_1$. Then $\varphi(g^{-1}F(g)) = \varphi(g)^{-1}F_1(\varphi(g)) = 1$ and so $g^{-1}F(g) \in Z$. Then we define $\alpha(g_1)$ to be the image of $g^{-1}F(g)$ in $L/L(Z)$. First note that this is well-defined. Indeed, if we also have $\varphi(g') = g_1$, then $g' = gz$ for some $z \in Z$. But then $(g')^{-1}F(g') = z^{-1}F(z)g^{-1}F(g)$, which is the same as $g^{-1}F(g)$ modulo $L(Z)$.

Next, we check that α is a group homomorphism. Let $g_1, g_1' \in G_1^{F_1}$, and choose $g, g' \in G$ such that $\varphi(g) = \varphi(g_1)$ and $\varphi(g') = g_1'$. Then we also have $\varphi(gg') = g_1 g_1'$ and $(gg')^{-1}F(gg') = (g')^{-1}g^{-1}F(g)F(g') = (g')^{-1}F(g')g^{-1}F(g)$, where the last equality holds since $g^{-1}F(g) \in \ker(\varphi)$ lies in the centre of G.

Now, α certainly is surjective. Indeed, consider any $z \in Z$. By the Lang–Steinberg theorem, we have $g^{-1}F(g) = z$ for some $g \in G$. Then $\varphi(g)^{-1}\varphi(F(g)) = 1$ and so $F_1(\varphi(g)) = \varphi(g)$. Thus, $\alpha(g_1) = zL(Z)$, where $g_1 := \varphi(g) \in G_1^{F_1}$.

Finally, it is clear that $\varphi(G^F) \subseteq \ker(\alpha)$. Conversely, assume that $\alpha(g_1) = 1$ for some $g_1 \in G_1^{F_1}$. This means that $g^{-1}F(g) = z^{-1}F(z)$ for some $z \in Z$ and some $g \in G$ such that $\varphi(g) = g_1$. But then $F(gz^{-1}) = gz^{-1}$, and this yields $g_1 = \varphi(gz^{-1}) \in \varphi(G^F)$.

Hence α induces an injective homomorphism $\bar\alpha\colon G_1^{F_1}/\varphi(G^F) \to Z/L(Z)$. We have already seen before that this is also surjective.

(c) We have $\ker(L|_Z) = \{z \in Z \mid z^{-1}F(z) = 1\} = Z^F = \ker(\varphi|_{G^F})$, and so $|Z/L(Z)| = |\ker(\varphi|_{G^F})|$. This yields $|G^F| = |\varphi(G^F)||Z/L(Z)|$. But, by (b), $Z/L(Z)$ and $G_1^{F_1}/\varphi(G^F)$ have the same order, and so we obtain $|G^F| = |G_1^{F_1}|$. $\qquad\square$

4.2.4 Theorem (Rosenlicht) *Let U be a connected affine algebraic group and $F\colon U \to U$ be a Frobenius map for an $\mathbb{F}_q$-rational* *structure on U. Then we have*

$$|U^F| = q^{\dim U} \qquad \text{if all elements of } U \text{ are unipotent.}$$

Proof We proceed in several steps.

Step 1. First we show that there exists a closed embedding $\varphi\colon U \to U_n(k)$ such that $\varphi \circ F = F_q \circ \varphi$, where $U_n(k)$ is the group of all upper unitriangular matrices and F_q is the standard Frobenius map on $\mathrm{GL}_n(k)$.

This is seen as follows. By Proposition 4.1.11, there exists a closed embedding $\varphi\colon G \to \mathrm{GL}_n(k)$ such that $\varphi \circ F = F_q \circ \varphi$. Furthermore, by Proposition 3.5.2, there exists some $P \in \mathrm{GL}_n(k)$ such that $P\varphi(U)P^{-1} \subseteq U_n(k)$. We must prove that P can be chosen to lie in $\mathrm{GL}_n(\mathbb{F}_q)$. For this purpose, we recall the proof of Proposition 3.5.2. The idea is to construct a new basis of k^n, such that the elements in U are given by upper unitriangular matrices with respect to that basis. The new basis is constructed inductively, where the crucial step is the existence of some non-zero $v \in k^n$ such that $\varphi(g).v = v$ for all $g \in U$. Thus, it remains to show that v can be chosen to have coordinates in $\mathbb{F}_q$. But this follows from Exercise 4.7.2(b): Indeed, by Proposition 3.5.2, we already know that

$$K := \{ v \in k^n \mid (\varphi(g) - I_n)v = v \text{ for all } g \in U \} \subseteq k^n$$

is a non-zero subspace. Now let us also consider the standard Frobenius map $F_q\colon k^n \to k^n$. This is compatible with $F_q\colon \mathrm{GL}_n(k) \to \mathrm{GL}_n(k)$, in the sense that $F_q(g \cdot v) = F_q(g) \cdot F_q(v)$ for all $g \in \mathrm{GL}_n(k)$ and $v \in k^n$. Now note that $F_q(K) \subseteq K$. Indeed, let $v \in k^n$ and $g \in U$. We can write $g = F(h)$ for some $h \in U$, since $F\colon U \to U$ is bijective; see Proposition 4.1.5. Then we obtain

$$\varphi(g) \cdot F_q(v) = \varphi(F(h)) \cdot F_q(v) = F_q(\varphi(h)) \cdot F_q(v) = F_q(\varphi(h) \cdot v) = F_q(v)$$

and so $F_q(v) \in K$. Thus, by Exercise 4.7.2(b), we have $K \cap \mathbb{F}_q^n \neq \{0\}$, as required. Now continue the argument as in the proof of Proposition 3.5.2.

Step 2. Assume $\dim U \geqslant 1$. We claim that there is a homomorphism of algebraic groups $\lambda\colon U \to \mathbb{G}_a(k)$ (where $\mathbb{G}_a(k)$ is the additive group of k) with

$$\lambda(U) = \mathbb{G}_a(k) \qquad \text{and} \qquad \lambda \circ F = F_q \circ \lambda,$$

where F_q denotes the standard Frobenius map on $\mathbb{G}_a(k)$.

This is proved as follows. Let $\varphi\colon U \to U_n(k)$ be a closed embedding as in Step 1. For $1 \leqslant i < j \leqslant n$, let $\lambda_{ij}\colon U \to k$ be the composition of φ with the projection onto the (i,j)-coordinate of a matrix in $U_n(k)$. Then λ_{ij} certainly is a morphism of affine varieties, and we have $\lambda_{ij} \circ F = F_q \circ \lambda_{ij}$. In general, λ_{ij} will not be a group homomorphism. However, let $r \geqslant 1$ be minimal such that $\varphi_{i,i+r}(U) \neq \{0\}$ for some i. (Such an r exists since $U \neq \{1\}$.) Then an easy matrix calculation shows that $\lambda_{i,i+r}\colon U \to k$ is a group homomorphism into the additive group of k. We set $\lambda := \lambda_{i,i+r}$. Since $\lambda(U) \neq \{0\}$, the image of λ is a non-trivial closed connected subgroup of $\mathbb{G}_a(k)$, and hence λ is surjective. Thus, the above claim is proved.

Step 3. Let $\lambda\colon U \to \mathbb{G}_a(k)$ be as in Step 2 and set $U_1 := \ker(\lambda)^\circ$. Then U_1 is an F-stable closed connected subgroup of U. We claim that there exists a homomorphism of algebraic groups $\gamma\colon U \to \mathrm{GL}_d(k)$ (for some $d \geqslant 1$) such that

$$U_1 = \ker(\gamma) \quad \text{and} \quad \gamma \circ F = F_q \circ \gamma.$$

This is proved as follows. By Proposition 3.3.9, there exists an algebraic representation $\rho\colon U \to \mathrm{GL}_n(k)$ and a non-zero $v \in k^n$ such that $U_1 = \mathrm{Stab}_U(\langle v \rangle)$. We claim that ρ can be chosen to be defined over $\mathbb{F}_q$. Indeed, as in the proof of Proposition 3.3.9, one chooses generators $f_1, \ldots, f_m$ for the vanishing ideal of U_1 in $A[U]$. By Corollary 4.1.5, the f_i can be chosen to lie in the subalgebra $A_0[U]$. Now carry out the construction in the proof of Proposition 3.3.9 with these generators $f_1, \ldots, f_m$. First, this leads to a representation of U on a certain finite-dimensional subspace $E \subseteq A[U]$. Then the required representation ρ is obtained by considering the induced representation on $V = \bigwedge^r E$ for a certain r. As in the proof of Proposition 4.1.11, one sees that the representation on E is defined over $\mathbb{F}_q$. Finally, taking an exterior power also leads to a reresentation defined over $\mathbb{F}_q$. Thus, there exists a homomorphism of algebraic groups $\rho\colon U \to \mathrm{GL}_n(k)$ and some non-zero $v \in k^n$ such that $\mathrm{Stab}_U(\langle v \rangle) = U_1$ and $\rho \circ F = F_q \circ \rho$. Now note that every $g \in U$ is unipotent, and hence all eigenvalues of $\rho(g)$ are 1. Thus, we actually have $\rho(g).v = v$ for all $g \in U_1$, and so the subspace

$$K := \{u \in k^n \mid \rho(g).u = u \text{ for all } g \in U_1\} \subseteq k^n$$

is non-zero. By a computation similar to that in Step 1, one checks that $F_q(K) \subseteq K$. Now consider the induced representation

$\gamma\colon U \to \mathrm{GL}(K)$. We claim that $U_1 = \ker(\gamma)$. Indeed, it is clear that $U_1 \subseteq \ker(\gamma)$. Conversely, let $g \in \ker(\gamma)$. Then $\rho(g).u = u$ for all $u \in K$. In particular, $\rho(g).v = v$ and so $g \in U_1$, as required. Finally, since K is F_q-stable, there exists a basis of K consisting of vectors in k^n which have all their coordinates in $\mathbb{F}_q$ (see Exercise 4.7.2). Then the resulting matrix representation has the required properties.

Step 4. Now we can prove that $|U^F| = q^{\dim U}$ by induction on $\dim U$. If $\dim U = 0$, the result is clear. Next assume that $\dim U = 1$. Let $\lambda\colon U \to \mathbb{G}_a(k)$ be as in Step 2. By Proposition 2.2.14, we have $\dim \ker(\lambda) = \dim U - 1 = 0$ and so λ is an isogeny. Now Proposition 4.2.3 implies that $|U^F| = |\mathbb{G}_a(k)^{F_q}| = |\mathbb{F}_q| = q$, as required. Finally, assume that $\dim U > 1$. Again, let $\lambda\colon U \to \mathbb{G}_a(k)$ be as in Step 2 and set $U_1 := \ker(\rho)^\circ$. Then U_1 is an F-stable closed connected subgroup, and we have $\dim U_1 = \dim U - 1$ by Proposition 2.2.14. Thus, by induction, we have $|U_1^F| = q^{\dim U - 1}$. Now let $\gamma\colon U \to \mathrm{GL}_d(k)$ be as in Step 3 and set $V := \gamma(U) \subseteq \mathrm{GL}_d(k)$. Then V is an F_q-stable closed connected subgroup of dimension 1 (see Proposition 2.2.14). Thus, we already know that $|V^{F_q}| = q$. Now γ restricts to a homomorphism $U^F \to V^{F_q}$, with kernel $\ker(\gamma) \cap U^F = U_1^F$. Hence, if we can show that this restriction is surjective, then $|U^F| = |V^{F_q}||U_1^F| = q^{\dim U}$ and we are done. The fact that $\gamma(U^F) = V^{F_q}$ is seen as follows. Let $h \in V^{F_q}$ and $g \in U$ be such that $\gamma(g) = h$. Then $\gamma(F(g)) = F_q(\gamma(g)) = F_q(h) = h$ and so $x := gF(g)^{-1} \in \ker(\gamma) = U_1$. Now F restricts to a Frobenius map on U_1 by Corollary 4.1.5. Since U_1 is connected, Theorem 4.1.12 yields that $x = y^{-1}F(y)$ for some $y \in U_1$. Now set $g_1 := yg$. Then $\gamma(g_1) = h$ and $F(g_1) = F(y)F(g) = yx \cdot x^{-1}g = yg = g_1$. So $U^F \to V^{F_q}$ is surjective, as required. $\qquad\square$

4.2.5 Corollary *Let G be a connected affine algebraic group defined over $\mathbb{F}_q$, with corresponding Frobenius map F. Assume that there exist F-stable closed subgroups $B, N \subseteq G$ which form a reductive BN-pair in G, with Weyl group W. Then the groups B^F, N^F form a split BN-pair in G^F, with Weyl group W^F. Now write $B = UH$ as usual and assume that $\dim U = l(w_0)$. Then*

$$|G^F| = q^{l(w_0)}|H^F| \sum_{w \in W^F} q^{l(w)}.$$

Note that, here, $l(w)$ is the length of w in terms of the original generators of W.

Proof We have already seen in §4.2.1 that the conditions in §1.7.1 are satisfied for the endomorphism $F\colon G \to G$. Now Proposition 1.7.2 shows that B^F, N^F form a split BN-pair in G^F, provided we can show that $U_{w_J}^F \neq \{1\}$ for every F-orbit J on S. But U_{w_J} is connected of dimension $\geqslant l(w_J)$ by Lemma 4.2.2. So we certainly have $U_{w_J}^F \neq \{1\}$ by Theorem 4.2.4. Finally, if $\dim U = l(w_0)$, then $\dim U_w = l(w)$ for all $w \in W$, and so the order formula in §4.2.1 can be rewritten as above, using Theorem 4.2.4. $\square$

4.2.6 Finite classical groups   Let $G \subseteq \mathrm{GL}_n(k)$ be one of the classical groups defined in §§1.3.15–16:

$$
G = \begin{cases}
\mathrm{GL}_n(k), & \text{any } n, \text{ any characteristic;} \\
\mathrm{SO}_{2m+1}(k), & n = 2m + 1,\ \mathrm{char}(k) \neq 2; \\
\mathrm{Sp}_{2m}(k), & n = 2m,\ \text{any characteristic;} \\
\mathrm{SO}_{2m}^+(k), & n = 2m,\ \text{any characteristic.}
\end{cases}
$$

By Proposition 1.6.10 and §1.7.9, we have a split BN-pair in G, where $B := B_n(k) \cap G$ and $N = N_n(k) \cap G$. Furthermore, we have $B = UH$ where $U := U_n(k) \cap G$ and $H := T_n(k) \cap G$. As pointed out in Definition 3.4.5, all these BN-pairs are reductive. Now let $F\colon G \to G$ be one of the Frobenius maps considered in Example 4.1.10. Then it is readily checked that B and N are F-stable. Furthermore, by §1.7.9(b), we have $\dim U = l(w_0)$. Hence, by Corollary 4.2.5, the groups B^F, N^F form a split BN-pair in G^F. Thus, we have established the existence of split BN-pairs in all the finite classical groups.

In order to evaluate the order formula in Corollary 4.2.5, it remains to compute $|H^F|$ and the sum $\sum_{w \in W^F} q^{l(w)}$. Now, by Exercise 1.8.24, we have $W^F = \langle w_J \mid J \in \bar{S} \rangle$. Furthermore, every $w \in W^F$ can be written in the form $w = w_{J_1} \cdots w_{J_r}$, where $J_i \in \bar{S}$ and $l(w) = l(w_{J_1}) + \cdots + l(w_{J_r})$. Thus, we need to know $l(w_J)$ for each $J \in \bar{S}$. In Table 4.1, we specify for each case a diagram for W^F with vertices corresponding to the generators $\{w_J \mid J \in \bar{S}\}$ (that is, one of the Coxeter graphs in Table 1.2), where the exponents of q indicate $l(w_J)$. The case where $G^F = \mathrm{GL}_n(\mathbb{F}_q)$ has been already dealt with in Example 4.1.10 and Exercise 1.8.21. Now let us consider the remaining cases.

Case 1. Assume that F is the restriction of the standard Frobenius map F_q on $\mathrm{GL}_n(k)$. Then the description of the diagonal matrices in Lemmas 1.5.9–11 shows that $|H^F| = (q-1)^m$ in each

Table 4.1 *BN*-pairs in finite classical groups

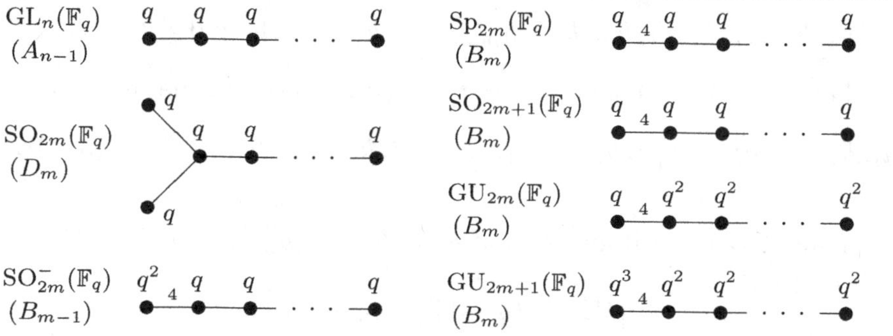

$$
\begin{aligned}
|\,\mathrm{GL}_n(\mathbb{F}_q)| &= q^{n(n-1)/2}(q-1)(q^2-1)(q^3-1)\cdots(q^n-1),\\
|\mathrm{GU}_n(\mathbb{F}_q)| &= q^{n(n-1)/2}(q+1)(q^2-1)(q^3+1)\cdots(q^n-(-1)^n),\\
|\mathrm{Sp}_{2m}(\mathbb{F}_q)| &= q^{m^2}(q^2-1)(q^4-1)\cdots(q^{2m}-1),\\
|\mathrm{SO}_{2m+1}(\mathbb{F}_q)| &= q^{m^2}(q^2-1)(q^4-1)\cdots(q^{2m}-1) \qquad (q \text{ odd}),\\
|\mathrm{SO}^+_{2m}(\mathbb{F}_q)| &= q^{m(m-1)}(q^2-1)(q^4-1)\cdots(q^{2m-2}-1)(q^m-1),\\
|\mathrm{SO}^-_{2m}(\mathbb{F}_q)| &= q^{m(m-1)}(q^2-1)(q^4-1)\cdots(q^{2m-2}-1)(q^m+1).
\end{aligned}
$$

of the above cases. Furthermore, F induces the identity automorphism on the Weyl group W. Thus, we have $J = \{s\}$ for each $J \in \bar{S}$, and so $l(s) = 1$ for each generator s of $W^F = W$.

Case 2. Let $G = \mathrm{GL}_n(k)$ and F be the Frobenius map described in Example 4.1.10(c), such that $G^F = \mathrm{GU}_n(\mathbb{F}_q)$ is the general unitary group. Then

$$
|H^F| = \begin{cases} (q^2-1)^m & \text{if } n = 2m,\\ (q^2-1)^m(q+1) & \text{if } n = 2m+1; \end{cases}
$$

see also Exercise 4.7.8. Furthermore, F induces on $W \cong \mathfrak{S}_n$ the automorphism given by conjugation with the longest element w_0. So, by Exercise 1.8.26, we have $W^F = \langle t, s_1, \ldots, s_{m-1}\rangle$, where $n = 2m$ (if n is even) or $n = 2m+1$ (if n is odd). The description of the generators of W^F shows that $l(s_i) = 2$ for $1 \leqslant i \leqslant m-1$, and $l(t) = 1$ (if $n = 2m$) or $l(t) = 3$ (if $n = 2m+1$).

Case 3. Let $G = \mathrm{SO}^+_{2m}(k)$, where $m \geqslant 2$, and F be the Frobenius map described in Example 4.1.10(d), such that $G^F = \mathrm{SO}^-_{2m}(\mathbb{F}_q)$ is the non-split orthogonal group. Let $h \in G$ be a diagonal matrix with diagonal entries

$$h_1, h_2, \ldots, h_{m-1}, h_m, h_m^{-1}, h_{m-1}^{-1}, \ldots, h_1^{-1}, \quad \text{where } 0 \neq h_i \in k;$$

see Lemma 1.5.11. Then $F(h) = t_m^{-1} F_q(h) t_m$ has diagonal entries

$$h_1^q, h_2^q, \ldots, h_{m-1}^q, h_m^{-q}, h_m^q, h_{m-1}^{-q}, \ldots, h_2^{-q}, h_1^{-q}, \quad \text{where } h_i^{-q} := (h_i^{-1})^q.$$

Thus, $F(h) = h$ if and only if $h_1^q = h_1, \ldots, h_{m-1}^q = h_{m-1}, h_m^{q+1} = 1$. So we obtain

$$|H^F| = (q-1)^{m-1}(q+1).$$

Furthermore, we have $W \cong W_m' \subseteq W_m = \langle t, s_1, \ldots, s_{m-1} \rangle$; see Exercise 1.8.26. Here, we have $W_m' = \langle u, s_1, \ldots, s_{m-1} \rangle$, where $u = ts_1t$. Now F induces on W the automorphism given by conjugation with t. Thus, we have $F(u) = s_1$, $F(s_1) = u$, and $F(s_i) = s_i$ for $2 \leqslant i \leqslant m-1$. Then we have $W^F = \langle t', s_2, \ldots, s_{m-1} \rangle$, where $t' := us_1$, and this is a group of type B_{m-1}. (This is verified by an argument analogous to that in Exercise 1.8.26; we leave the details to the reader.) We have $l(t') = 2$ and $l(s_i) = 1$ for $2 \leqslant i \leqslant m-1$.

Now there is a beautiful and general theory for the evaluation of the sum in Corollary 4.2.5; but we will not go into this here. The result is that there exist certain roots of unity ε_j, ε_{0j} in $\mathbb{C}$ and integers $d(j) \geqslant 1$ such that

$$\sum_{w \in W^F} q^{l(w)} = \prod_{j=1}^{|S|} \frac{1 - \varepsilon_j q^{d(j)}}{1 - \varepsilon_{0j} q}; \qquad \text{see §2 of Steinberg (1968).}$$

For the finite classical groups, this yields the formulas in Table 4.1. In these cases, a more direct proof (based on an induction) for the factorization of $\sum_{w \in W^F} q^{l(w)}$ can also be found in §10.5.1 of Geck and Pfeiffer (2000), for example.

4.2.7 Finite reductive groups Let G be a connected affine algebraic group over $\overline{\mathbb{F}}_p$ and $F \colon G \to G$ be a generalized Frobenius map. Assume that there are F-stable closed subgroups B, N which form a reductive BN-pair in G. Then G^F is called a *finite group of Lie type* or a *finite reductive group*. By Proposition 3.4.6, G is a connected reductive group and $G' := [G, G]$ is a *semisimple group*, where $F(G') \subseteq G'$.

Now, by the basic classification result of Chevalley (1956–8), it is known that G' is a group arising from a semisimple complex Lie algebra, by the construction sketched in Example 1.5.15. Therefore the structure of G' is known in great detail; in particular, all homomorphisms of G' into itself can be described explicitly. By a detailed study of the various possibilities, Steinberg has shown the following results. Let us write $B = UH$ as usual and recall that $U \subseteq G'$. Then there exists a real number $q_0 > 1$ such that

(a) $q_0^d = q$ if F^d is the Frobenius map for an $\mathbb{F}_q$-rational structure on G;

(b) $|U_w^F| = q_0^{l(w)}$ for all $w \in W^F$;

see §11.15 and §11.17 of Steinberg (1968). We have established these facts in Corollary 4.2.5 in the case where F is a Frobenius map for an $\mathbb{F}_q$-rational structure on G. (In these cases, $q_0 = q$.) In Section 4.6, we will see an example of a pair (G, F) where q_0 is an odd power of $\sqrt{2}$.

4.3 Further applications of the Lang–Steinberg theorem

We now discuss a number of further applications of the Lang–Steinberg theorem. Let G be an affine algebraic group over $k = \overline{\mathbb{F}}_p$ and $F\colon G \to G$ a generalized Frobenius map. The first series of applications arises from an axiomatization of the argument that we already used in §4.2.1, for example. This will provide a tool for transferring results on G to the finite group G^F. In another direction, we will introduce Deligne–Lusztig varieties in finite groups of Lie type.

4.3.1 Transitive action Assume that G acts *transitively* on a set $X \neq \varnothing$, where we write the action as $(g, x) \mapsto g.x$ ($g \in G$, $x \in X$), and that there is a map $F'\colon X \to X$ such that the following conditions hold:

(a) The map F' is compatible with F and the action of G on X; that is, we have $F'(g.x) = F(g).F'(x)$ for all $g \in G$ and $x \in X$.
(b) The stabilizer of a point in X is a closed subgroup of G.

Note that X is just a set (finite or infinite); in particular, we do not assume that X is an affine variety. Let $X^{F'} := \{x \in X \mid F'(x) = x\}$. By (a), we know that the finite group G^F acts on the set $X^{F'}$ (but this may no longer be transitive).

4.3.2 Proposition *In the above set-up, assume that G is connected.*

(a) The set $X^{F'} = \{x \in X \mid F'(x) = x\}$ is non-empty.

(b) Let $x_0 \in X^{F'}$ and assume, moreover, that the stabilizer $\mathrm{Stab}_G(x_0)$ is a connected subgroup of G. Then G^F acts transitively on $X^{F'}$.

Proof (a) We begin by taking any $x \in X$. Since G acts transitively on X, we have $F'(x) = g^{-1}.x$ for some $g \in G$. By the Lang–Steinberg theorem, we can write $g = h^{-1}F(h)$ for some $h \in G$. Then we have $F'(h.x) = F(h).F'(x) = hg.F'(x) = h.x$ and so $h.x \in X^{F'}$.

(b) Let $x_0 \in X^{F'}$ and $H := \mathrm{Stab}_G(x_0)$. By §4.3.1(a), this is a closed subgroup of G. Since $F'(x_0) = x_0$, we have $F(H) \subseteq H$, and so $F|_H$ is a generalized Frobenius map on H; see Corollary 4.1.5. Now let $x \in X^{F'}$ and write $x = g.x_0$ for some $g \in G$. Then $g.x_0 = x = F'(x) = F'(g.x_0) = F(g).F'(x_0) = F(g).x_0$ and so $g^{-1}F(g) \in H$. Since H is connected, we can apply the Lang–Steinberg theorem and conclude that $g^{-1}F(g) = h^{-1}F(h)$ for some $h \in H$. Then $gh^{-1} \in G^F$ and $x = g.x_0 = gh^{-1}.x_0$. Thus, x and x_0 lie in the same G^F-orbit. $\square$

4.3.3 Example Assume that G is connected and has a BN-pair where B is a Borel subgroup. Let $\mathfrak{B}$ be the set of all Borel subgroups in G. By Theorem 3.4.3 and Corollary 1.6.4, we have

$$\mathfrak{B} = \{xBx^{-1} \mid x \in G\} \quad \text{and} \quad \mathrm{N}_G(B) = B.$$

We claim that the map $F': \mathfrak{B} \to \mathfrak{B}$, $B' \mapsto F(B')$, satisfies the requirements in §4.3.1. Indeed, let B' be any Borel subgroup in G. Then, since F is a homomorphism of algebraic groups, $F(B') \subseteq G$ is a closed connected solvable subgroup; see Proposition 2.2.14. Furthermore, F is injective and so $\dim F(B') = \dim B'$; see Corollary 2.2.9. Consequently, $F(B')$ is a Borel subgroup and so $F(B') \in \mathfrak{B}$. Also note that $\mathrm{N}_G(B')$ is a closed subgroup by Example 2.5.4; so the condition in §4.3.1(b) holds. Finally, the compatibility condition in §4.3.1(a) certainly is satisfied. So we can apply Proposition 4.3.2(a), and hence there exists some $B_0 \in \mathfrak{B}^{F'}$. Now, since $\mathrm{N}_G(B_0) = B_0$ is connected, we can also apply Proposition 4.3.2(b), and this yields that

$$\mathfrak{B}^{F'} = \{xB_0x^{-1} \mid x \in G^F\}; \quad \text{(in particular, } \mathfrak{B}^{F'} \text{ is a finite set).}$$

We can also consider the set of all pairs (B_1, T_1), where $B_1 \subseteq G$ is a Borel subgroup and $T_1 \subseteq B_1$ is a maximal torus. We claim that:

(a) there exists a pair (B_1, T_1) such that $F(B_1) = B_1$ and $F(T_1) = T_1$;

(b) if (B_2, T_2) is another pair as in (a), then there exists some $g \in G^F$ such that $B_2 = gB_1g^{-1}$ and $T_2 = gT_1g^{-1}$.

To see this, we apply again Proposition 4.3.2. Just note that G acts transitively by conjugation on the set of all pairs (B_1, T_1) as above (by Theorem 3.4.3 and Theorem 3.5.6), that set is stable under F, and that we have

$$\mathrm{Stab}_G(B_1, T_1) = \mathrm{N}_G(B_1) \cap \mathrm{N}_G(T_1) = B_1 \cap \mathrm{N}_G(T_1)$$

$$= \mathrm{N}_{B_1}(T_1) = \mathrm{C}_{B_1}(T_1) \qquad \text{(by Exercise 3.6.10)};$$

but $\mathrm{C}_{B_1}(T_1)$ is connected by Theorem 3.5.9.

Now we turn to the question of what happens if $\mathrm{Stab}_G(x_0)$ is not connected.

4.3.4 Twisted conjugation   We keep the assumptions of §4.3.1. By Proposition 4.3.2(a), we know that $X^{F'} \neq \varnothing$. Now let us fix some $x_0 \in X^{F'}$ and consider its stabilizer $H := \mathrm{Stab}_G(x_0)$. By §4.3.1(a), this is a closed subgroup of G. Denote by $A(x_0)$ the quotient of H by the connected component H° of the identity (see Proposition 1.3.13). Since x_0 is fixed by F', we have $F(H) \subseteq H$ and then also $F(H^\circ) \subseteq H^\circ$. Thus, F induces a group automorphism of the finite group $A(x_0)$ which we denote by the same symbol. We say that $a, a' \in A(x_0)$ are *F-conjugate* if there exists some $c \in A(x_0)$ such that $a' = c^{-1}aF(c)$. This defines an equivalence relation on $A(x_0)$. If F acts as the identity on $A(x_0)$, then this reduces to the usual conjugation action of a group on itself.

We shall denote the canonical map $H \to A(x_0)$ by $h \mapsto \bar{h}$. Let us begin with the following observation. Let $x \in X^{F'}$. Then we can write $x = g.x_0$ for some $g \in G$. We have $g.x_0 = x = F'(x) = F'(g.x_0) = F(g).F'(x_0) = F(g).x_0$, and so $g^{-1}F(g) \in \mathrm{Stab}_G(x_0) = H$. Thus, we have $\overline{g^{-1}F(g)} \in A(x_0)$.

Given $x \in X^{F'}$ and $a \in A(x_0)$, we write $x \sim a$ if there exists some $g \in G$ such that $x = g.x_0$ and $a = \overline{g^{-1}F(g)}$.

4.3.5 Theorem *In the above set-up, assume that G is connected. Then the relation $\sim$ induces a bijective correspondence between the G^F-orbits on $X^{F'}$ and the F-conjugacy classes of $A(x_0)$.*

Proof We proceed in several steps.

Step 1. Let $x_1, x_2 \in X^{F'}$ and $a_1, a_2 \in A(x_0)$ be such that $x_i \sim a_i$ for $i = 1, 2$. We claim that, if x_1, x_2 lie in the same G^F-orbit, then a_1, a_2 are F-conjugate.

Indeed, by assumption, there exist $g_1, g_2 \in G$ such that $x_i = g_i.x_0$ and $a_i = \overline{g_i^{-1}F(g_i)}$ for $i = 1, 2$. If x_1, x_2 are in the same G^F-orbit, we have $x_2 = g.x_1$ for some $g \in G^F$. This yields $gg_1.x_0 = g.x_1 = x_2 = g_2.x_0$ and so $h := g_2^{-1}gg_1 \in H$. A straightforward computation, using $F(g) = g$, shows that $h^{-1}g_2^{-1}F(g_2)F(h) = g_1^{-1}F(g_1)$. Hence we obtain $\bar{h}^{-1}a_2F(\bar{h}) = a_1$, as required.

The result of step 1 shows that $\sim$ induces a well-defined map from the set of G^F-orbits on $X^{F'}$ to the set of F-conjugacy classes of $A(x_0)$. It remains to show that this map is bijective.

Step 2. Surjectivity. We must show that, for any given $a \in A(x_0)$, there exists some $x \in X^{F'}$ such that $x \sim a$. To do this, let $h \in H$ be such that $\bar{h} = a$. By the Lang–Steinberg theorem, there exists some $g \in G$ such that $g^{-1}F(g) = h$. Then we have $F'(g.x_0) = F(g).x_0 = gh.x_0 = g.x_0$ and so $x := g.x_0 \in X^{F'}$. By construction, this means that $x \sim a$.

Step 3. Injectivity. Let $x_1, x_2 \in X^{F'}$ and $a_1, a_2 \in A(x_0)$ be such that $x_i \sim a_i$ for $i = 1, 2$. We must show that if a_1, a_2 are F-conjugate, then x_1, x_2 are in the same G^F-orbit on $X^{F'}$. Now, by assumption, there exist $g_i \in G$ such that $x_i = g_i.x_0$ and $\overline{g_i^{-1}F(g_i)} = a_i$ for $i = 1, 2$. Assume that a_1, a_2 are F-conjugate and write $a_2 = \bar{h}^{-1}a_1F(\bar{h})$ for some $h \in H$. Then we have

$$h'g_2^{-1}F(g_2) = h^{-1}g_1^{-1}F(g_1)F(h) \in H^\circ, \quad \text{where } h' \in H^\circ.$$

Now H° is F-stable and $g_2^{-1}F(g_2) \in H$ normalizes H°. Consequently, we have $F(g_2H^\circ g_2^{-1}) = F(g_2)H^\circ F(g_2)^{-1} = g_2H^\circ g_2^{-1}$; that is, the restriction of F to $g_2H^\circ g_2^{-1}$ is a generalized Frobenius map by Corollary 4.1.5. Applying the Lang–Steinberg theorem to $g_2h'g_2^{-1} \in g_2H^\circ g_2^{-1}$, there exists some $h_0 \in H^\circ$ such that

$$g_2h^{-1}g_1^{-1}F(g_1)F(h)F(g_2)^{-1} = g_2h'g_2^{-1} = (g_2h_0g_2^{-1})^{-1}F(g_2h_0g_2^{-1})$$

$$= g_2h_0^{-1}g_2^{-1}F(g_2)F(h_0)F(g_2)^{-1},$$

which yields $g_2 h_0 h^{-1} g_1^{-1} = F(g_2)F(h_0)F(h)^{-1}F(g_1)^{-1} = F(g_2 h_0 h^{-1} g_1^{-1})$. So $\tilde{g} := g_2 h_0 h^{-1} g_1^{-1} \in G^F$ and, moreover, $\tilde{g}.x_1 = \tilde{g}g_1.x_0 = g_2 h_0 h^{-1}.x_0 = g_2.x_0 = x_2$. Thus, we have shown that x_1, x_2 lie in the same G^F-orbit, as desired. $\qquad\square$

4.3.6 Example Assume that G is connected, and let C be an F-stable conjugacy class in G. Denote by $F' \colon C \to C$ the restriction of F to C. Then the conditions in §4.3.1 are satisfied; we have $\mathrm{Stab}_G(x) = \mathrm{C}_G(x) := \{y \in G \mid xy = yx\}$ for $x \in C$, and this is a closed subgroup. So there exists some $x_0 \in C^F$; moreover, the G^F-classes contained in C^F are in bijection to the F-conjugacy classes of $A(x_0) = \mathrm{C}_G(x_0)/\mathrm{C}_G(x_0)^\circ$. In particular, this shows that, if $\mathrm{C}_G(x_0)$ is connected, then C^F is a single G^F-conjugacy class.

This situation occurs, for example, in $G = \mathrm{GL}_n(k)$, where the centralizer of any element is connected. Indeed, let $g \in G$. Then $K(g) := \{A \in M_n(k) \mid Ag = gA\}$ is a linear subspace of $M_n(k)$ and, hence, irreducible. Consequently, $\mathrm{C}_G(g) = K(g) \cap \mathrm{GL}_n(k)$ is a principal open set in $K(g)$ and, hence, irreducible.

However, this is a rather exceptional example. Non-connected centralizers already occur in $\mathrm{SL}_2(k)$ where the characteristic of k is not 2; see Exercise 2.7.12. Examples of unipotent elements in algebraic groups where $A(x_0)$ is an elementary abelian group of order 2^e (for some $e \geqslant 1$) or a symmetric group of degree $3, 4$, or 5 can be found in the tables in §13.1 of Carter (1985).

4.3.7 Maximal tori Assume that there are closed F-stable subgroups B, N which form a reductive BN-pair in G. Note that this implies, in particular, that G itself is connected. Write $B = UH$ as usual, where H is an F-stable maximal torus. Then we have $H = \mathrm{C}_G(H)$, $N = \mathrm{N}_G(H)$ and the Weyl group W of G is given by N/H. Recall that G acts transitively by conjugation on the set of all maximal tori and that this set is stable under F; see the arguments in §4.3.3. Hence, since $A(H) = N/H = W$, we deduce from Theorem 4.3.5 that the G^F-conjugacy classes of F-stable maximal tori in G are parametrized by the F-conjugacy classes of W. For any $w \in W$, let us denote by $T_w \subseteq G$ an F-stable maximal torus in the G^F-conjugacy class parametrized by w; we shall say that

$$T_w \text{ is obtained from } H \text{ by twisting with } w.$$

Explicitly, we have $T_w = xHx^{-1}$, where $x \in G$ is such that $x^{-1}F(x) = n_w$. For any $t \in H$, we have $F(xtx^{-1}) = xtx^{-1}$ if and only if $F(t) = n_w^{-1} t n_w$, and so

$$T_w^F \cong H^{[w]} := \{t \in H \mid F(t) = n_w^{-1} t n_w\}.$$

Similarly, we have $\mathrm{N}_G(T_w) = xNx^{-1}$ and so, using Exercise 4.7.6,

$$W(T_w)^F \cong \mathrm{C}_{W,F}(w) := \{x \in W \mid x^{-1} w F(x) = w\},$$

where we denote $W(T_w) := \mathrm{N}_G(T_w)/T_w$. Note also that, by Exercise 4.7.5, the map $F_w \colon g \mapsto n_w F(g) n_w^{-1}$ is a generalized Frobenius map on G, and $F_w(H) \subseteq H$.

4.3.8 Proposition *In the setting of §4.3.7, every semisimple element in G^F is contained in an F-stable maximal torus of G.*

Proof Let $s \in G^F$ be semisimple and set $C := \mathrm{C}_G(s)^\circ$. By Exercise 3.6.11, we have $s \in C$. Furthermore, since $s \in G^F$, we certainly have $F(C) \subseteq C$. Thus, F restricts to a generalized Frobenius map on C. On the other hand, by Exercise 3.6.12, s is contained in some maximal torus T of G. Then $T \subseteq \mathrm{C}_G(s)$, and so $T \subseteq C$ since T is connected. Now, by Corollary 3.5.7, all maximal tori in C are conjugate. Using Proposition 4.3.2, we conclude that there exists some F-stable maximal torus $T' \subseteq C$. Since T' is conjugate to T by an element of C and s lies in the centre of C, we conclude that $s \in T'$, as desired. $\qquad\square$

4.3.9 Example Consider the group $G = \mathrm{GL}_n(k)$ with its standard BN-pair (see Proposition 1.6.10), and let $F = F_q$ be the standard Frobenius map on G; see Example 4.1.10(a). Here, $H = B \cap N$ is the maximal torus consisting of all diagonal matrices in G, and we can identify $W = \mathfrak{S}_n$, where F acts as the identity. Thus, the G^F-conjugacy classes of F-stable maximal tori are parametrized by the conjugacy classes of $\mathfrak{S}_n$ and, hence, by the partitions of n. We claim that

$$|T_\lambda^F| = \prod_{i=1}^{r}(q^{\lambda_i} - 1) \qquad \text{for } \lambda = (\lambda_1 \geqslant \lambda_2 \geqslant \cdots \geqslant \lambda_r > 0).$$

To prove this formula, we first consider the case where $\lambda = (n)$. Let w_n be the permutation matrix corresponding to the n-cycle $(1, 2, \ldots, n)$; thus, w_n has coefficient 1 at positions $(2, 1)$, $(3, 2)$, $\ldots$, $(n, n-1)$, $(1, n)$, and 0 otherwise. Now let $t \in H$ be a diagonal matrix with diagonal entries $t_1, \ldots, t_n$. Then $w_n^{-1} t w_n$ has diagonal entries $t_2, t_3, \ldots, t_n, t_1$, and so $F(t) = w_n^{-1} t w_n$ if and only if

$$t_1^q = t_2, t_2^q = t_3, \ldots, t_{n-1}^q = t_n, t_n^q = t_1,$$

or, equivalently, $t_1^{q^n} = t_1$ and $t_{i+1} = t_1^{q^i}$ for $1 \leqslant i \leqslant n - 1$. Thus we have

$$H^{[w_n]} = \{\operatorname{diag}(\xi, \xi^q, \xi^{q^2}, \ldots, \xi^{q^{n-1}}) \mid \xi \in \mathbb{F}_{q^n}^\times\} \cong \mathbb{F}_{q^n}^\times,$$

and this yields the above formula in this case. For a general partition λ with non-zero parts $\lambda_1, \ldots, \lambda_r$, we can take an element $w_\lambda = w_{\lambda_1} \cdots w_{\lambda_r}$, where the w_{λ_i} correspond to disjoint cycles. We can apply the above discussion cycle by cycle, which yields the above formula in general.

We now introduce a notion which leads to the definition of the so-called Deligne–Lusztig varieties, which play a fundamental role in the representation theory of the finite groups G^F.

4.3.10 Relative position of Borel subgroups   Assume that there are closed F-stable subgroups B, N which form a reductive BN-pair in G. As in §4.3.3, let $\mathfrak{B}$ be the set of all Borel subgroups of G. Recall that all Borel subgroups are conjugate (see Theorem 3.4.3) and that $\mathrm{N}_G(B) = B$ (see Corollary 1.6.4). These two facts imply that the map

$$\pi \colon \mathfrak{B} = \{gBg^{-1} \mid g \in G\} \to G/B = \{gB \mid g \in G\}, \quad gBg^{-1} \mapsto gB,$$

is a well-defined G-equivariant bijection, where G acts by conjugation on $\mathfrak{B}$ and by left multiplication on G/B. Now consider the cartesian product $\mathfrak{B} \times \mathfrak{B}$, where G acts diagonally, by $g.(B', B'') = (gB'g^{-1}, gB''g^{-1})$. We also have a diagonal action of G on $G/B \times G/B$, by $g.(g'B, g''B) = (gg'B, gg''B)$. Since $G = \coprod_{w \in W} B n_w B$, the orbits of G on $G/B \times G/B$ are in bijective correspondence with W. Explicitly, the maps

$$
\begin{array}{ccccc}
W & \to & \{G\text{-orbits on } G/B \times G/B\} & \to & \{G\text{-orbits on } \mathfrak{B} \times \mathfrak{B}\}, \\
w & \mapsto & G\text{-orbit of } (B, n_w B) & \mapsto & G\text{-orbit of } (B, n_w B n_w^{-1}),
\end{array}
$$

are bijections. (This is true for any group and any subgroup which equals its own normalizer, and where W is replaced by a set of double coset representatives.) For $w \in W$, we denote by $\mathcal{O}(w)$ the G-orbit in $\mathfrak{B} \times \mathfrak{B}$ containing the pair $(B, n_w B n_w^{-1})$. Now an arbitrary pair (B', B'') of Borel subgroups is said to be in *relative position* w if $(B', B'') \in \mathcal{O}(w)$; in this case, we write

$$B' \xrightarrow{\ w\ } B''.$$

Thus, we have $B' \xrightarrow{\ w\ } B''$ if and only if there exists some $g \in G$ such that $B' = gBg^{-1}$ and $B'' = g n_w B n_w^{-1} g^{-1}$.

4.3.11 Definition (Deligne–Lusztig) Let $w \in W$. Then the *Deligne–Lusztig variety*[1] associated with w is defined as

$$X_w := \{B' \in \mathfrak{B} \mid B' \xrightarrow{\ w\ } F(B')\}.$$

Under the bijection $\pi \colon \mathfrak{B} \to G/B$, $gBg^{-1} \mapsto gB$, we have

$$\pi(X_w) = \{gB \in G/B \mid g^{-1}F(g) \in Bn_wB\}.$$

Furthermore, the Bruhat decomposition shows that $\mathfrak{B} = \coprod_{w \in W} X_w$. The finite group G^F acts on each X_w by conjugation: if $g \in G^F$ and $B' \in X_w$, it is clear that $gB'g^{-1} \xrightarrow{\ w\ } F(gB'g^{-1}) = gF(B')g^{-1}$. Under the bijection $\pi \colon \mathfrak{B} \to G/B$, this action corresponds to the natural action of G^F on G/B by left multiplication.

4.3.12 Example (a) Let $w = 1$. Then we have

$$X_1 = \{B' \in \mathfrak{B} \mid B' = gBg^{-1} \text{ and } F(B') = gBg^{-1} \text{ for some } g \in G\}$$
$$= \{B' \in \mathfrak{B} \mid F(B') = B'\}.$$

Thus, X_1 consists precisely of all F-stable Borel subgroups and we have $\pi(X_1) = \{gB \in G/B \mid g^{-1}F(g) \in B\} = (G/B)^F \cong G^F/B^F$ (see Exercise 4.7.6). Thus, we have a G^F-equivariant bijection $X_1 \cong G^F/B^F$.

(b) Consider the *longest element* $w_0 \in W$. We claim that

$$X_{w_0} = \{B' \in \mathfrak{B} \mid B' \cap F(B') \text{ is a maximal torus in } G\}.$$

Indeed, let $B' \in X_w$ for some $w \in W$. Then $B' = gBg^{-1}$ and $F(B') = g n_w B n_w^{-1} g^{-1}$ for some $g \in G$, and so $B' \cap F(B') =$

---

[1] We have not introduced the general notion of quasi-projective varieties. If we had done so, then one could show that G/B is a projective variety and each X_w is a locally closed subset of G/B. For the time being, we just consider X_w as an abstract set. In §4.3.15, we will obtain a description of X_w as an orbit space of an affine variety under an algebraic group action, which will be sufficient for all our applications. Note also that the definition of X_w only makes sense over a field of positive characteristic: there is no analogue of X_w in characteristic zero!

$g(B \cap n_w B n_w^{-1}) g^{-1}$. Now note that $B \cap n_w B n_w^{-1} = (U \cap n_w U n_w^{-1}) H$. So this intersection is a maximal torus if and only if $U_{w_0 w^{-1}} = U \cap n_w U n_w^{-1} = \{1\}$. Using Theorem 1.6.19, we see that $U_{w_0 w^{-1}} = \{1\} \Leftrightarrow w = w_0$, as claimed.

4.3.13 The zeta function of X_w Recall that some power of F is a Frobenius map and that F itself induces an automorphism of finite order on W. So there exists some $\delta \geqslant 1$ such that F^δ is a Frobenius map (for an $\mathbb{F}_q$-rational structure on G) and F^δ acts as the identity on W. Let us fix such a $\delta \geqslant 1$.

First note that X_w is stable under F^δ. (Indeed, if $B' \in X_w$, then there exists some $g \in G$ such that $B' = gBg^{-1}$ and $F(B') = g n_w B n_w^{-1} g^{-1}$; applying F^δ yields a similar relation between $F^\delta(B')$ and $F(F^\delta(B'))$.)

Now let $g \in G^F$; then g acts on X_w by conjugation. Hence X_w is stable under the composition $F^\delta g = g F^\delta$. Since $g F^\delta$ is a Frobenius map (see Exercise 4.7.4), we see that the fixed-point set of $\mathfrak{B}$ under $g F^\delta$ is finite; see Example 4.3.3. Consequently, the fixed set of $X_w \subseteq \mathfrak{B}$ under $g F^\delta$ is finite also. More generally, the above discussion works if we replace F by F^n for any $n \geqslant 1$. Thus, we have

$$|X_w^{g F^{\delta n}}| < \infty \qquad \text{for any } g \in G^F \text{ and any } n \geqslant 1.$$

We define a formal power series $Z(X_w, g; t)$ (where t is an indeterminate) by

$$Z(X_w, g; t) = \exp\left(\sum_{n=1}^{\infty} \left|X_w^{g F^{\delta n}}\right| \frac{t^n}{n}\right) \in \mathbb{Q}[[t]].$$

Here, exp is given by $\exp(t) = \sum_{n=0}^{\infty} t^n / n!$, as usual. (For basic properties of formal power series, see Chapter V, §11, of Lang (1984). This is called the *zeta function* of X_w (with respect to g). Formally, we have $Z(X_w, g; 0) = 1$ and

$$\frac{\mathrm{d}}{\mathrm{d}t} Z(X_w, g; t) = \left(\sum_{n=1}^{\infty} \left|X_w^{g F^{\delta n}}\right| t^{n-1}\right) Z(X_w, g; t). \qquad (*)$$

Note that the above two properties uniquely determine the zeta function. Thus, the important point is to compute the power series

$$P(X_w, g; t) := \sum_{n=1}^{\infty} \left|X_w^{g F^{\delta n}}\right| t^{n-1},$$

which is nothing but the generating function of the sequence $\{|X_w^{gF^{\delta n}}|\}_{n \geqslant 1}$. For example, for $w = 1$, we have $X_1 \cong G^F/B^F$ by Example 4.3.12(a) and so

$$|X_1^{gF^{\delta n}}| = |(G^F/B^F)^g| \qquad \text{for all } n \geqslant 1.$$

(Indeed, X_1 is a finite set, and F^δ acts as the identity on it.)

We will see in the following sections that $Z(X_w, g; t)$ contains some rather subtle information about X_w. Eventually, following Deligne and Lusztig, we will show in §4.4.17 that $Z(X_w, g; t)$ also has a meaning in the representation theory of the finite group G^F.

We close this section by proving the promised description of X_w as an orbit space. Recall that $B = UH$, where $U = R_{\mathrm{u}}(B)$ and $H = B \cap N$ is a torus.

4.3.14 Lemma *Let $w \in W$ and set $H^{[w]} := \{t \in H \mid F(t) = n_w^{-1} t n_w\}$. Furthermore, set $U^{[w]} := U \cap n_w U n_w^{-1}$. Then the following hold.*

(a) The group $U^{[w]} H^{[w]}$ acts by right multiplication on $L^{-1}(n_w U)$.

(b) The map $g \mapsto gBg^{-1}$ induces a surjective map $L^{-1}(n_w U) \to X_w$.

(c) The fibres of the map in (b) are the orbits of $U^{[w]} H^{[w]}$ on $L^{-1}(n_w U)$.

Proof First note that $H^{[w]}$ normalizes $U^{[w]}$, and so $U^{[w]} H^{[w]}$ is a group.

(a) Let us show that $U^{[w]} H^{[w]}$ acts on $L^{-1}(n_w U)$. So let $g \in L^{-1}(n_w U)$, $u \in U^{[w]}$, and $t \in H^{[w]}$. We must show that $gut \in L^{-1}(n_w U)$. By assumption, we have $F(t) = n_w^{-1} t n_w$ and $g^{-1} F(g) = n_w u_0$ for some $u_0 \in U$. Hence we have

$$(gut)^{-1} F(gut) = t^{-1} u^{-1} g^{-1} F(g) F(u) F(t)$$

$$= t^{-1} u^{-1} n_w u_0 F(u) n_w^{-1} t n_w$$

$$= n_w (n_w^{-1} t^{-1} n_w)(n_w^{-1} u^{-1} n_w) u_0 F(u) n_w^{-1} t n_w.$$

Now note that $n_w^{-1} u^{-1} n_w \in U$ by assumption on u. Hence the above expression lies in $n_w U$, as required.

(b) Let $g \in G$ be such that $L(g) \in n_w U$. Then we have $F(g) = g n_w u$ for some $u \in U$, and so $F(gBg^{-1}) = F(g) B F(g)^{-1} = g n_w B n_w^{-1} g^{-1}$. By definition, this means that $gBg^{-1} \in X_w$.

Now we prove that this map is surjective. For this purpose, let $g \in G$ be such that $gBg^{-1} \in X_w$. Then $F(gBg^{-1}) = gn_w Bn_w^{-1} g^{-1}$ and so $n_w^{-1} g^{-1} F(g) \in N_G(B) = B$. Thus, we have $g^{-1}F(g) = n_w u_0 t_0$ for some $u_0 \in U$ and $t_0 \in H$. In order to see that g can be chosen such that $t_0 = 1$, we replace g by gt (for some $t \in H$) in the above computation. Then we still have $(gt)B(gt)^{-1} = gBg^{-1}$, but we also obtain

$$L(gt) = tn_w u_0 t_0 F(t)^{-1} = n_w(n_w^{-1} tn_w)u_0 t_0 F(t)^{-1}$$
$$= n_w u_0'(n_w^{-1} tn_w)t_0 F(t)^{-1},$$

where $u_0' \in U$. We would like to choose t such that $(n_w^{-1} tn_w)t_0 F(t)^{-1} = 1$ or, equivalently, $t_0 = n_w t^{-1} n_w^{-1} F(t)$. Thus, it is enough to show that the map $H \to H$, $t \mapsto n_w^{-1} t^{-1} n_w F(t)$, is surjective. But, by Exercise 4.7.5, the map $t \mapsto F(n_w t n_w^{-1})$ is a generalized Frobenius map on H. Hence the assertion follows from the Lang–Steinberg theorem.

(c) Let $g, g' \in L^{-1}(n_w U)$ be such that $gBg^{-1} = g'B(g')^{-1}$. Then we have $g^{-1}g' \in B$. Since B is F-stable, we also have $F(g)^{-1}F(g') \in B$. On the other hand, we can write $F(g) = gn_w u$ and $F(g') = g'n_w u'$ for some $u, u' \in U$. This yields $F(g)^{-1}F(g') = u^{-1} n_w^{-1} g^{-1} g' n_w u' \in B$ and so $n_w^{-1} g^{-1} g' n_w \in B$. Thus, $g^{-1}g' \in B \cap n_w Bn_w^{-1} = U^{[w]} H^{[w]}$, and we can write $g' = gu_0 t_0$, where $u_0 \in U^{[w]}$ and $t_0 \in H$. It remains to show that $t_0 \in H^{[w]}$. To see this, we consider

$$L(g') = L(gu_0 t_0) = t_0^{-1} u_0^{-1} g^{-1} F(g) F(u_0) F(t_0)$$
$$= t_0^{-1} u_0^{-1} n_w u_0 F(u_0) F(t_0).$$

We have $n_w^{-1} L(g') \in U$ and so $F(t_0)n_w^{-1} L(g')F(t_0)^{-1} \in U$. This yields

$$F(t_0)(n_w^{-1} t_0^{-1} n_w)(n_w^{-1} u_0^{-1} n_w)u_0 F(u_0) \in U.$$

Since we also have $n_w^{-1} u_0^{-1} n_w u_0 F(u_0) \in U$ (note that $u_0 \in U^{[w]}$), we deduce that $F(t_0)n_w^{-1} t_0^{-1} n_w \in U \cap H = \{1\}$ and so $F(t_0) = n_w^{-1} t_0 n_w^{-1}$. Conversely, it is easily checked that $L^{-1}(n_w U) \to X_w$ is constant on the orbits of $U^{[w]} H^{[w]}$. $\square$

4.3.15 The varieties Y_{n_w}   In the above setting, since $H^{[w]}$ is a finite group, the *affine quotient*

$$Y_{n_w} := L^{-1}(n_w U)/H^{[w]}$$

exists (see §2.5.8). Furthermore, we have an induced surjective map $Y_w \to X_w$, $g \mapsto gBg^{-1}$, whose fibres are the orbits of the action of $U^{[w]}$ by right translation.

Indeed, if $g \in L^{-1}(n_w U)$ and $t \in H^{[w]}$, then we have $tBt^{-1} = B$ and so $L(gt) = L(g)$. Thus, we have a well-defined map $Y_{n_w} \to X_w$ as required. Furthermore, by Lemma 4.3.14(c), the fibres of that map are the orbits of $U^{[w]}$.

Now the finite group G^F acts on $L^{-1}(n_w U)$ by left multiplication. (Indeed, for $g \in G^F$ and $x \in G$ such that $L(x) \in n_w U$, we have $L(gx) = x^{-1}gF(g)F(x) = L(x) \in n_w U$, as required.) Thus, we also has an induced action of G^F on Y_{n_w} by left multiplication. Furthermore, $L^{-1}(n_w U)$ is invariant under F^δ, since F^δ acts as the identity on W. Hence, gF^δ restricts to a Frobenius map on $L^{-1}(n_w U)$. Since the action of $H^{[w]}$ commutes with gF^δ, we obtain an induced Frobenius map on Y_{n_w} which we denote by the same symbol (see Exercise 4.7.9).

4.3.16 Lemma *Let us set $F' := F^{\delta n}$, where $n \geqslant 1$. Recall that F^δ is a Frobenius map for an $\mathbb{F}_q$-rational structure on G. Then, for any $g \in G^F$, we have*

$$|Y_{n_w}^{gF'}| = q^{nd}\,|X_w^{gF'}|, \quad \text{where } d = \dim U^{[w]}.$$

Proof Consider the map $Y_{n_w} \to X_w$, $y \mapsto yBy^{-1}$, defined in §4.3.15. First note that this map commutes with the action of gF', and so we obtain a map

$$Y_{n_w}^{gF'} \to X_w^{gF'}, \qquad y \mapsto yBy^{-1}.$$

Furthermore, the finite group $(U^{[w]})^{F'}$ acts on each fibre of that map by right multiplication. We now claim that the above map is surjective. To see this, let y be such that yBy^{-1} is fixed by gF'. Then $(gF')(y)$ and y lie in the same fibre, and so $(gF')(y) = yt_0u_0$ for some $t_0 \in H^{[w]}$ and $u_0 \in U^{[w]}$; see Lemma 4.3.14.

Now F' is a Frobenius map on $U^{[w]}$. Hence, so is the map $u \mapsto t_0F'(u)t_0^{-1}$; see Exercise 4.7.5. So, by the Lang–Steinberg theorem, there exists some $u \in U^{[w]}$ such that $t_0u_0t_0^{-1} = ut_0F'(u)^{-1}t_0^{-1}$. Then we obtain $(gF')(yu) = yt_0u_0F'(u) = yut_0$. Thus, we have found an element in $Y_{n_w}^{gF'}$ which maps to yBy^{-1}. So the above map is surjective, as claimed. This yields

$$|Y_{n_w}^{gF'}| = |(U^{[w]})^{F'}|\,|X_w^{gF'}|.$$

Now, $F' = F^{\delta n}$ corresponds to an $\mathbb{F}_{q^n}$-rational structure on $U^{[w]}$; see Exercise 4.7.3(a). Further note that $U^{[w]} = U_{w_0 w^{-1}}$ in the notation of Theorem 1.6.19. So Theorem 4.2.4 and Lemma 4.2.2 show that $|(U^{[w]})^{F'}| = q^{nd}$.
$\qquad\square$

4.4 Counting points on varieties over finite fields

In the previous section, we introduced the zeta function of a Deligne–Lusztig variety. This involved the consideration of a sequence like

$$|X^{F^n}| \qquad \text{for } n = 1, 2, 3, \ldots,$$

where X is a variety defined over $\mathbb{F}_q$, with corresponding Frobenius map F. Note that $X = \bigcup_{n \geqslant 1} X^{F^n}$, and so it is conceivable that properties of X might be detected by carefully looking at the collection of the above finite subsets of X. It is, perhaps, suprising that this is a very difficult subject, where any result of substance will involve some rather deep mathematics. The purpose of this section is to give an idea of the complexity of these matters, and to explain without proof some results. Throughout, let $k = \mathbb{F}_p$ be an algebraic closure of the field with p elements, where $p > 0$ is a prime number.

4.4.1 Points over finite fields Let (X, A) be an affine variety over k, and assume that X is defined over $\mathbb{F}_q$ (where q is some power of p), with corresponding Frobenius map F. By Exercise 4.7.3, if $n \geqslant 1$, then $F^n \colon X \to X$ is the Frobenius map for an $\mathbb{F}_{q^n}$-rational structure. In particular, by Proposition 4.1.4, we have

$$N_n := |X^{F^n}| < \infty \qquad \text{for all } n \geqslant 1.$$

In order to understand the behaviour of the function $n \mapsto N_n$, we consider the generating function of the sequence $\{N_n \mid n = 1, 2, 3, \ldots\}$. We set

$$P(X/\mathbb{F}_q; t) := \sum_{n=1}^{\infty} N_n\, t^{n-1},$$

where t is an indeterminate and we regard this as a formal power series in $\mathbb{Q}[[t]]$. (For basic properties of formal power series, see Chapter V, §11, of Lang (1984).)

4.4.2 Remark For some purposes, it may also be useful to know whether $P(X/\mathbb{F}_q; t)$ is a convergent power series, where we consider t as a complex variable. This is indeed so. To see this, we note that, under an embedding $X \subseteq k^m$ as in Proposition 4.1.4, we have $X^{F^n} \subseteq \mathbb{F}_{q^n}^m$ and so $0 \leqslant N_n \leqslant q^{mn}$. Thus, the series $P(X/\mathbb{F}_q; t)$ is absolutely convergent for all $t \in \mathbb{C}$ such that $|t| < 1/q^m$.

We begin with the question of obtaining approximations to the numbers N_n. It seems natural that the function $n \mapsto N_n$ should be related in some way to the dimension of X. The following result provides a first indication in this direction.

4.4.3 Lemma *Let $d = \dim X$. Then there exists a constant $C \geqslant 1$ such that*

$$N_n \leqslant Cq^{dn} \qquad \text{for all } n \geqslant 1.$$

Proof Let $A = A[X]$. By Lemma 4.1.3, there exists a finitely generated $\mathbb{F}_q$-subalgebra $A_0 \subseteq A$ such that the natural map $k \otimes_{\mathbb{F}_q} A_0 \to A$ given by multiplication is an isomorphism. Let $\mathbb{F}_q[a_1, \ldots, a_r] \subseteq A_0$ be a Noether normalization of A_0, where $a_1, \ldots, a_r$ are algebraically independent. Then the inclusion $\mathbb{F}_q[a_1, \ldots, a_r] \subseteq A_0$ induces an inclusion $k[a_1, \ldots, a_r] \subseteq A$ which is still an integral extension of rings (and where $a_1, \ldots, a_r$ remain algebraically independent in $k[a_1, \ldots, a_r]$). This inclusion corresponds to a finite morphism $\varphi \colon X \to k^r$, which is defined over $\mathbb{F}_q$ by construction. It now also follows that $r = d = \dim X$.

Now, by Lemma 2.2.3, there exists some $C \geqslant 1$ such that $|\varphi^{-1}(v)| \leqslant C$ for all $v \in k^d$. Consequently, we have $|N_n| \leqslant C |(k^d)^{F_q^n}| = Cq^{dn}$, as required. $\square$

It is much more difficult to obtain a lower bound for the numbers N_n. We state without proof the following result (which will not be used in this book).

4.4.4 Theorem (Lang–Weil) *Assume that X is irreducible. Then we have*

$$\lim_{n \to \infty} \frac{N_n}{q^{dn}} = 1, \qquad \text{where } d = \dim X.$$

The reference for the original article is Lang and Weil (1954). An elementary exposition can be found in Chapter V of Schmidt (1976).

The deepest facts about the numbers N_n are yet to come! To state them, we now introduce zeta functions in general.

4.4.5 Definition (Weil) Let (X, A) be an affine variety defined over $\mathbb{F}_q$, with corresponding Frobenius map $F\colon X \to X$. The *zeta function* of X is defined by

$$Z(X/\mathbb{F}_q; t) := \exp\left(\sum_{n=1}^{\infty} N_n \frac{t^n}{n}\right) \in \mathbb{Q}[[t]].$$

Here, exp is given by $\exp(t) = \sum_{n=0}^{\infty} t^n/n!$, as usual. Formally, we have

$$Z(X/\mathbb{F}_q; 0) = 1 \quad \text{and} \quad \frac{d}{dt} Z(X/\mathbb{F}_q; t) = \left(\sum_{n=1}^{\infty} N_n\, t^{n-1}\right) Z(X/\mathbb{F}_q; t).$$

$$(*)$$

The above two properties uniquely determine the zeta function.

4.4.6 Example (a) Let $X = k^d$ with its standard $\mathbb{F}_q$-rational structure (with Frobenius map given by F_q). Then $N_n = q^{dn}$ for all $n \geqslant 1$, and so

$$\sum_{n=1}^{\infty} q^{dn} t^{n-1} = q^d \sum_{n=0}^{\infty} (q^d t)^n = \frac{q^d}{1 - q^d t}.$$

This implies that $Z(k^d/\mathbb{F}_q; t) = 1/(1 - q^d t)$.

(b) Let $X = \{(x, y) \in k^2 \mid x^2 + y^2 = 1\}$ with its standard $\mathbb{F}_q$-rational structure, where q is assumed to be odd. Then one can show that

$$Z(X/\mathbb{F}_q; t) = \frac{1 - \varepsilon t}{1 - qt}, \qquad \text{where } \varepsilon \in \{\pm 1\} \text{ is such that } q \equiv \varepsilon \bmod 4.$$

For the details, see p. 112 of Koblitz (1984).

In the above examples, $Z(X/\mathbb{F}_q; t)$ is a rational function, that is, a quotient of two polynomials. Weil conjectured that this should be true in general and predicted further properties of the polynomials in the numerator and the denominator; see Weil (1949). The first step in the proof of these conjectures is given by the following fundamental result.

4.4.7 Theorem (Dwork)   *Let X be an affine variety defined over $\mathbb{F}_q$. Then there exist polynomials $P, Q \in \mathbb{Z}[t]$ with constant term equal to 1 such that $Z(X/\mathbb{F}_q; t) = P/Q$.*

The reference for the original article is Dwork (1960). An elementary exposition of the proof, which uses methods of p-adic analysis, can be found in Chapter 5 of Koblitz (1984). For a completely different proof, see §4.4.9 below.

4.4.8 Corollary   *There exist non-zero complex numbers $\alpha_1, \ldots, \alpha_a$, $\beta_1, \ldots, \beta_b$ (where $a + b \geqslant 1$) such that*

$$N_n = \sum_{i=1}^{a} \alpha_i^n - \sum_{j=1}^{b} \beta_j^n \qquad \text{for all } n \geqslant 1.$$

Proof   Let P, Q be as in Theorem 4.4.7. Since P and Q have constant term 1, there exist non-zero $\alpha_i, \beta_j \in \mathbb{C}$ ($1 \leqslant i \leqslant a$, $1 \leqslant j \leqslant b$) such that

$$P = \prod_{j=1}^{b} (1 - \beta_j t) \quad \text{and} \quad Q = \prod_{i=1}^{a} (1 - \alpha_i t);$$

note, however, that at this stage it is not yet clear that $a + b \geqslant 1$. In order to obtain the desired expression for N_n, we use the relation 4.4.5($*$) and the formula for the derivative of the quotient of two polynomials. This yields

$$\frac{Q}{P} \frac{\mathrm{d}}{\mathrm{d}t} \left(\frac{P}{Q} \right) = \sum_{i=1}^{a} \frac{\alpha_i}{1 - \alpha_i t} - \sum_{j=1}^{b} \frac{\beta_j}{1 - \beta_j t}$$

$$= \sum_{n=1}^{\infty} \left(\underbrace{\sum_{i=1}^{a} \alpha_i^n - \sum_{j=1}^{b} \beta_j^n}_{=N_n} \right) t^{n-1},$$

where we used the expansion $1/(1 - at) = \sum_{i=0}^{\infty} a^i t^i$. Thus, we have obtained the desired expression of N_n. Since $N_n > 0$ for some n, we now see that $a + b \geqslant 1$. $\qquad\square$

Now Weil himself predicted that a proof of his conjectures could be obtained by using a suitable cohomological interpretation of the numbers N_n, in the spirit of the *Lefschetz fixed-point theorem* in algebraic topology (see, for example, Chapter 4, §7, of Spanier (1966)). Such a cohomological interpretation was found by M. Artin, A. Grothendieck *et al.*, in the Paris seminars [SGA] (Séminaire de Géométrie Algébrique). It has been outstandingly successful for many purposes; see Appendix C of Hartshorne (1977) for further references and historical remarks. By an approach similar to that of Carter (1985), we shall formulate a number of precise properties concerning that cohomology theory and use these as axioms on which the subsequent development is based.

4.4.9 Cohomological interpretation To any affine algebraic variety X over k, one can attach canonically a family of finite-dimensional $\mathbb{Q}_\ell$-vector spaces denoted by $H_c^i(X, \mathbb{Q}_\ell)$, where $i \in \mathbb{Z}$. (Here, $\mathbb{Q}_\ell$ is the field of ℓ-adic numbers, for a fixed prime $\ell \neq p$.) For the construction and basic properties of $H_c^i(X, \mathbb{Q}_\ell)$, see the appendix of Carter (1985) or Chapter V of Srinivasan (1979). We have

$$H_c^i(X, \mathbb{Q}_\ell) = \{0\} \quad \text{unless } 0 \leqslant i \leqslant 2\dim X.$$

A finite morphism $\varphi \colon X \to Y$ between affine varieties induces a linear map

$$\varphi^* \colon H_c^i(Y, \mathbb{Q}_\ell) \to H_c^i(X, \mathbb{Q}_\ell),$$

and if $\psi \colon Y \to Z$ is a further finite morphism, we have $(\psi \circ \varphi)^* = \varphi^* \circ \psi^*$.

Now assume that X is defined over $\mathbb{F}_q$, with corresponding Frobenius map F. Then $F \colon X \to X$ is a finite morphism, and so we have induced linear maps

$$F^* \colon H_c^i(X, \mathbb{Q}_\ell) \to H_c^i(X, \mathbb{Q}_\ell).$$

The required analogue of the Lefschetz fixed-point theorem states that F^* is a linear isomorphism on each $H_c^i(X, \mathbb{Q}_\ell)$ and that

$$|X^F| = \sum_i (-1)^i \mathrm{trace}(F^*, H_c^i(X, \mathbb{Q}_\ell)).$$

This is called *Grothendieck's trace formula*. For a proof see Deligne (1977).

In order to illustrate the power of these methods, let us prove Corollary 4.4.8. Let $d_i := \dim H_c^i(X, \mathbb{Q}_\ell)$ and $\lambda_{ij} \in \bar{\mathbb{Q}}_\ell$ $(1 \leqslant j \leqslant d_i)$ be the eigenvalues of F^* on $H_c^i(X, \mathbb{Q}_\ell)$. Here, $\bar{\mathbb{Q}}_\ell$ is an algebraic closure of $\mathbb{Q}_\ell$. Then λ_{ij}^n are the eigenvalues of $(F^*)^n = (F^n)^*$ on $H_c^i(X, \mathbb{Q}_\ell)$ for $n \geqslant 1$. So Grothendieck's trace formula yields

$$|X^{F^n}| = \sum_i (-1)^i \operatorname{trace}((F^*)^n, H_c^i(X, \mathbb{Q}_\ell)) = \sum_i (-1)^i \sum_{j=1}^{d_i} \lambda_{ij}^n.$$

Thus, the formula in Corollary 4.4.8 holds with $\{\alpha_1, \ldots, \alpha_a\} = \{\lambda_{ij} \mid i \text{ even}\}$ and $\{\beta_1, \ldots, \beta_b\} = \{\lambda_{ij} \mid i \text{ odd}\}$. Note that $\lambda_{ij} \neq 0$ since F^* is a linear isomorphism. Running the argument in Corollary 4.4.8 backwards, we also obtain an expression of $Z(X/\mathbb{F}_q; t)$ as a quotient of two polynomials as in Theorem 4.4.7.

4.4.10 Definition Let X be an affine variety over k and $g\colon X \to X$ be an automorphism of finite order. Then the *Lefschetz number* of g is defined by

$$\mathcal{L}(g, X) := \sum_i (-1)^i \operatorname{trace}(g^*, H_c^i(X)).$$

Note that, a priori, $\mathcal{L}(g, X) \in \mathbb{Q}_\ell$ for a prime $\ell \neq p$. In Theorem 4.4.12 we will show that $\mathcal{L}(g, X)$ in fact is an integer independent of ℓ. For the proof, we need the following elementary result on power series.

4.4.11 Lemma *Let K be any field and $R = \sum_{n=0}^\infty a_n t^n \in K[[t]]$ be a formal power series. For any $m, s \geqslant 0$, we consider the matrix*

$$A_{s,m} := \begin{bmatrix} a_s & a_{s+1} & a_{s+2} & \cdots & a_{s+m} \\ a_{s+1} & a_{s+2} & a_{s+3} & \cdots & a_{s+m+1} \\ a_{s+2} & a_{s+3} & a_{s+4} & \cdots & a_{s+m+2} \\ \vdots & \vdots & \vdots & & \vdots \\ a_{s+m} & a_{s+m+1} & a_{s+m+2} & \cdots & a_{s+2m} \end{bmatrix} \in M_{m+1}(K).$$

Then we have $R = P/Q$ for some polynomials $P, Q \in K[t]$, if and only if there exist integers $m \geqslant 0$ and s_0 such that $\det(A_{s,m}) = 0$ for all $s \geqslant s_0$.

Proof See Lemma 5, §5.5 of Koblitz (1984). $\qquad\qquad\square$

4.4.12 Theorem *Let X be an affine variety over k and $g\colon X \to X$ be an automorphism of finite order. Then there exists a Frobenius* *map $F\colon X \to X$ such that $Fg = gF$. For any such F, there are polynomials $P, Q \in \mathbb{Q}[t]$ such that*

$$-\sum_{n=1}^{\infty} |X^{gF^n}| \, t^n = P/Q \qquad \text{and} \qquad \mathcal{L}(g, X) = \lim_{t \to \infty} P(t)/Q(t).$$

We have $\mathcal{L}(g, X) = \mathcal{L}(g^{-1}, X) \in \mathbb{Z}$, and this is independent of $\ell \neq p$.

Proof We may assume that X is a closed subset of k^m for some $m \geqslant 1$. Then X and g are defined by (finitely many) polynomials in m variables over k. The coefficients of all these polynomials will lie in some finite subfield $\mathbb{F}_q \subseteq k$. Let F_q be the standard Frobenius map on k^m. Then $F_q(X) \subseteq X_q$ and $gF_q = F_q g$. Thus, $F := F_q|_X \colon X \to X$ is a Frobenius map as required.

Now consider the cohomology spaces $H_c^i(X, \mathbb{Q}_\ell)$. Let $d_i :=$ $\dim H_c^i(X, \mathbb{Q}_\ell)$ and $\lambda_{ij} \in \bar{\mathbb{Q}}_\ell$ $(1 \leqslant j \leqslant n_i)$ be the eigenvalues of F^* on $H_c^i(X, \mathbb{Q}_\ell)$. Then λ_{ij}^n are the eigenvalues of $(F^*)^n = (F^*)^n$ on $H_c^i(X, \mathbb{Q}_\ell)$ for any $n \geqslant 1$. Furthermore, let $\alpha_{ij} \in \bar{\mathbb{Q}}_\ell$ $(1 \leqslant j \leqslant d_i)$ be the eigenvalues of g^* on $H_c^i(X, \mathbb{Q}_\ell)$. Since F and g commute with each other, so do g^* and $(F^*)^n$. Consequently, $\alpha_{ij} \lambda_{ij}^n$ are the eigenvalues of $(gF^n)^*$ on $H_c^i(X, \mathbb{Q}_\ell)$, and so

$$|X^{gF^n}| = \sum_i (-1)^i \operatorname{trace}((gF^n)^*, H_c^i(X, \mathbb{Q}_\ell)) = \sum_i (-1)^i \sum_{j=1}^{d_i} \alpha_{ij} \lambda_{ij}^n.$$

Using the expansion $\sum_{n \geqslant 1} t^n = t/(1-t)$, we obtain

$$-\sum_{n=1}^{\infty} |X^{gF^n}| \, t^n = \sum_i (-1)^i \sum_{j=1}^{d_i} \alpha_{ij} \lambda_{ij}^n t^n = \sum_i (-1)^i \sum_{j=1}^{d_i} \alpha_{ij} \frac{\lambda_{ij} t}{\lambda_{ij} t - 1}.$$

$$(*)$$

Thus, we see that the formal power series on the left-hand side has an expression as a quotient of two polynomials, where the degree of the numerator is less than or equal to the degree of the denominator. Consequently, the limit as $t \to \infty$ exists. Since all λ_{ij} are non-zero, the limit of the above expression is given by

$$\sum_i (-1)^i \sum_{j=1}^{d_i} \alpha_{ij} = \sum_i (-1)^i \operatorname{trace}(g^*, H_c^i(X, \mathbb{Q}_\ell)),$$

and this equals $\mathcal{L}(g, X)$. Thus, $(*)$ yields polynomials $P', Q' \in \bar{\mathbb{Q}}_\ell[t]$ such that

$$-\sum_{n=1}^{\infty} |X^{gF^n}| t^n = P'/Q' \qquad \text{and} \qquad \mathcal{L}(g, X) = \lim_{t \to \infty} P'(t)/Q'(t).$$

Now regard the power series on the left hand side as a series in $\bar{\mathbb{Q}}_\ell[[t]]$. Then Lemma 4.4.11 shows that the coefficients satisfy a certain set of determinantal conditions. But these coefficients are in $\mathbb{Q}$, and so the determinantal conditions are valid when we regard that power series in $\mathbb{Q}[[t]]$. Thus, another application of Lemma 4.4.11 shows that there exist $P, Q \in \mathbb{Q}[t]$ such that $P'/Q' = P/Q$. This certainly implies that $\mathcal{L}(g, X) \in \mathbb{Q}$ and that this number is independent of ℓ. Finally, since the eigenvalues of g^* are roots of unity, the defining formula in Definition 4.4.10 shows that $\mathcal{L}(g, X)$ is an algebraic integer. Being a rational number and an algebraic integer, we conclude that $\mathcal{L}(g, X) \in \mathbb{Z}$, as required. This also implies the equality $\mathcal{L}(g, X) = \mathcal{L}(g^{-1}, X)$, using the fact that α_{ij}^{-1} are the eigenvalues of $(g^{-1})^*$ and that there is a field automorphism of $\bar{\mathbb{Q}}_\ell$ which sends roots of unity to their inverses. $\square$

In the following discussion, we shall simply write

$$\mathcal{L}(g, X) = -\lim_{t \to \infty} \sum_{n=1}^{\infty} |X^{gF^n}| t^n,$$

without referring explicity to the polynomials P and Q.

4.4.13 Example Let X be a 0-dimensional affine variety over k (a finite set). Let $g \colon X \to X$ be an automorphism of finite order. Then

$$\mathcal{L}(g, X) = |X^g| = |\{x \in X \mid g(x) = x\}|.$$

This is seen as follows. Let $F \colon X \to X$ be a Frobenius map such that F commutes with g. Since X is a finite set and F is a bijection, some power of F is the identity on X. Hence, replacing F by that power, we can assume that F is the identity on X. Then $|X^{gF^n}| = |X^g|$ for all $n \geqslant 1$. Thus, by Theorem 4.4.12, have

$$\mathcal{L}(g, X) = -\lim_{t \to \infty} \sum_{n=1}^{\infty} |X^{gF^n}| t^n = -|X^g| \lim_{t \to \infty} \sum_{n=1}^{\infty} t^n$$

$$= |X^g| \lim_{t \to \infty} \frac{t}{t-1} = |X^g|.$$

4.4.14 Example Let $\varphi\colon X \to X_1$ be a bijective morphism of affine varieties. Let $g\colon X \to X$ and $g_1\colon X_1 \to X_1$ be automorphisms of finite order such that $g_1 \circ \varphi = \varphi \circ g$. Then we have

$$\mathcal{L}(g, X) = \mathcal{L}(g_1, X_1).$$

Indeed, choose Frobenius maps $F\colon X \to X$, $F_1\colon X_1 \to X_1$ such that $Fg = gF$, $F_1 g_1 = g_1 F_1$, $\varphi F = F_1 \varphi$. Then φ induces a bijection $X^{gF^n} \to X_1^{g_1 F_1^n}$, and so

$$|X^{gF^n}| = |X_1^{g_1 F_1^n}| \quad \text{for all } n \geqslant 1.$$

Consequently, we have $\mathcal{L}(g, X) = \mathcal{L}(g_1, X_1)$ by Theorem 4.4.12.

4.4.15 Example Let U be a connected unipotent group over k, and $g\colon U \to U$ be an automorphism of finite order. Then we have

$$\mathcal{L}(g, U) = 1.$$

Indeed, choose a Frobenius map $F\colon U \to U$ for some $\mathbb{F}_q$-rational structure on U which commutes with g. Then F^n is the Frobenius map for an $\mathbb{F}_{q^n}$-rational structure on U by Exericse 4.7.3(a), and so is gF^n by Exercise 4.7.4(a). Hence, by Theorem 4.2.4, we have $|U^{gF^n}| = q^{dn}$ for all $n \geqslant 1$, where $d = \dim U$. This implies the above claim, since

$$\mathcal{L}(g, U) = -\lim_{t \to \infty} \sum_{n=1}^{\infty} |U^{gF^s}| \, t^n = -\lim_{t \to \infty} \sum_{n=1}^{\infty} (q^d t)^n = \lim_{t \to \infty} \frac{q^d t}{q^d t - 1} = 1.$$

Finally, we discuss some applications of the above constructions to representation theory. We assume that the reader has some basic familiarity with the character theory of finite groups. (Part I of Serre (1977) would be sufficient.)

4.4.16 Example Let X be an affine variety over k and assume that Γ is a finite group acting on X. Then $g \in \Gamma$ acts on $H_c^i(X, \mathbb{Q}_\ell)$ by $(g^*)^{-1}$, and this gives a representation of Γ. By Theorem 4.4.12, we have

$$\mathcal{L}(g, X) = \mathcal{L}(g^{-1}, X) = \sum_i (-1)^i \operatorname{trace}((g^*)^{-1}, H_c^i(X, \mathbb{Q}_\ell)) \quad \text{for } g \in \Gamma.$$

Thus, $g \mapsto \mathcal{L}(g, X)$ is the alternating sum of the characters of the representations of Γ on the spaces $H_c^i(X, \mathbb{Q}_\ell)$. Now note the following fact. Given a function $f\colon \Gamma \to \mathbb{Z}$ and a field $K \supseteq \mathbb{Q}$ such that f is an integral linear combination of characters over K, then f will also be an integral linear combination of the complex irreducible characters of Γ. (Indeed, let $\bar{K}$ be an algebraic closure of K and $\bar{\mathbb{Q}}_K$ be the algebraic closure of $\mathbb{Q}$ in $\bar{K}$. Then the irreducible representations of Γ over $\bar{K}$ are obtained by scalar extension from the irreducible representations over $\bar{\mathbb{Q}}_K$. Hence, every representation of Γ over K is equivalent to a representation over $\bar{\mathbb{Q}}_K$. Now embed $\bar{\mathbb{Q}}_K$ into $\mathbb{C}$.) Thus, we have:

$$\text{The map } g \mapsto \mathcal{L}(g, X) \text{ is a virtual character of } \Gamma.$$

A *virtual character* of a finite group is, by definition, an integral linear combination of the complex irreducible characters of that group.

4.4.17 The virtual characters of Deligne and Lusztig, I

Let G be a connected affine algebraic group and $F\colon G \to G$ a generalized Frobenius map. Assume that there are closed F-stable subgroups B, N which form a reductive BN-pair in G. Recall the definition of the Deligne–Lusztig varieties from Section 4.3. Let $w \in W$. We have seen in §4.3.15 that G^F acts by left multiplication on the affine variety Y_{n_w}. Thus, by Example 4.4.16, the map

$$R_w\colon G^F \to \mathbb{Z}, \qquad g \mapsto \mathcal{L}(g, Y_{n_w}),$$

is a virtual character of G^F. A priori, R_w depends on the choice of $n_w \in N$. However, we will now obtain an alternative expression which shows the independence of n_w. By Theorem 4.4.12, there exist $P, Q \in \mathbb{Q}[t]$ such that

$$-\sum_{n=1}^{\infty} |Y_{n_w}^{gF^{\delta n}}|\, t^n = P/Q \quad \text{and} \quad \mathcal{L}(g, Y_{n_w}) = \lim_{t \to \infty} P(t)/Q(t).$$

Now, using Lemma 4.3.16, we obtain

$$-\sum_{n=1}^{\infty} |X_w^{gF^{\delta n}}|\, t^n = -\sum_{n=1}^{\infty} |Y_{n_w}^{gF^{\delta n}}|\, (t/q^d)^n = \frac{P(t/q^d)}{Q(t/q^d)}.$$

Since $\displaystyle \lim_{t \to \infty} \frac{P(t/q^d)}{Q(t/q^d)} = \lim_{t \to \infty} \frac{P(t)}{Q(t)}$, we conclude that

$$\mathcal{L}(g, Y_{n_w}) = -\lim_{t \to \infty} \sum_{n=1}^{\infty} |X_w^{gF^{\delta n}}|\, t^n.$$

Thus, the above formal power series can also be expressed as a rational function, and this rational function has a limit as $t \to \infty$, where the limit equals $\mathcal{L}(g, Y_{n_w})$.

4.4.18 Definition (Deligne–Lusztig) Let χ be a complex irreducible character of G^F. Then χ is called a *unipotent character* if $\langle R_w, \chi \rangle_{G^F} \neq 0$ for some $w \in W$, where $\langle, \rangle_{G^F}$ denotes the usual hermitian scalar product on the space of complex-valued class functions on G^F. Thus, the unipotent characters of G^F form a certain subset of the set of irreducible characters of G^F. We remark that no elementary description of that subset is known!

4.4.19 Example Let $w = 1$. We have seen in Example 4.3.12 that X_1 is a finite set and that we have a G^F-equivariant bijection $X_1 \cong G^F/B^F$. So we can choose $\delta \geqslant 1$ such that F^δ acts as the identity on X_1. Then we have

$$|X_1^{gF^{\delta n}}| = |(G^F/B^F)^g| \qquad \text{for all } n \geqslant 1.$$

Arguing as in Example 4.4.13 yields that the limit of the power series in §4.4.17 is $|(G^F/B^F)^g|$. Thus, we have

$$\mathcal{L}(g, Y_{n_1}) = |(G^F/B^F)^g| \qquad \text{for all } g \in G^F.$$

In other words, this means that the virtual character R_1 is nothing but the character of the permutation repesentation of G^F on the cosets of B^F. This shows, for example, that the *unit character* $1_{G^F}\colon g \mapsto 1$ is always unipotent.

4.5 The virtual characters of Deligne and Lusztig

Let G be a connected affine algebraic group over $k = \overline{\mathbb{F}}_p$, and $F\colon G \to G$ be a generalized Frobenius map. Assume that there are closed F-stable subgroups B, N which form a reductive BN-pair in G.

In this section, we describe a generalization of the construction of the virtual characters R_w introduced in §4.4.17. This yields virtual characters $R_{T,\theta}$ for each F-stable maximal torus T in G and each irreducible character θ of T^F.

We assume that the reader is familiar with the character theory of finite groups; see, for example, Part I of Serre 1977. For any finite group Γ, we denote by $\mathrm{Irr}(\Gamma)$ the set of complex irreducible characters of Γ and by $\langle , \rangle_\Gamma$ the usual hermitian scalar product on the space of complex-valued class functions on Γ. A virtual character of Γ is an integral linear combination of $\mathrm{Irr}(\Gamma)$.

4.5.1 The virtual characters of Deligne and Lusztig, II Let $T \subseteq G$ be an F-stable maximal torus in G, and let $\theta \in \mathrm{Irr}(T^F)$. We wish to associate with the pair (T, θ) a virtual character of G^F. This is done as follows. Recall that $B = UH$ where $U = R_u(B)$ and $H = B \cap N$ is a maximal torus. Since T is conjugate to H, there exists some Borel subgroup $\tilde{B} \subseteq G$ containing T. Let us write $\tilde{B} = \tilde{U}T$, where $\tilde{U} = R_{\mathrm{u}}(\tilde{B})$. Note that, in general, there will be several choices of $\tilde{B}$ and none of these may be F-stable! As before, let $L \colon G \to G$, $g \mapsto g^{-1}F(g)$, be the Lang–Steinberg map. Then $L^{-1}(\tilde{U})$ is stable under left multiplication with elements of G^F and under right multiplication with elements of T^F. Indeed, let $g \in G^F$, $t \in T^F$, and $x \in L^{-1}(\tilde{U})$. Then we have

$$L(gxt) = (gxt)^{-1}F(gxt) = t^{-1}x^{-1}g^{-1}F(g)F(x)F(t)$$

$$= t^{-1}x^{-1}F(x)t \in \tilde{U},$$

since $x^{-1}F(x) \in \tilde{U}$ and $\tilde{U}$ is normalized by T. Thus, the map

$$(G^F \times T^F) \times L^{-1}(\tilde{U}) \to L^{-1}(\tilde{U}), \qquad ((g,t),x) \mapsto (g \cdot t) \cdot x := gxt^{-1},$$

defines an action of the finite group $G^F \times T^F$ on $L^{-1}(\tilde{U})$. We shall set

$$R_{T,\tilde{U},\theta}(g) := \frac{1}{|T^F|} \sum_{t \in T^F} \mathcal{L}((g,t), L^{-1}(\tilde{U}))\, \theta(t) \qquad \text{for any } g \in G^F,$$

where $\mathcal{L}((g,t), L^{-1}(\tilde{U}))$ is the *Lefschetz number*. Then $R_{T,\tilde{U},\theta}$ is a complex-valued function on G^F. (We will see later that it does not depend on $\tilde{U}$.)

4.5.2 Proposition $R_{T,\tilde{U},\theta}$ *is a virtual character of G^F. In particular, the values of $R_{T,\tilde{U},\theta}$ are algebraic integers.*

Proof We begin with the following general remarks. Consider a $(G^F \times T^F)$-module V. Let us fix some $\theta \in \mathrm{Irr}(T^F)$ and define $e :=$ $|T^F|^{-1} \sum_{t \in T^F} \theta(t^{-1})\, t$. Then e is an idempotent in the group algebra of T^F and we have a direct-sum decomposition $V = eV \oplus (1-e)V$.

Since the action of G^F commutes with that of T^F, the subspace eV is a G^F-module and, for any $g \in G^F$, we have

$$\mathrm{trace}(g, eV) = \mathrm{trace}((g, e), V) = \frac{1}{|T^F|} \sum_{t \in T^F} \mathrm{trace}((g, t), V)\, \theta(t^{-1}).$$

Consequently, if χ is a virtual character of $G^F \times T^F$, then the function

$$g \mapsto \frac{1}{|T^F|} \sum_{t \in T^F} \chi(g, t)\, \theta(t^{-1})$$

is a virtual character of G^F. Thus, the desired assertion follows by applying the above discussion to the virtual character given by $(g, t) \mapsto \mathcal{L}((g, t), L^{-1}(\tilde{U})) = \mathcal{L}((g^{-1}, t^{-1}), L^{-1}(\tilde{U}))$ of $G^F \times T^F$. $\qquad\square$

Our aim will be to establish at least some properties of the characters $R_{T,\tilde{U},\theta}$. The subsequent discussion will rely on the following two basic results from Deligne and Lusztig (1976) which we state here without proof.

4.5.3 Theorem (Deligne–Lusztig) *Let X be an affine variety over k and let $g\colon X \to X$ be an automorphism of finite order. Write $g = us = su$, where $s\colon X \to X$ is of order prime to p and $u\colon X \to X$ is of order a power of p. Then*

$$\mathcal{L}(g, X) = \mathcal{L}(u, X^s), \quad \text{where } X^s := \{x \in X \mid s(x) = x\}.$$

For the proof, see §3 of Deligne and Lusztig (1976) and also Theorem 6.3 of Srinivasan (1979). It does not seem to be possible to prove this statement by just using the elementary characterization of Lefschetz numbers in Theorem 4.4.12. Instead, one has to use the cohomological interpretation in §4.4.9.

The next result gives an orthogonality relation for the characters $R_{T,\tilde{U},\theta}$. Given F-stable maximal tori T, T' in G, we set

$$\mathrm{N}_G(T, T') := \{n \in G \mid n^{-1}Tn = T'\} \quad (= \mathrm{Tran}_G(T, T')).$$

Note that $\mathrm{N}_G(T, T')^F$ is empty unless T, T' are G^F-conjugate, in which case we have $|T^F| = |T'^F|$. Any $n \in \mathrm{N}_G(T, T')^F$ defines a bijection $\mathrm{Irr}(T^F) \to \mathrm{Irr}(T'^F)$, $\theta \mapsto {}^n\theta$, where ${}^n\theta(t') := \theta(nt'n^{-1})$ $(t' \in T'^F)$.

4.5.4 Theorem (Deligne–Lusztig)   *Using the above notation, we have*

$$\langle R_{T,\tilde{U},\theta}, R_{T',\tilde{U}',\theta'}\rangle_{G^F} = |\{n \in \mathrm{N}_G(T,T')^F \mid {}^n\theta = \theta'\}|/|T^F|,$$

where $\tilde{U}$ and $\tilde{U}'$ are the unipotent radicals of Borel subgroups containing T and T', respectively.

This is a deep result whose proof is based on a detailed study of the variety $L^{-1}(\tilde{U}) \times L^{-1}(\tilde{U}')$. The original proof of Deligne and Lusztig (1976) is considerably simplified in §2.3 of Lusztig (1977). A detailed and perfectly readable exposition of that proof can be found on pp. 214–19 of Carter (1985). The material that we have developed so far provides all the ingredients used in that proof. (For example, it uses Theorem 3.4.8.) The cohomological interpretation of Lefschetz numbers is only used via reference to Theorem 4.5.3.

4.5.5 Corollary $R_{T,\tilde{U},\theta}$ *is independent of $\tilde{U}$; it will henceforth be denoted by $R_{T,\theta}$. Furthermore, the following hold.*

(a) If T, T' are F-stable maximal tori, then $R_{T,\theta}$ and $R_{T',\theta'}$ are either equal or orthogonal to each other.

(b) If $\theta \in \mathrm{Irr}(T^F)$ is in general position, that is, $\{n \in \mathrm{N}_G(T)^F \mid {}^n\theta = \theta\} = T^F$, then $\pm R_{T,\theta} \in \mathrm{Irr}(G^F)$.

Proof   Let $\tilde{U}'$ be the unipotent radical of another Borel subgroup containing T. Then Theorem 4.5.4 shows that we have

$$\langle R_{T,\tilde{U},\theta}, R_{T,\tilde{U},\theta}\rangle_{G^F} = \langle R_{T,\tilde{U},\theta}, R_{T,\tilde{U}',\theta}\rangle_{G^F} = \langle R_{T,\tilde{U}',\theta}, R_{T,\tilde{U}',\theta}\rangle_{G^F}.$$

It follows that the scalar product of $R_{T,\tilde{U},\theta} - R_{T,\tilde{U}',\theta}$ with itself is zero and so $R_{T,\tilde{U},\theta} = R_{T,\tilde{U}',\theta}$. Thus, $R_{T,\tilde{U},\theta}$ is independent of $\tilde{U}$. The assertions in (a) and (b) are proved similarly. $\square$

Now we can show that the construction of $R_{T,\theta}$ indeed is a generalization of the construction of the virtual characters R_w in §4.4.17.

4.5.6 Proposition   *Let $T \subseteq G$ be an F-stable maximal torus, and assume that T is obtained from H by twisting with $w \in W$; see §4.3.7. Then we have $R_w = R_{T,1}$, where 1 stands for the unit character of T^F.*

Proof By §4.3.7, we have $T = xHx^{-1}$, where $x \in G$ is such that $x^{-1}F(x) = n_w$. Then we also have $T = F(x)HF(x)^{-1}$, and so $F(x)BF(x)^{-1}$ is a Borel subgroup containing T. Consequently, we have

$$R_{T,1}(g) = \frac{1}{|T^F|} \sum_{t \in T^F} \mathcal{L}((g,t), L^{-1}(F(x)UF(x)^{-1})).$$

We have seen in Lemma 4.3.14 that $L^{-1}(n_w U)$ is stable under right multiplication with elements of the finite group $H^{[w]} = x^{-1}T^F x$. Now the map

$$L^{-1}(F(x)UF(x)^{-1}) \to L^{-1}(n_w U), \quad y \mapsto yx,$$

is an isomorphism which commutes with the action of G^F (by left multiplication on both sides) and transforms the action of T^F on $L^{-1}(F(x)UF(x)^{-1})$ into the action of $H^{[w]}$ on $L^{-1}(n_w U)$. (Indeed, for $y \in G$, we have $L(y) \in F(x)UF(x)^{-1} = xn_w U n_w^{-1} x^{-1}$ if and only if $L(yx) = x^{-1}L(y)F(x) \in x^{-1}F(x)U = n_w U$.) Hence, by Example 4.4.14 and Exercise 4.7.10, we have

$$R_{T,1}(g) = \frac{1}{|H^{[w]}|} \sum_{t \in H^{[w]}} \mathcal{L}((g,t), L^{-1}(n_w U)) = \mathcal{L}(g, L^{-1}(n_w U)/H^{[w]}),$$

and this equals $\mathcal{L}(g, Y_{n_w})$, as required. $\qquad\square$

The orthogonality relations in Theorem 4.5.4 show that $\pm R_{T,\theta}$ is irreducible whenever θ is in general position. At the other extreme, the character $R_{T,1}$ is not irreducible, where 1 stands for the unit character of T^F. The following result gives at least a piece in the decomposition of $R_{T,1}$ into irreducible characters.

4.5.7 Lemma $\langle R_{T,\theta}, 1_{G^F} \rangle_{G^F}$ *equals* 1 *if* $\theta = 1$, *and is* 0 *otherwise.*

Proof By definition, the scalar product $\langle R_{T,\theta}, 1_{G^F} \rangle_{G^F}$ equals

$$\frac{1}{|G^F|} \sum_{g \in G^F} R_{T,\theta}(g) = \frac{1}{|G^F|} \frac{1}{|T^F|} \sum_{g \in G^F} \sum_{t \in T^F} \mathcal{L}((g,t), L^{-1}(\tilde{U}))\, \theta(t).$$

Now recall that G^F acts by left multiplication on $L^{-1}(\tilde{U})$. Hence, by Exercise 4.7.10, the right-hand side of the above expression equals

$$\frac{1}{|T^F|} \sum_{t \subset T^F} \mathcal{L}(t, L^{-1}(\tilde{U})/G^F)\, \theta(t).$$

Now note that the natural map $L^{-1}(\tilde{U})/G^F \to \tilde{U}, y \mapsto y^{-1}F(y)$, is a bijective morphism of affine varieties which commutes with the action of T^F by right multiplication on $L^{-1}(\tilde{U})/G^F$ and by conjugation on $\tilde{U}$. Hence we have $\mathcal{L}(t, L^{-1}(\tilde{U})/G^F) = \mathcal{L}(t, \tilde{U}) = 1$ by Examples 4.4.14 and 4.4.15. So we finally obtain $\langle R_{T,\theta}, 1_{G^F} \rangle_{G^F} = |T^F|^{-1} \sum_{t \in T^F} \theta(t)$, and this equals 1 if $\theta = 1$, and is 0 otherwise, as required. $\qquad\square$

We now address the question of determining the values of $R_{T,\theta}$. In general, this is an extremely difficult problem, which is not yet completely solved. (The remaining problems arise in groups over a field of characteristic 2 or 3, as far as the values on unipotent elements are concerned.) If $g \in G^F$ is semisimple, then the value $R_{T,\theta}(g)$ can be determined explicitly; see the formula in §2.13 of Lusztig (1977). We only treat here the following special case.

4.5.8 Proposition *Let $s \in G^F$ be a regular semisimple element; that is, $\mathrm{C}_G(s)$ is a maximal torus. Then we have*

$$R_{T,\theta}(s) = \sum_{w \in W(T)^F} \theta(\dot{w}s\dot{w}^{-1}), \quad where \ W(T)^F = \mathrm{N}_G(T)^F/T^F,$$

where $\dot{w}$ denotes a representative of w in $\mathrm{N}_G(T)^F$. In particular, $R_{T,\theta}(s) = 0$ if s is not G^F-conjugate to an element in T^F.

Proof Let $\tilde{U}$ be the unipotent radical of a Borel subgroup containing T. By Theorem 4.5.3, we have $\mathcal{L}((s,t), L^{-1}(\tilde{U})) = \mathcal{L}(1, L^{-1}(\tilde{U})^{(s,t)})$. We claim that

$$L^{-1}(\tilde{U})^{(s,t)} = \{g \in G^F \mid s = gtg^{-1}\}.$$

Indeed, let $g \in L^{-1}(\tilde{U})$ be fixed by (s,t). This means that $sgt^{-1} = g$, and so s, t are conjugate; in particular, t also is a regular semisimple element. Furthermore, since s, t are fixed by F, we have $g^{-1}F(g) \in \mathrm{C}_G(t) \cap \tilde{U} = \{1\}$. Thus, $g \in G^F$ and $s = gtg^{-1}$. Conversely, it is easily checked that all g with these properties lie in $L^{-1}(\tilde{U})^{(s,t)}$. Thus, the above claim is established. First of all, this shows that $L^{-1}(\tilde{U})^{(s,t)}$ is a finite set and so

$$\mathcal{L}((s,t), L^{-1}(\tilde{U})) = |L^{-1}(\tilde{U})^{(s,t)}| = |\{g \in G^F \mid s = gtg^{-1}\}|,$$

by Example 4.4.13. Furthermore, the cardinality of the set on the right-hand side equals $|T^F| = |\mathrm{C}_G(s)^F|$ if s, t are G^F-conjugate and 0 otherwise. So, by the defining formula for $R_{T,\theta}(s)$, we already see that $R_{T,\theta}(s) = 0$ if s is not G^F-conjugate to an element in T^F. Finally, assume that $s \in T^F$. Then we have

$$R_{T,\theta}(s) = \frac{1}{|T^F|} \sum_{t \in T^F} |\{g \in G^F \mid s = gtg^{-1}\}| \, \theta(t) = \sum_t \theta(t),$$

where the second sum runs over all $t \in T^F$ which are G^F-conjugate to s. Now let $t \in T^F$, and assume that $gtg^{-1} = s$ for some $g \in G^F$. Then g also conjugates the centralizer of t onto the centralizer of s. Since s is regular and semisimple, we conclude that $g \in N_G(T)^F$. Hence the G^F-conjugates of s in T^F are the elements $\dot{w}s\dot{w}^{-1}$, where $\dot{w}$ runs over a set of representatives of $N_G(T)/T^F$. $\qquad\square$

4.5.9 Proposition *Let $u \in G^F$ be unipotent. Then $R_{T,\theta}(u) \in \mathbb{Z}$ is independent of θ and will be denoted by $Q_T(u)$. We have the following orthogonality relations.*

$$\frac{1}{|G^F|} \sum_{\substack{g \in G^F \\ \text{unipotent}}} Q_T(g) \, Q_{T'}(g) = \frac{|N(T,T')^F|}{|T^F||T'^F|}.$$

Proof For any $t \in T^F$, we have $\mathcal{L}((u,t), L^{-1}(U)) = \mathcal{L}((u,1), L^{-1}(\tilde{U})^t)$ by Theorem 4.5.3. But $g \in L^{-1}(\tilde{U})$ can only be fixed by t if $t = 1$. Thus, the defining formula for $R_{T,\theta}$ shows that

$$R_{T,\theta}(u) = \frac{1}{|T^F|} \sum_{t \in T^F} \mathcal{L}((u,t), L^{-1}(\tilde{U})) \, \theta(t) = \frac{1}{|T^F|} \mathcal{L}((u,1), L^{-1}(\tilde{U}));$$

hence, this value does not depend on θ. The above formula also shows that $R_{T,\theta}(u) \in \mathbb{Q}$. Then the fact that $R_{T,\theta}(u)$ is an integer is a consequence of Proposition 4.5.2. Next we claim that

$$\sum_{\theta \in \mathrm{Irr}(T^F)} R_{T,\theta}(g) = \begin{cases} |T^F| \, Q_T(g) & \text{if } g \in G^F \text{ is unipotent,} \\ 0 & \text{otherwise.} \end{cases} \tag{$*$}$$

Indeed, using the defining formula, we see that the left-hand side equals

$$\frac{1}{|T^F|} \sum_{t \in T^F} \mathcal{L}((g,t), L^{-1}(\tilde{U})) \sum_{\theta \in \mathrm{Irr}(T^F)} \theta(t) = \mathcal{L}((g,1), L^{-1}(\tilde{U})),$$

where the equality results from the fact that $\sum_\theta \theta$ is the character of the regular representation of T^F. Now let us write $g = su = us$, where $s \in G^F$ is semisimple and $u \in G^F$ is unipotent. By Theorem 4.5.3, we have

$$\mathcal{L}((g,1), L^{-1}(\tilde{U})) = \mathcal{L}((u,1), L^{-1}(\tilde{U})^s).$$

Now note that an element in $L^{-1}(\tilde{U})$ can only be fixed by s if $s = 1$. Thus, we have $L^{-1}(\tilde{U})^s = \varnothing$ and $\sum_\theta R_{T,\theta}(g) = 0$ if $s \neq 1$. If $s = 1$, then g is unipotent, and the assertion is clear by the defining property of Q_T. Now the required orthogonality relations are seen to be a consequence of Theorem 4.5.4. Indeed, by $(*)$, the expression

$$\frac{1}{|G^F|} \sum_{\substack{g \in G^F \\ \text{unipotent}}} Q_T(g)\, Q_{T'}(g)$$

can be rewritten as follows, where we also use that $Q_{T'}(g) = Q_{T'}(g^{-1})$.

$$\frac{1}{|G^F||T^F||T'^F|} \sum_{\theta \in \mathrm{Irr}(T^F)} \sum_{\theta' \in \mathrm{Irr}(T'^F)} \sum_{g \in G^F} R_{T,\theta}(g)\, R_{T',\theta'}(g^{-1})$$

$$= \frac{1}{|T^F||T'^F|} \sum_{\theta \in \mathrm{Irr}(T^F)} \sum_{\theta' \in \mathrm{Irr}(T'^F)} \langle R_{T,\theta}, R_{T',\theta'} \rangle_{G^F}.$$

Using the formula in Theorem 4.5.4, the expression on the right-hand side is easily seen to be equal to $|N_G(T,T')^F||T^F|^{-1}|T'^F|^{-1}$, as required. $\qquad\square$

4.5.10 Definition Let $w \in W$ and T_w be an F-stable maximal torus obtained from H by twisting with w; see §4.3.7. Then the function

$$Q_w \colon \{g \in G^F \mid g \text{ unipotent}\} \to \mathbb{Z}, \quad g \mapsto R_{T_w,\theta}(g),$$

is called the *Green function*[2] associated with w. By Corollary 4.5.5(a) and §4.3.7, we have $Q_w = Q_{w'}$ if $w, w' \in W$ are F-conjugate. (We will see examples of Green functions in Example 4.5.12 below.)

Now there is also a general formula for the value of $R_{T,\theta}$ at the identity element of G^F. In fact, let T be an F-stable maximal torus

---

[2]These functions were first investigated, using completely different (and purely combinatorial) methods by Green (1955) in the case when $G^F = \mathrm{GL}_n(\mathbb{F}_q)$.

in G and assume that T is obtained by twisting H with w; see §4.3.7. Then we have

$$R_{T,\theta}(1) = (-1)^{l(w)}[G^F : T^F]_{p'},$$

where m_p, for any integer $m \geqslant 1$, denotes the largest power of p dividing m and $m_{p'} = m/m_p$. The proof relies on some properties of the *Steinberg character*, which can be constructed as in Exercise 4.7.11; see §2.9 of Lusztig (1977) or §7.5.1 of Carter (1985) for the details. With the methods that we have developed so far, we can establish this formula at least in the following special case. (This will be required in Section 4.6; see also Example 4.5.12 below.)

4.5.11 Lemma *Assume that W^F has order 2, and let T be an F-stable maximal torus which is not G^F-conjugate to H. Then the following hold.*

(a) We have $R_{H,1} = 1_G + \chi$ where $\chi \in \mathrm{Irr}(G^F)$ is such that $\chi(1) = |U^F|$, and $\chi(u) = 0$ for all unipotent elements $u \neq 1$ in G^F.

(b) We have $R_{T,\theta}(1) = -[G^F : T^F]_{p'}$ and $\langle R_{T,1}, \chi \rangle_{G^F} = -1$.

In the above situation, we set $\mathrm{St}_{G^F} := \chi$ and call this the *Steinberg character* of G^F; see Exercise 4.7.11 for a general construction of this character.

Proof (a) First note that we have $R_{H,1}(g) = |(G^F/B^F)^g)|$ for all $g \in G^F$, by Proposition 4.5.6 and Example 4.4.19. Furthermore, since $|W^F| = 2$, we have $\langle R_{H,1}, R_{H,1} \rangle_{G^F} = 2$ by Theorem 4.5.4. Since $R_{H,1}$ is the character of a transitive permutation representation, the trivial character of G^F occurs with multiplicity 1 in $R_{H,1}$. Thus, we must have

$$R_{H,1} = 1_{G^F} + \chi, \quad \text{where } \chi \in \mathrm{Irr}(G^F).$$

Now, the sharp form of the Bruhat decomposition reads $G^F = B^F \amalg B^F n_{w_0} U^F$ in our case. This yields $[G^F : B^F] = 1 + |U^F|$. Thus, we must have $\chi(1) = |U^F|$. It remains to show $\chi(u) = 0$ for all unipotent non-unit $u \in G^F$. This is seen as follows. By §4.2.7(b), we have $U^F \neq \{1\}$, and so U^F is a Sylow p-subgroup of G^F. Thus, we may assume that $u \in U^F$. We have already remarked that $R_{H,1}(u) = |(G^F/B^F)^u)|$. So let $g \in G^F$ be such that $ugB^F = gB^F$. Then $g^{-1}ug \in B^F$ and so $u \in gB^Fg^{-1}$. Now, if $g \notin B^F$, then we can write $g = bn_{w_0}u'$, where $b \in B^F$ and $u' \in U^F$. Thus, $gB^Fg^{-1} = bn_{w_0}B^Fn_{w_0}^{-1}b^{-1}$, and so $u = bn_{w_0}b'n_{w_0}^{-1}b^{-1}$ for some

$b' \in B^F$. This yields $b^{-1}ub = n_{w_0}b'n_{w_0}^{-1} \in U^F \cap n_{w_0}B^Fn_{w_0}^{-1} = \{1\}$ by Lemma 1.6.16, contrary to our assumption. Thus, we must have $g \in B^F$. So there is precisely one coset of B^F which is fixed by u (namely, B^F). This shows that $R_{H,1}(u) = 1$ and so $\chi(u) = 0$.

(b) By Proposition 4.5.9, the value $R_{T,\theta}(1)$ is independent of θ. So it will be enough to determine $R_{T,1}(1)$. Combining the fact that χ is zero on unipotent elements $\neq 1$ with the relation $(*)$ in the proof of Proposition 4.5.9, we see that

$$\sum_{\theta \in \mathrm{Irr}(T^F)} \langle R_{T,\theta}, \chi \rangle_{G^F} = \frac{|T^F|R_{T,1}(1)\chi(1)}{|G^F|} = R_{T,1}(1)[G^F : T^F]_{p'}^{-1}.$$

Then we obtain $\langle R_{T,\theta}, \chi \rangle_{G^F} = \langle R_{T,\theta}, R_{H,1} - 1_G \rangle_{G^F} = -\langle R_{T,\theta}, 1_T \rangle_{G^F}$ using Theorem 4.5.4. The result is -1 if $\theta = 1$ and 0 otherwise; see Lemma 4.5.7. This yields all the desired assertions. $\square$

4.5.12 Example Assume that W is irreducible and that W^F has order 2. Then the Green functions of G^F can be computed explicitly. By the classification of finite Coxeter groups (see Table 1.2), we have to consider the following cases:

(a) $W = \langle s_1 \rangle$ and F acts as the identity. This situation occurs for $G = \mathrm{GL}_2(k)$ with the standard Frobenius map; we have

$$G^F = \mathrm{GL}_2(\mathbb{F}_q) \quad \text{and} \quad |G^F| = q(q-1)^2(q+1).$$

The orders of the maximal tori are computed in Example 4.3.9. The unipotent classes are specified by their Jordan type, as in Section 2.6. The Green functions are given in Table 4.2. For the complete character table, see Chapter 15 of Digne and Michel (1991) or §I.5 of Fulton and Harris (1991).

(b) $W = \langle s_1, s_2 \rangle$, where $s_1 s_2$ has order 3 and $F(s_1) = s_2$. This situation occurs for $G = \mathrm{GL}_3(k)$, where F is given as in Example 4.1.10(c); we have

$$G^F = \mathrm{GU}_3(\mathbb{F}_q) \quad \text{and} \quad |G^F| = q^3(q+1)^3(q-1)(q^2-q+1).$$

The F-conjugacy classes are given by $\{1, s_1s_2, s_2s_1\}$, $\{s_1, s_2\}$, and $\{w_0 = s_1s_2s_1\}$. The orders of the maximal tori are computed as in Exercise 4.7.8. The three unipotent classes in G (specified by their Jordan type) are all F-stable, as is easily checked. The Green functions are given in Table 4.2. The complete character table can be found in Ennola's (1963) paper.

Table 4.2 Green functions for $\mathrm{GL}_2(\mathbb{F}_q)$, $\mathrm{GU}_3(\mathbb{F}_q)$, and $^2G_2(q^2)$

$\mathrm{GL}_2(\mathbb{F}_q)$	(1^2)	(2)
$|C_G(u)^F|$ $|G^F|$		$q(q-1)$
Q_1	$q+1$	1
Q_{s_1}	$-(q-1)$	1

$\mathrm{GU}_3(\mathbb{F}_q)$	(1^3)	$(2,1)$	(3)
$|C_G(u)^F|$ $|G^F|$		$q^3(q+1)^2$	$q^2(q+1)$
Q_1	q^3+1	1	1
Q_{w_0}	$-(q-1)(q^2-q+1)$	$-2q+1$	1
Q_{s_1}	$-(q+1)^2(q-1)$	$q+1$	1

$^2G_2(q^2)$	1	X	T	T^{-1}	Y	YT	YT^{-1}
$|C_G(u)^F|$ $|G^F|$		q^6	$2q^4$	$2q^4$	$3q^2$	$3q^2$	$3q^2$
Q_1	$(q^2+1)(q^4-q^2+1)$	1	1	1	1	1	1
$Q_{s_\alpha s_\beta s_\alpha}$	$-(q^2-1)(q^4-q^2+1)$	$-2q^2+1$	1	1	1	1	1
$Q_{s_\alpha w_0}$	$-(q^4-1)(q^2-\sqrt{3}q+1)$	$q^2-\sqrt{3}q+1$	$-\sqrt{3}q+1$	$-\sqrt{3}q+1$	1	1	1
Q_{s_α}	$-(q^4-1)(q^2+\sqrt{3}q+1)$	$q^2+\sqrt{3}q+1$	$\sqrt{3}q+1$	$\sqrt{3}q+1$	1	1	1

The notation for the elements in $^2G_2(q^2)$ is taken from Table III of Ward (1966).

(c) $W = \langle s_\alpha, s_\beta \rangle$, where $s_\alpha s_\beta$ has order 4 and $F(s_\alpha) = s_\beta$. This case corresponds to the finite *Suzuki groups* denoted $^2B_2(q^2)$ (where q^2 is an odd power of 2) and will be studied in detail in Section 4.6.

(d) $W = \langle s_\alpha, s_\beta \rangle$ where $s_\alpha s_\beta$ has order 6 and $F(s_\alpha) = s_\beta$. This case corresponds to the finite *Ree group* denoted $^2G_2(q^2)$, where q^2 is an odd power of 3; see Chapter 14 of Carter (1972). (Here, α corresponds to a short root and β corresponds to a long root in a root system of type G_2.) We have

$$|^2G_2(q^2)| = q^6(q^2 - 1)(q^2 + 1)(q^4 - q^2 + 1).$$

Furthermore, $1, s_\alpha s_\beta s_\alpha, s_\alpha w_0$, and s_α form a complete set of representatives for the F-conjugacy classes of W. The Green functions are given in Table 4.2. The complete character table can be found in the paper of Ward (1966).

The results in Table 4.2 are computed as follows. By Lemma 4.5.11, we know the values of all Green functions at the identity, and we know that $Q_1(u) = 1$ for $u \neq 1$; furthermore, all values must be integers by Proposition 4.5.9. Then the values that have not yet been determined are obtained by solving the system of equations arising from the orthogonality relations in Proposition 4.5.9. We leave the details as an exercise to the reader; similar computations will also be performed in the proof of Proposition 4.6.8 below. For the Ree groups, however, one needs one additional relation: By §9.16 of Deligne and Lusztig (1976), the values at Y, YT, and YT^{-1} must all be 1 (since these are regular unipotent elements). We omit further details.

4.6 An example: the characters of the Suzuki groups

Throughout this section, let k be an algebraic closure of $\mathbb{F}_2$ and

$$G = \mathrm{Sp}_4(k) = \{A \in M_4(k) \mid A^{\mathrm{tr}} JA = J\}, \quad \text{where} \quad J = \begin{bmatrix} 0 & 0 & 0 & 1 \\ 0 & 0 & 1 & 0 \\ 0 & 1 & 0 & 0 \\ 1 & 0 & 0 & 0 \end{bmatrix},$$

the 4-dimensional symplectic group over k. The Suzuki group is defined as the finite group G^F, where $F \colon G \to G$ is a certain endomorphism such that F^2 is a Frobenius map. The aim of this section is to study this group using the techniques developed so far. In particular, we will determine the conjugacy classes, the Green functions and the complex irreducible characters of G^F.

4.6.1 The BN-pair in G By Theorem 1.7.4, we know that G can be obtained as the fixed-point set of $\mathrm{GL}_4(k)$ under the automorphism $\varphi \colon A \to J(A^{\mathrm{tr}})^{-1} J$. Furthermore, G has a reductive BN-pair, where B and N are obtained by taking fixed points under φ of the corresponding subgroups in $\mathrm{GL}_4(k)$. In particular, writing $B = UH$, then U is the group of all upper unitriangular matrices which are contained in G and H is the group of all diagonal matrices contained in G. Explicitly, we obtain

$$H = \{h(t, u) \mid t, u \in k^{\times}\}, \quad \text{where} \quad h(t, u) := \begin{bmatrix} t & 0 & 0 & 0 \\ 0 & u & 0 & 0 \\ 0 & 0 & u^{-1} & 0 \\ 0 & 0 & 0 & t^{-1} \end{bmatrix}.$$

Furthermore, $N = H.W$ where the Weyl group W is generated by

$$s_\alpha := \begin{bmatrix} 0 & 1 & 0 & 0 \\ 1 & 0 & 0 & 0 \\ 0 & 0 & 0 & 1 \\ 0 & 0 & 1 & 0 \end{bmatrix} \quad \text{and} \quad s_\beta := \begin{bmatrix} 1 & 0 & 0 & 0 \\ 0 & 0 & 1 & 0 \\ 0 & 1 & 0 & 0 \\ 0 & 0 & 0 & 1 \end{bmatrix}.$$

Since $s_\alpha s_\beta$ has order 4, we see that $W = \langle s_\alpha, s_\beta \rangle$ is a dihedral group of order 8:

$$W = \{1, s_\alpha, s_\beta, s_\alpha s_\beta, s_\beta s_\alpha, s_\alpha s_\beta s_\alpha, s_\beta s_\alpha s_\beta, w_0\}$$

where $w_0 = s_\alpha s_\beta s_\alpha s_\beta = s_\beta s_\alpha s_\beta s_\alpha$ is the unique longest element. We now define some elements in G (see also Lemma 1.5.9):

$$x_\alpha(t) = \begin{bmatrix} 1 & t & 0 & 0 \\ 0 & 1 & 0 & 0 \\ 0 & 0 & 1 & t \\ 0 & 0 & 0 & 1 \end{bmatrix}, \qquad x_\beta(t) = \begin{bmatrix} 1 & 0 & 0 & 0 \\ 0 & 1 & t & 0 \\ 0 & 0 & 1 & 0 \\ 0 & 0 & 0 & 1 \end{bmatrix},$$

$$x_{\alpha+\beta}(t) = \begin{bmatrix} 1 & 0 & t & 0 \\ 0 & 1 & 0 & t \\ 0 & 0 & 1 & 0 \\ 0 & 0 & 0 & 1 \end{bmatrix}, \qquad x_{2\alpha+\beta}(t) = \begin{bmatrix} 1 & 0 & 0 & t \\ 0 & 1 & 0 & 0 \\ 0 & 0 & 1 & 0 \\ 0 & 0 & 0 & 1 \end{bmatrix},$$

where $t \in k$. Let $x_{-r}(t) = x_r(t)^{\mathrm{tr}}$ for $r \in \{\alpha, \beta, \alpha+\beta, 2\alpha+\beta\}$. Finally, we set[3]

$$\Phi = \{\pm\alpha, \pm\beta, \pm(\alpha + \beta), \pm(2\alpha + \beta)\}.$$

The labelling is chosen such that $U_{s_\alpha} = \{x_\alpha(t) \mid t \in k\}$ and $U_{s_\beta} = \{x_\beta(t) \mid t \in k\}$ (notation of Theorem 1.6.19); the subgroup U consists of the elements

$$x_\alpha(t_1)x_\beta(t_2)x_{\alpha+\beta}(t_3)x_{2\alpha+\beta}(t_4) = \begin{bmatrix} 1 & t_1 & t_3+t_1t_2 & t_4+t_1t_3 \\ 0 & 1 & t_2 & t_3 \\ 0 & 0 & 1 & t_1 \\ 0 & 0 & 0 & 1 \end{bmatrix},$$

where $t_i \in k$. We also note that, if we define

$$n_\alpha(t) := x_\alpha(t)x_{-\alpha}(t^{-1})x_\alpha(t) \quad \text{and} \quad n_\beta(t) := x_\beta(t)x_{-\beta}(t^{-1})x_\beta(t)$$

for $t \in k^\times$, then $s_\alpha = n_\alpha(1)$, $s_\beta = n_\beta(1)$, and $h(t, u) = n_\alpha(t)n_\alpha(1)n_\beta(tu)n_\beta(1)$. In particular, this shows that $G = \langle x_r(t) \mid r \in \Phi, t \in k \rangle$.

---

[3]Here, Φ is just an abstract indexing set for the elements $x_r(t)$. In Lie notation, Φ may be regarded as a root system of type $B_2 = C_2$, where α, β are fundamental roots; furthermore, $\pm\alpha, \pm(\alpha + \beta)$ are so-called *short roots* and $\pm\beta, \pm(2\alpha + \beta)$ are *long roots*.

Now, by Theorem 3.3.6, there is a unique bijective homomorphism of algebraic groups $\theta \colon G \to G$ such that

$$\theta(x_\alpha(t)) = x_\beta(t^2), \qquad\qquad \theta(x_{-\alpha}(t)) = x_{-\beta}(t^2),$$
$$\theta(x_\beta(t)) = x_\alpha(t), \qquad\qquad \theta(x_{-\beta}(t)) = x_{-\alpha}(t),$$
$$\theta(x_{\alpha+\beta}(t)) = x_{2\alpha+\beta}(t^2), \qquad \theta(x_{-\alpha-\beta}(t)) = x_{-2\alpha-\beta}(t^2),$$
$$\theta(x_{2\alpha+\beta}(t)) = x_{\alpha+\beta}(t), \qquad \theta(x_{-2\alpha-\beta}(t)) = x_{-\alpha-\beta}(t).$$

It is readily checked that $\theta \circ F_2 = F_2 \circ \theta$ and $\theta^2 = F_2$, where $F_2 \colon G \to G$ is the standard Frobenius map which raises every matrix entry to its square.

4.6.2 Definition Let $r = 2^f$ $(f \geqslant 0)$ and consider the standard Frobenius map $F_r \colon G \to G$ (where $F_1 = \mathrm{id}$ by convention). Then θ commutes with F_r and, setting $F := \theta \circ F_r$, we have

$$F^2 = \theta^2 \circ F_r^2 = F_2 \circ F_{r^2} = F_{2r^2} = F_{2^{2f+1}}.$$

Thus, F is a *generalized Frobenius map* on G. The finite group G^F is called a *Suzuki group* and denoted by ${}^2B_2(q^2)$, where $q = \sqrt{2}r$. (See also the bibliographic remarks at the end of this chapter for further historical comments.)

Note that the case $r = 1$ is explicitly allowed here; then we have $q^2 = 2$ and so $|{}^2B_2(2)| = 20$, by the order formula in the following result.

4.6.3 Proposition *The groups B and N are F-stable, and B^F, N^F form a split BN-pair in G^F. We have $B^F = U^F H^F$ where*

$$U^F = \{x_\alpha(t^r)x_\beta(t)x_{\alpha+\beta}(t^{r+1} + u^r)x_{2\alpha+\beta}(u) \mid t, u \in \mathbb{F}_{q^2}\},$$

$$H^F = \{h(t^{2r+1}, t) \mid t \in \mathbb{F}_{q^2}^\times\} \cong \mathbb{Z}/(q^2 - 1)\mathbb{Z},$$

$$W^F = \{1, w_0\}, \quad \text{where } w_0 = s_\alpha s_\beta s_\alpha s_\beta.$$

We have $G^F = B^F \amalg B^F w_0 U^F$ with uniqueness of expressions, and so

$$|G^F| = q^4(q^2 - 1)(q^4 + 1).$$

Note that the number $q = \sqrt{2}r$ indeed is the positive real number q_0 in §4.2.7(b), since $U = U_{w_0}$ and $l(w_0) = 4$.

Proof It is readily checked that B and N are F-stable. So we are in the setting of §4.2.1. Furthermore, one easily checks that

Table 4.3 Unipotent classes of $\mathrm{Sp}_4(\overline{\mathbb{F}}_2)$

Jordan type		representatives $u \in G^F$	$\dim \mathrm{C}_G(u)$	$A(u)$
C_1	(1^4)	id	10	$\{1\}$
C_2	$(2,1^2)$	$x_\beta(1)$	6	$\{1\}$
C_3	$(2,2)$	$x_\alpha(1)$	6	$\{1\}$
C_4	$(2,2)$	$x_{\alpha+\beta}(1)x_{2\alpha+\beta}(1)$	4	$\{1\}$
C_5	(4)	$\left\{\begin{array}{c} x_\alpha(1)x_\beta(1)x_{\alpha+\beta}(1) \\ (x_\alpha(1)x_\beta(1)x_{\alpha+\beta}(1))^{-1} \end{array}\right\}$	2	$\mathbb{Z}/2\mathbb{Z}$

$W^F = \{1, w_0\}$. The above descriptions of U^F and H^F are obtained by an explicit computation, using the defining equations for θ. This yields that $|U^F| = q^4$ and $|H^F| = q^2 - 1$. In particular, this shows that the conditions in Proposition 1.7.2 are trivially satisfied for $U_{w_0}^F \neq \{1\}$. Consequently, the groups B^F and N^F do form a split BN-pair in G^F, and we have $G^F = B^F \amalg B^F w_0 U^F$ with uniqueness of expressions. This also yields the formula for $|G^F|$. $\qquad\square$

We shall now determine the F-stable conjugacy classes of G.

4.6.4 Proposition *There are 5 unipotent classes in G, denoted by $C_i(1 \leqslant i \leqslant 5)$; see Table 4.3. The classes C_1, C_4, and C_5 are F-stable, while $F(C_2) = C_3$.*

(a) The set C_4^F is a single conjugacy class in G^F, of size $|G^F|/q^4$ and with representative as specified in Table 4.3.

(b) The set C_5^F splits into two conjugacy classes in G^F, of size $|G^F|/2q^2$ each and with representatives as specified in Table 4.3.

Thus, the finite group G^F has 4 unipotent classes.

Proof First note that any unipotent element of G is contained in a Borel subgroup and, consequently, in U; see Propositions 3.4.9 and 3.5.4. We shall now determine the intersections of the unipotent classes of G with U.

First, let C_5 be the conjugacy class of $x_\alpha(1)x_\beta(1)$ in G. We claim that

$$C_5 \cap U = \{x_\alpha(t_1)x_\beta(t_2)x_{\alpha+\beta}(t_3)x_{2\alpha+\beta}(t_4) \mid t_1 \neq 0, t_2 \neq 0\}.$$

To see this, we begin by noting that $x_\alpha(1)x_\beta(1) - $ id has rank 3 and so the Jordan normal form of $x_\alpha(1)x_\beta(1)$ just consists of one

Jordan block of size 4. Hence, if $x \in C_5 \cap U$, then $x - \mathrm{id}$ also must have rank 3. Considering the matrix of an element $x = x_\alpha(t_1)x_\beta(t_2)x_{\alpha+\beta}(t_3)x_{2\alpha+\beta}(t_4)$ as in §4.6.1, we see that we must have $t_1 \neq 0$ and $t_2 \neq 0$. Conversely, assume that we have an element x as above, where $t_1 \neq 0$ and $t_2 \neq 0$. We must show that x is conjugate to $x_\alpha(1)x_\beta(1)$. To see this, let $u \in k^\times$ be such that $u^2 = t_2^{-1}$, and set $t := ut_1^{-1} \in k^\times$. Then we have

$$x' := h(t, u)xh(t, u)^{-1} = x_\alpha(1)x_\beta(1)x_{\alpha+\beta}(t_3')x_{2\alpha+\beta}(t_4')$$

for some $t_3', t_4' \in k$. Now let $r, s \in k$. A straightforward computation shows that

$$x_\alpha(r)^{-1}x_\beta(s)^{-1}x'x_\beta(s)x_\alpha(r)$$
$$= x_\alpha(1)x_\beta(1)x_{\alpha+\beta}(r+s+t_3')x_{2\alpha+\beta}(r^2+s+t_4').$$

Thus, choosing r such that $r^2+r+t_3'+t_4' = 0$, and taking $s := t_3'+r$, we see that x is conjugate to $x_\alpha(1)x_\beta(1)$, as claimed.

Now we can also see that the set $C_5 \cap U$ is invariant under F; furthermore, one immediately checks that $u_5 := x_\alpha(1)x_\beta(1)x_{\alpha+\beta}(1) \in C_5^F$. Now consider the centralizer of u_5. The Jordan normal form of this element consists of just one Jordan block of size 4. Hence, by Lemma 2.6.1, the centralizer of that element in $\mathrm{GL}_4(k)$ is connected of dimension 4; in fact that centralizer is contained in the Borel subgroup of $\mathrm{GL}_4(k)$. Consequently, we also have

$$\dim \mathrm{C}_G(u_5) \subseteq B.$$

Now a straightforward computation shows that

$$\mathrm{C}_G(u_5) = \{x_\alpha(\varepsilon)x_\beta(\varepsilon)x_{\alpha+\beta}(t_3)x_{2\alpha+\beta}(t_4) \mid \varepsilon \in \{0, 1\}, t_3, t_4 \in k\}.$$

Thus, we have $\dim \mathrm{C}_G(u_5) = 2$ and $A(u_5) \cong \mathbb{Z}/2\mathbb{Z}$, as claimed; more precisely, $A(u_5)$ is generated by the image of u_5 in $A(u_5)$. By Example 4.3.6, the set C_5^F splits into two conjugacy classes in G^F. To find a second representative, we just have to write u_5 in the form $g^{-1}F(g)$ (since the image of u_5 generates $A(u_5)$). Let $g := x_\alpha(1)x_{\alpha+\beta}(t)x_{2\alpha+\beta}(t')$ where $t, t' \in k$ are such that $t^2 + t + 1 = 0$ and $t' = t$ (if f is odd) or $t' = t + 1$ (if f is even). Then we have

$$g^{-1}F(g) = x_\alpha(1)x_{\alpha+\beta}(t)x_{2\alpha+\beta}(t')x_\beta(1)x_{\alpha+\beta}(t'^{2^f})x_{2\alpha+\beta}(t^{2^{f+1}})$$
$$= x_\alpha(1)x_\beta(1)x_{\alpha+\beta}(t + t'^{2^f})x_{2\alpha+\beta}(t' + t^{2^{f+1}}) = u_5$$

as required. Then $u_5' := g u_5 g^{-1}$ is the required second representative; see Theorem 4.3.5. We compute that

$$g u_5 g^{-1} = x_\alpha(1) u_5 x_\alpha(1)^{-1} = x_\alpha(1) x_\beta(1) x_{2\alpha+\beta}(1) = u_5^{-1},$$

as claimed. Thus, all the assertions concerning C_5 are established.

Next, let C_2 be the conjugacy class of $x_\beta(1)$ in G. By a similar computation as above, one shows that

$$C_2 \cap U = \{ x_\beta(t_2) x_{\alpha+\beta}(t_3) x_{2\alpha+\beta}(t_4) \mid t_3^2 = t_2 t_4, (t_2, t_3, t_4) \neq (0,0,0) \}.$$

Note that all elements $u \in C_2$ have Jordan type $(2, 1^2)$, which means that $u - \mathrm{id}$ has rank 1. Furthermore, a straightforward computation yields that any element in C_2 has a connected centralizer of dimension 6, as claimed.

Since $F(x_\beta(1)) = x_\alpha(1)$ does not have Jordan type $(2, 1^2)$, we conclude that $x_\alpha(1)$ is not conjugate to $x_\beta(1)$. Hence, denoting by C_3 the conjugacy class of $x_\alpha(1)$, we have $F(C_2) = C_3 \neq C_2$. A similar computation as above shows that

$$C_3 \cap U = \{ x_\alpha(t_1) x_{\alpha+\beta}(t_3) x_{2\alpha+\beta}(t_4) \mid t_1 t_3 = t_4, (t_1, t_3, t_4) \neq (0,0,0) \}.$$

The description of the above sets shows that $U \setminus (C_1 \cup C_2 \cup C_3 \cup C_5)$ is not empty; for example, the element $u_4 := x_{\alpha+\beta}(1) x_{2\alpha+\beta}(1) \in G^F$ lies in that set. Let C_4 be the conjugacy class of u_4; then C_4 is F-stable. Computing the centralizer of u_4, we find that $\mathbf{C}_G(u_4) = U$. Hence we have $A(u_4) = \{1\}$ and $\dim \mathbf{C}_G(u_4) = 4$. Finally, one shows by a direct computation that any element in $U \setminus (C_1 \cup C_2 \cup C_3 \cup C_5)$ is conjugate to u_4.

It remains to determine the cardinalities of the centralizers of the above elements in G^F. We have $\mathbf{C}_G(u_4) = U$ and so $|\mathbf{C}_G(u_4)^F| = |U^F| = q^4$, using the description of U^F in Proposition 4.6.3. That description also shows that

$$x_\alpha(\varepsilon) x_\beta(\varepsilon) x_{\alpha+\beta}(t_3) x_{2\alpha+\beta}(t_4) \in \mathbf{C}_G(u_5)^F$$

if and only if $t_3 = \varepsilon + t_4^r$ and $t_4 \in \mathbb{F}_{q^2}$. Thus, $|\mathbf{C}_G(u_5)^F| = 2q^2$, as claimed. $\qquad\qquad\square$

4.6.5 Proposition *Let T be an F-stable maximal torus in G. Then T^F is a cyclic group, and we have $\mathbf{C}_G(g) = T$ for every non-unit $g \in T^F$. The structure of T_w^F for w in the various F-conjugacy classes of W is specified in Table 4.4.*

Table 4.4 F-stable maximal tori in $\mathrm{Sp}_4(\overline{\mathbb{F}}_2)$

F-conj. class	$\{1, s_\alpha s_\beta, s_\beta s_\alpha, w_0\}$	$\{s_\alpha s_\beta s_\alpha, s_\beta s_\alpha s_\beta\}$	$\{s_\alpha, s_\beta\}$
$|T_w^F|$	q^2-1	$q^2+\sqrt{2}q+1$	$q^2-\sqrt{2}q+1$
Generator	π_0	π_1	π_2
$|W(T_w)^F|$	2	4	4
$W(T_w)^F$-orbit	$\pi_0^{\pm 1}$	$\pi_1^{\pm 1}, \pi_1^{\pm q^2}$	$\pi_2^{\pm 1}, \pi_2^{\pm q^2}$

[Here, the last row contains the $W(T_w)^F$-orbit of a generator of T_w^F.]

Proof Let T be an F-stable maximal torus and $g \in T^F$. Let C be the G- conjugacy class of g. Then $C \cap H$ is F-stable and a single orbit under the action of W on H; see Exercise 3.6.12. Now the action of F on H is given by

$$F(h(t,u)) = h(t^r u^r, t^r u^{-r}) \qquad \text{for any } h(t,u) \in H.$$

The following table describes the action of $w \in W$ on $h(t,u) \in H$:

1	s_α	s_β	$s_\alpha s_\beta$	$s_\beta s_\alpha$	$s_\alpha s_\beta s_\alpha$	$s_\beta s_\alpha s_\beta$	w_0
(t,u)	(u,t)	(t,u^{-1})	(u^{-1},t)	(u,t^{-1})	(t^{-1},u)	(u^{-1},t^{-1})	(t^{-1},u^{-1})

Thus, if $t = 1$ or $u = 1$, then the W-orbit of $h(t,u)$ has length 4 and consists of $(t,1), (1,t), (1,t^{-1}), (t^{-1},1)$. Assume, for example, that the orbit of $h(t,1)$ is F-stable, where $t \in k^\times$. Then $F(h(t,1)) = h(t^r, t^r)$ must be contained in that orbit. By inspection of the above list, we see that this is only possible for $t = 1$. Thus, there are no non-identity F-stable orbits of this type. Similarly, if $u = t$ or $u = t^{-1}$, then the orbit has length 4 and consists of $(t,t), (t,t^{-1}), (t^{-1},t), (t^{-1},t^{-1})$. Arguing as above, we see that there are no non-identity F-stable orbits of this type. Thus, only W-orbits of length 8 have a chance to be F-stable. But, if the orbit of $h(t,u)$ has length 8, then $t \neq 1, u \neq 1, t \neq u, t \neq u^{-1}$. Such an element is represented by a diagonal matrix with distinct entries on the diagonal and so the centralizer of that element is just H. This already proves that $C_G(g) = T$ for every non-unit $g \in T^F$, as claimed.

Next we determine the structure of $T_w^F \cong H^{[w]}$ for any $w \in W$, where $H^{[w]}$ is defined as in §4.3.7. The F-conjugacy classes of W are easily determined:

$$\{1, s_\alpha s_\beta, s_\beta s_\alpha, w_0\}, \{s_\alpha, s_\beta\}, \{s_\alpha s_\beta s_\alpha, s_\beta s_\alpha s_\beta\}.$$

First let $w = 1$. Then we have $F(h(t, u)) = h(t, u)$ for some $t, u \in k^\times$ if and only if $t^r u^r = t$ and $t^r u^{-r} = u$. Thus, we see that

$$H^{[1]} = \{h(u^{2r+1}, u) \mid u^{2r^2-1} = 1\} \cong \mathbb{Z}/(q^2 - 1)\mathbb{Z}.$$

Next, let $w = s_\alpha s_\beta s_\alpha$. Then $F(h(t, u)) = wh(t, u)w^{-1}$ if and only if $t^r u^r = t^{-1}$ and $t^r u^{-r} = u$. Thus, we see that

$$H^{[s_\alpha s_\beta s_\alpha]} = \{h(u^{2r+1}, u) \mid u^{2r^2+2r+1} = 1\} \cong \mathbb{Z}/(q^2 + \sqrt{2}q + 1)\mathbb{Z}.$$

Finally, if $w = s_\alpha$, then $F(h(t, u)) = s_\alpha h(t, u)s_\alpha$ holds if and only if $t^r u^r = u$ and $t^r u^{-r} = t$. This yields

$$H^{[s_\alpha]} = \{h(u^{2r-1}, u) \mid u^{2r^2-2r+1} = 1\} \cong \mathbb{Z}/(q^2 - \sqrt{2}q + 1)\mathbb{Z}.$$

In all cases, we see that T_w^F is a cyclic group. Finally, a direct computation also yields the F-centralizers as defined in §4.3.7:

$$C_{W,F}(1) = \{1, w_0\} \quad \text{and} \quad C_{W,F}(s_\alpha s_\beta s_\alpha) = C_{W,F}(s_\alpha)$$
$$= \{1, s_\alpha s_\beta, s_\beta s_\alpha, w_0\}.$$

Applying an element of $C_{W,F}(w)$ to $h(t, u) \in H^{[w]}$, we obtain the last row in Table 4.4. For example, let $w = s_\alpha$ and $\pi_2 = h(u^{2r-1}, u)$ be a generator of $H^{[w]}$, for a suitable $u \in k^\times$ such that $u^{2r^2-2r+1} = 1$. Then $s_\alpha s_\beta \pi_2 s_\beta s_\alpha = h(u^{-1}, u^{2r-1}) = \pi_2^{-q^2}$ since $u^{2r-1} = u^{-2r^2}$; the images of π_2 under the remaining elements of $C_{W,F}(w)$ are computed similarly. $\qquad\square$

4.6.6 Corollary   *The group G^F has $q^2 + 3$ conjugacy classes, four of which consist of unipotent elements and q^2 of which consist of semisimple elements. If a non-unit $s \in G^F$ is semisimple, then $C_G(s)$ is an F-stable maximal torus in G.*

Proof   The conjugacy classes of unipotent elements are determined in Proposition 4.6.4. Now let the non-unit $g \in G^F$ be semisimple. Then, by Proposition 4.3.8, g is contained in some F-stable maximal torus $T \subseteq G$. Since $C_G(g) = T$ by Proposition 4.6.5, we conclude that T is the unique maximal torus such that $g \in T$. Consequently, g is G^F-conjugate to an element in exactly one of the three tori in Table 4.4. A similar argument also shows that $g, g' \in T^F$ are conjugate in G^F if and only if g, g' are conjugate in $N_G(T)^F$. Indeed, if $g' = xgx^{-1}$ for some $x \in G^F$, then we also have $xC_G(g)x^{-1} = C_G(g')$, and so $x \in N_G(T)^F$ since $C_G(g) = C_G(g') = T$. Thus, we obtain a complete list of representatives of the G-conjugacy classes of semisimple elements in G^F by taking complete lists of representatives of the $N_G(T_w)^F/T_w^F$-classes on T_w^F for $w \in \{1, s_\alpha s_\beta s_\alpha, s_\alpha\}$. Using the information in the last row of Table 4.4, we obtain $(q^2 - 2)/2 + (q^2 + \sqrt{2}q)/4 + (q^2 - \sqrt{2}q)/4 = q^2 - 1$ non-trivial semisimple classes.

Finally, note that every element in G^F is unipotent or semisimple. Indeed, let $g \in G^F$ and write $g = su = us$, where $s \in G^F$ is semisimple and $u \in G^F$ is unipotent. Assume that $s \neq 1$. As we have seen, $C_G(s)$ is a maximal torus, and so we necessarily have $u = 1$, as required. $\qquad\qquad\square$

4.6.7 Proposition *Every irreducible character of G^F occurs with non-zero multiplicity in some virtual character $R_{T,\theta}$. We have*

$$
\begin{aligned}
R_{T_1,1} &= 1_{G^F} & & & & + & \mathrm{St}_{G^F}, \\
R_{T_{s_\alpha s_\beta s_\alpha},1} &= 1_{G^F} &+ \mathcal{W} &+ \overline{\mathcal{W}} &- & & \mathrm{St}_{G^F}, \\
R_{T_{s_\alpha},1} &= 1_{G^F} &- \mathcal{W} &- \overline{\mathcal{W}} &- & & \mathrm{St}_{G^F}.
\end{aligned}
$$

where $\mathcal{W}$ is an irreducible character of degree $\frac{1}{\sqrt{2}}q(q^2-1)$ and the bar denotes complex conjugation. Furthermore, if $\theta \neq 1$, then $\pm R_{T,\theta} \in \mathrm{Irr}(G^F)$.

Proof Let $w \in \{1, s_\alpha, s_\alpha s_\beta s_\alpha\}$. Then one immediately checks that every non-trivial irreducible character of T_w^F is in general position. Thus, we have $\pm R_{T_w,\theta} \in \mathrm{Irr}(G^F)$ whenever $\theta \neq 1$; see Theorem 4.5.4. Taking into account the action of $N_G(T_w)^F/T_w^F$, we obtain $(q^2 - 2)/2$ irreducible characters arising from T_1^F, $(q^2 + \sqrt{2}q)/4$ irreducible characters arising from $T_{s_\alpha s_\beta s_\alpha}^F$, and $(q^2 - \sqrt{2}q)/4$ irreducible characters arising from $T_{s_\alpha}^F$. Thus, we have already obtained

$$
\frac{1}{2}(q^2 - 2) + \frac{1}{4}(q^2 + \sqrt{2}q) + \frac{1}{4}(q^2 - \sqrt{2}q) = q^2 - 1
$$

irreducible characters of G^F. So, by Corollary 4.6.6, it remains to find 4 more irreducible characters. Now, we certainly have the unit character 1_{G^F} and the Steinberg character St_{G^F} which, in our case, is given by the relation $R_{T_1,1} = 1_{G^F} + \mathrm{St}_{G^F}$; see Lemma 4.5.11.

So it remains to find two irreducible characters of G^F. Now, all the irreducible characters listed above have the same value on ρ and ρ^{-1} (where ρ is a representative in C_5, as in Table 4.3.) Hence one of the missing characters must have different values on ρ and ρ^{-1}. In fact, denoting such a character by $\mathcal{W}$, we must have $\mathcal{W}(\rho^{-1}) = \overline{\mathcal{W}}(\rho)$, where the bar denotes complex conjugation. Thus, $\overline{\mathcal{W}} \neq \mathcal{W}$ is the second missing character.

We now consider $R_{T_{s_\alpha},1}$. By Theorem 4.5.4, this character has norm 4 and it is orthogonal to all $\pm R_{T,\theta}$ with θ in general position. Thus, the only characters of G^F that might possibly occur in $R_{T_{s_\alpha},1}$ are 1_{G^F}, St_{G^F}, $\mathcal{W}$, and $\overline{\mathcal{W}}$. Furthermore, note that $\mathcal{W}$ and $\overline{\mathcal{W}}$ must occur with the same multiplicity since $R_{T_{s_\alpha},1}$ is rational-valued; see §4.5.1. Since 1_{G^F} occurs with multiplicity 1 and St_{G^F} occurs with multiplicity -1 (see Lemmas 4.5.7 and 4.5.11), we conclude that

$$R_{T_{s_\alpha},1} = 1_{G^F} \pm (\mathcal{W} + \overline{\mathcal{W}}) - \mathrm{St}_{G^F}.$$

Evaluating at 1 yields $-(q^2-1)(q^2+\sqrt{2}q+1) = 1 \pm 2\mathcal{W}(1) - q^4$, and so $\mathcal{W}(1) = \mp\frac{1}{\sqrt{2}}q(q^2-1)$. Since $\mathcal{W}(1) > 0$, the sign must be $+1$. By a similar argument, we also find the decomposition of $R_{T_{s_\alpha s_\beta s_\alpha},1}$. $\quad\square$

4.6.8 Proposition *The Green functions of G^F are given as follows.*

	1		σ	ρ	ρ^{-1}
Q_1	q^4+1		1	1	1
$Q_{s_\alpha s_\beta s_\alpha}$	$-(q^2-1)(q^2-\sqrt{2}q+1)$		$-\sqrt{2}q+1$	1	1
Q_{s_α}	$-(q^2-1)(q^2+\sqrt{2}q+1)$		$\sqrt{2}q+1$	1	1

where $\sigma := x_{\alpha+\beta}(1)x_{2\alpha+\beta}(1)$ and $\rho := x_\alpha(1)x_\beta(1)x_{\alpha+\beta}(1)$ as in Table 4.3.

Proof Firstly, we know the values of the Green functions at 1 by Lemma 4.5.11; this also yields $Q_1(u) = 1$ for $u \neq 1$. Secondly, we also know that all the values at ρ and ρ^{-1} must be equal. Thus, we already have the following table of values:

	1		σ	ρ	ρ^{-1}
Q_1	q^4+1		1	1	1
$Q_{s_\alpha s_\beta s_\alpha}$	$-(q^2-1)(q^2-\sqrt{2}q+1)$		a	b	b
Q_{s_α}	$-(q^2-1)(q^2+\sqrt{2}q+1)$		c	d	d

where a, b, c, d are unknown integers. We now evaluate the orthogonality relations in Proposition 4.5.9 to obtain further conditions on a, b, c, d. Evaluating the orthogonality relation between Q_1 and

Table 4.5 The character table of G^F

χ	1	σ	ρ	ρ^{-1}	π_0^a	π_1^b	π_2^c
1_{G^F}	1	1	1	1	1	1	1
$\mathcal{W}$	$r(q^2-1)$	$-r$	ir	$-ir$	0	1	-1
$\overline{\mathcal{W}}$	$r(q^2-1)$	$-r$	$-ir$	ir	0	1	-1
St_{G^F}	q^4	0	0	0	1	-1	-1
$\mathcal{X}_l$	q^4+1	1	1	1	$\varepsilon_0^l(\pi^a)$	0	0
$\mathcal{Y}_m$	$(q^2-1)(q^2-2r+1)$	$2r-1$	-1	-1	0	$\varepsilon_1^m(\pi^b)$	0
$\mathcal{Z}_n$	$(q^2-1)(q^2+2r+1)$	$-2r-1$	-1	-1	0	0	$\varepsilon_2^n(\pi_2^c)$

See Theorem 4.6.9 for notation.

$Q_{s_\alpha s_\beta s_\alpha}$, we obtain the condition $a = q^2 - \sqrt{2}q + 1 - q^2 b$. Evaluating the orthogonality relation of $Q_{s_\alpha s_\beta s_\alpha}$ with itself, we obtain a quadratic relation for b which has two solutions:

$$b = 1 \quad \text{or} \quad b = \frac{q^2 - 2\sqrt{2}q + 1}{q^2 + 1}.$$

The second solution is not an integer. Hence we must have $b = 1$ and so $a = -\sqrt{2}q + 1$. The orthogonality relation between Q_1 and Q_{s_α} now yields $c = -q^2 d + q^2 + \sqrt{2}q + 1$, while the orthogonality relation between $Q_{s_\alpha s_\beta s_\alpha}$ and Q_{s_α} yields $d = 1$ and so $c = \sqrt{2}q + 1$, as required. $\square$

4.6.9 Theorem (Suzuki) *Recall that $q = \sqrt{2}r$, where $r = 2^f$ ($f \geqslant 0$). Let $i = \sqrt{-1}$ and $\varepsilon_0, \varepsilon_1, \varepsilon_2 \in \mathbb{C}$ be primitive roots of unity of order $2r^2 - 1$, $2r^2 + 2r + 1$ and $2r^2 - 2r + 1$, respectively. Then the character table of G^F is given as in Table 4.5.*

In the table, $1_{G^F}, \mathrm{St}_{G^F}, \mathcal{W}$, and $\overline{\mathcal{W}}$ are the unipotent characters of G^F, and

$$\varepsilon_0^l(\pi_0^a) = \varepsilon_0^{al} + \varepsilon_0^{-al},$$

$$\varepsilon_1^m(\pi_1^b) = \varepsilon_1^{bm} + \varepsilon_1^{bmq^2} + \varepsilon_1^{-bm} + \varepsilon_1^{-bmq^2},$$

$$\varepsilon_2^n(\pi_2^c) = \varepsilon_2^{cn} + \varepsilon_2^{cnq^2} + \varepsilon_2^{-cn} + \varepsilon_2^{-cnq^2}.$$

Here, the exponents a, b, c have to be chosen such that $\pi_0^a, \pi_1^b, \pi_2^c$ run over a complete set of representatives of the semisimple classes in G^F; furthermore, the exponent l runs over the same range as

a, m runs over the same range as b, and n runs over the same range as c.

(Note that the relation between q and r is not the same as that of Suzuki (1962).)

Proof By Corollary 4.6.6, every element in G^F is either unipotent or regular and semisimple. Hence we obtain all values of $R_{T,\theta}$ by Proposition 4.6.8 and Proposition 4.5.8, where we use the information in the last row of Table 4.4 to evaluate the sum in Proposition 4.5.8. In particular, this yields the values of $\mathcal{X}_l, \mathcal{Y}_m$, and $\mathcal{Z}_n$, which are characters of the form $\pm R_{T_w,\theta}$ with θ in general position, for $w = 1$, $w = s_\alpha s_\beta s_\alpha$, and $w = s_\alpha$, respectively. It also yields the values of St_{G^F}. Now, from the decomposition of $R_{T_{s_\alpha},1}$ in Proposition 4.6.7, we obtain the values of $\mathcal{W} + \overline{\mathcal{W}}$. Consequently, we know the values of $\mathcal{W}$ on all elements $g \in G^F$ such that g is conjugate to g^{-1}. Thus, the only unknown values of $\mathcal{W}$ are those on ρ and ρ^{-1}. Let us write $\mathcal{W}(\rho) = x + y\mathrm{i}$, where $x, y \in \mathbb{R}$. Now, using the relation $\sum_\chi \chi(\rho) = 0$ (sum over all $\chi \in \mathrm{Irr}(G^F)$), we obtain $2x = (\mathcal{W} + \overline{\mathcal{W}})(\rho) = 0$ and so $x = 0$. Thus, we have $\mathcal{W}(\rho) = y\mathrm{i}$ and so $\mathcal{W}(\rho^{-1}) = -y\mathrm{i}$. Now the orthogonality relation $\langle \mathcal{W}, \mathcal{W} \rangle_{G^F} = 1$ will yield a quadratic equation for y; inserting all the known information, we obtain $y^2 = r^2$ and so $y = \pm r$, as required. $\square$

4.6.10 Remark The characters $\mathcal{W}, \overline{\mathcal{W}}$ are so-called *cuspidal unipotent* characters. In the paper of Lusztig (1976b), they are denoted by $^2B_2[\pm\mathrm{i}]$ and constructed explicitly as subspaces of certain cohomology spaces $H^i_{\mathrm{c}}(X_{s_\alpha}, \overline{\mathbb{Q}}_\ell)$.

4.7 Bibliographic remarks and exercises

The discussion of Frobenius maps follows Chapter II of Srinivasan (1979). The proof of the Lang–Steinberg theorem is taken from Steinberg (1977). In fact, Steinberg proves a more general result: If G is a connected algebraic group and ϕ is any endomorphism of G such that G^ϕ is finite, then the map $x \mapsto x^{-1}\phi(x)(x \in G)$ is surjective. The original theorem of S. Lang states that, if G is a connected affine algebraic group defined over $\mathbb{F}_q$, with corresponding Frobenius map F, then the map $G \to G, g \mapsto g^{-1}F(g)$, is surjective. The original proof of Lang can be found in §4.4.17 of Springer (1998) and §V.16 of Borel (1991). For a further discussion of the finite Chevalley groups as well as the twisted groups of type $^2E_6, ^3D_4, ^2G_2$, and $^2F_4$, see Carter (1972) and Steinberg (1967).

Rosenlicht's Theorem 4.2.4 is just a special case of more general results concerning fields of definition of algebraic groups; see §V.15 of Borel (1991) or §14.2.7 of Springer (1998). The proof of Proposition 4.2.3 is an adaptation of the argument in §3 of Lusztig (1988). For a completely different proof of Proposition 4.2.3(c), see §16.8 of Borel (1991). The discussion of applications of the Lang–Steinberg theorem (in the general setting of §4.3.1) follows Chapter I, §2 of Springer and Steinberg (1970). The computation in Example 4.3.9 can be formalized and this leads to a general formula for the order of T^F, where T is an F-stable maximal torus; see §2.10 of Srinivasan (1979) or §3.3 of Carter (1985).

The conjectures on the zeta function of an algebraic variety appeared in Weil's (1949) paper. Dwork's Theorem 4.4.7 was one of the first steps towards the proof of these conjectures. The deepest parts of these conjectures, which are concerned with the eigenvalues of F^*, have been proved by Deligne; for more details, historical remarks and further references, see Appendix C of Hartshorne (1977). Theorem 4.4.12 and the examples following it can be found in Part 1 of Lusztig (1977). The argument proving that $\mathcal{L}(g, X) \in \mathbb{Z}$ is mentioned on p. 455 of Hartshorne (1977). The original argument in Proposition 3.3 of Deligne and Lusztig (1976) is different. Zeta functions of Deligne–Lusztig varieties are studied in detail by Digne and Michel (1985); see also the survey article by Curtis (1988).

The construction of the Deligne–Lusztig characters has turned out to be the key to the problem of determining the character tables of the finite groups of Lie type. First of all, a detailed study of the multiplicities by which the irreducible characters of G^F occur in the various $R_{T,\theta}$ leads to a classification of $\mathrm{Irr}(G^F)$. Thus, it is possible to attach a 'natural' label to each $\chi \in \mathrm{Irr}(G^F)$ and to compute $\chi(1)$ in a purely combinatorial fashion from that label; see Main Theorem 4.23 of Lusztig (1984) and Lusztig (1988). Furthermore, Lusztig has developed a theory of so-called *character sheaves* which eventually leads to an algorithm for computing the character table of G^F; see Lusztig (1992), Shoji (1995), and the references there. For further recent developments, see the survey articles of Carter and Geck (1998), Lusztig (1991), and Lusztig (1999). The known character tables of finite groups of Lie type (and computer programs for working with them) are available through the CHEVIE system Geck *et al.* (1996).

Lusztig (1976a) uses a generalization of the construction of $R_{T,\theta}$ to show that the number of unipotent classes in a connected reductive algebraic over $\overline{\mathbb{F}}_p$ is always finite. The variety of all unipotent elements in G still poses a number of problems. For example, the values of the Green functions are not yet completely known for $p = 2$. See also Geck and Malle (1999) for a conjecture on the number of $\mathbb{F}_q$-rational points in certain subsets of the *unipotent variety* of G, which were called 'special pieces' by Lusztig.

Let us add some comments on the Suzuki groups in Section 4.6. These groups were discovered by Suzuki (1962), as an exceptional case in a classification problem. A finite group is called a *(ZT)-group* if it has the following properties.

(a) The group acts doubly transitive on a set X where $|X|$ is odd;

(b) the identity is the only element which fixes three distinct points of X;

(c) there are no normal subgroups of index $|X|$.

The (ZT)-groups were classified by Suzuki (1962). That classification led to the discovery of a family of finite simple groups whose members have order $q^4(q^2 - 1)(q^4 + 1)$, where q^2 is any odd power of 2; see Theorem 8 of Suzuki (1962). These groups can be identified with the groups G^F in Definition 4.6.2.

Indeed, G^F has the correct order by Proposition 4.6.3; furthermore, let us consider the natural action of G^F on $X = G^F/B^F$. The character of the corresponding permutation representation is given by $R_{T_1,1} = 1_{G^F} + \mathrm{St}_{G^F}$. Then we see that $R_{T_1,1}(1) = q^4 + 1$ is odd and that $R_{T_1,1}(g) \leqslant 2$ for any non-unit $g \in G^F$. Thus, (a) and (b) are satisfied. Furthermore, G^F is simple if $r \geqslant 2$ (since the kernel of any irreducible character is trivial). Thus, (c) also holds and so G^F is a (ZT)-group. If $r = 1$, then $|G^F| = 20$ and condition (c) is also easily checked.

The character table in Theorem 4.6.9 has been determined by Suzuki (1962) using completely different methods (since the Deligne–Lusztig theory was not yet available at that time).

4.7.1 Exercise Let (X, A) be an affine variety over $k = \mathbb{F}_p$. Assume that $A_0 \subseteq A$ is a finitely generated $\mathbb{F}_q$-subalgebra (where q is a power of p) such that the natural map $k \otimes_{\mathbb{F}_q} A_0 \to A$ given by multiplication is an isomorphism. Show that there exists a unique

Frobenius map $F \colon X \to X$ such that

$$A_0 = \{f \in A \mid F^*(f) = f^q\}.$$

[*Hint.* Identify $A = k \otimes_{\mathbb{F}_q} A_0$ and show that there is a unique k-linear map $\gamma \colon A \to A$ such that $\gamma(\xi \otimes f) = \xi \otimes f^q$ for all $\xi \in k$ and $f \in A_0$. Then check that γ satisfies the conditions in Definition 4.1.1.]

4.7.2 Exercise   Consider the standard Frobenius map $F_q \colon k \to k$, where $k = \overline{\mathbb{F}}_p$ and q is some power of p. Let $V \subseteq k^n$ be an F_q-stable closed subset.

(a) Assume that $\mathbf{I}(V) = (S)$, where $S \subseteq k[X_1, \ldots, X_n]$ is a finite set. Choose some $m \geqslant 1$ such that all $f \in S$ have their coefficients in the field $\mathbb{F}_{q^m}$. Let $\{\xi_0, \ldots, \xi_{m-1}\}$ be an $\mathbb{F}_q$-basis of $\mathbb{F}_{q^m}$ and set

$$\tilde{f}_i := \sum_{j=0}^{m-1} \xi_i^{q^j} \sigma^j(f) \qquad \text{for } f \in S \text{ and } 0 \leqslant i \leqslant m - 1.$$

Then show that $\tilde{f}_i \in \mathbb{F}_q[X_1, \ldots, X_n]$ and $V = \mathbf{V}(\{\tilde{f}_i \mid f \in S, 0 \leqslant i \leqslant m - 1\}$.

(b) Assume that V is a linear subspace. Then show that V has a basis consisting of vectors which have all their coordinates in $\mathbb{F}_q$.

[*Hint.* (a) See the proof of Lemma 4.1.3 and the implication '(b)$\Rightarrow$ (c)' in the proof of Corollary 4.1.5. (b) Use Exercise 1.8.6 and apply (a).]

4.7.3 Exercise   Let $F \colon X \to X$ be a Frobenius map for an $\mathbb{F}_q$-rational structure on X. Then show the following statements.

(a) For any $m \geqslant 1, F^m$ is the Frobenius map for an $\mathbb{F}_{q^m}$-structure on X.

(b) For each $x \in X$, there exists some $m \geqslant 1$ such that $F^m(x) = x$. In particular, the set $\{x, F(x), F^2(x), \ldots\} \subseteq X$ is finite.

(c) Let $F' \colon X \to X$ be another Frobenius map for an $\mathbb{F}_{q'}$-rational structure on X. Then there exist some $n, n' \geqslant 1$ such that $F^{n'} = (F')^n$.

[*Hint.* (a) This is a straightforward verification. (b) Use Proposition 4.1.4. (c) Write $A[X] = k[f_1, \ldots, f_r]$, where $f_i \in A[X]$. Then one can find some $m \geqslant 1$ such that $(F^*)^m(f_i) = f_i^{q^m}$ and

$(F'^*)^m(f_i) = f_i^{(q')^m}$ for all i. Now q and q' are powers of p; let us write $q = p^s$ and $q' = p^{s'}$. Then show that $F^{ms'} = (F')^{ms}$.]

4.7.4 Exercise Let (X, A) be an affine variety defined over $\mathbb{F}_q$, with corresponding Frobenius map F.

(a) Let $g\colon X \to X$ be a morphism such that $g^e = \mathrm{id}_X$, for some $e \geqslant 1$, and $g \circ F = F \circ g$. Show that the morphism $F_g = F \circ g$ is also a Frobenius map for an $\mathbb{F}_q$-rational structure on X.

(b) Let $n \geqslant 2$ and consider the affine variety $X^n := X \times \cdots \times X$ (n factors). Then show that the map $\tilde{F}\colon X^n \to X^n$ defined by

$$(x_1, \ldots, x_n) \mapsto (F(x_n), F(x_1), \ldots, F(x_{n-1})),$$

is a Frobenius map with respect to $\mathbb{F}_{q^n}$ and that the projection onto the first factor defines a bijection $(X^n)^{\tilde{F}} \xrightarrow{\sim} X^{F^n}$.

[*Hint.* (a) We set $g' := F^* \circ g^*$. Then g' is a k-algebra homomorphism and, hence, there exists a well-defined morphism $F'\colon X \to X$ such that $(F')^* = g'$. Now check that the two conditions in Definition 4.1.1 are satisfied for g.]

4.7.5 Exercise Let G be a connected affine algebraic group and $F\colon G \to G$ be a generalized Frobenius map such that F^d (where $d \geqslant 1$) is the Frobenius map for some $\mathbb{F}_q$-rational structure on G. Fix an element $x \in G$. Show that the map

$$F'\colon G \to G, \qquad g \mapsto xF(g)x^{-1},$$

also is a generalized Frobenius map such that $(F')^d$ is the Frobenius map for some $\mathbb{F}_q$-rational structure on G. Show that $G^{F'} \cong G^F$.

[*Hint.* The following argument can be found in §10.9 of Steinberg (1968). Write $x = y^{-1}F(y)$ and let c_y be the automorphism of G given by conjugation with y. Then check that $F' = c_y^{-1} \circ F \circ c_y$. This yields all the required conditions; the isomorphism $G^{F'} \cong G^F$ is given by conjugation with y.]

4.7.6 Exercise Let G be an affine algebraic group and $F\colon G \to G$ be a generalized Frobenius map on G. Let $H \subseteq G$ be a closed *connected* subgroup of G such that $F(H) \subseteq H$.

(a) Show that, if $F(Hg) \subseteq Hg \, (g \in G)$, then $F(hg) = hg$ for some $h \in H$.

(b) Show that the natural map $\pi\colon G \to G/H$ induces a bijection $G^F/H^F \cong (G/H)^F$. Here, we also denote by F the induced map on G/H. Show that $G^F/H^F \cong (G/H)^F$ is an isomorphism of abstract groups if H is a normal subgroup.

[*Hint.* (a) Let H act on the coset Hg by left multiplication and check that the conditions in §4.3.1 are satisfied. (b) By restriction, π gives rise to a map $G^F \to (G/H)^F$ whose fibres are the cosets of H^F.]

4.7.7 Exercise   Let G be a connected affine algebraic group over $\mathbb{F}_p$. Let Z be the centre of G and assume that Z is finite of order prime to p. The purpose of this exercise is to show that G can always be embedded as a closed subgroup into a connected affine algebraic group with a connected centre. (This construction can be found in §1.21 of Deligne and Lusztig (1976).)

(a) Show that there exists a maximal torus T in G such that $Z \subseteq T$. Let $\tilde{Z} := \{(z, z^{-1}) \mid z \in Z\} \subseteq G \times T$ and show that this is closed normal subgroup of $G \times T$.

Now let us set $\tilde{G} := (G \times T)/\tilde{Z}$ and let $\tilde{T}$ be the image of $\{1\} \times T$ in $\tilde{G}$.

(b) Show that $\tilde{T}$ is the centre of $\tilde{G}$ and that $\tilde{G}/\tilde{T} \cong G/Z$ (as abstract groups). Thus, by §2.5.12, $\tilde{G}$ is a connected affine algebraic group whose centre is connected.

(c) Show that the map $\iota\colon G \to \tilde{G}$, $g \mapsto (g, 1)\tilde{Z}$, is an isomorphism of G onto a closed normal subgroup and that $\iota([G, G]) = [\tilde{G}, \tilde{G}]$.

[*Hint.* (a) Let B be a Borel subgroup of G and write $B = UT$ as in Theorem 3.5.6. Let $z \in Z$. Then z lies in some Borel subgroup and, hence, $z \in B$, since all Borel subgroups are conjugate. By a similar argument, $z \in T$. (c) Use the differential criterion in Example 2.3.16; further note that $\tilde{G} = \iota(G).\tilde{T}$.]

4.7.8 Exercise   Let $G = \mathrm{GL}_n(k)$ and consider the Frobenius map F' described in Example 4.2.6, such that $G^{F'} = \mathrm{GU}_n(\mathbb{F}_q)$ is the general unitary group. Consider the standard BN-pair in G, where B is the group of all upper triangular matrices and N is the group of all monomial matrices in G; see Example 1.6.10. The Weyl group of G will be identified with $\mathfrak{S}_n$; let $w_0 \in \mathfrak{S}_n$ be defined as in Exercise 1.8.26.

(a) Show that B and N are F'-stable. Show that F' induces φ on $\mathfrak{S}_n$, where φ is the automorphism given by conjugation with w_0. Furthermore, show that the map $w \mapsto ww_0$ induces a bijection between the F'-conjugacy classes and the usual conjugacy classes of $\mathfrak{S}_n$.

(b) Let $\lambda = (\lambda_1 \geqslant \lambda_2 \geqslant \cdots \geqslant \lambda_r > 0)$ be a partition of n and $w_\lambda \in W$ be an element of cycle type λ. Show that

$$|T_{w_\lambda w_0}^{F'}| = \prod_{i=1}^{r}(q^{\lambda_i} - (-1)^{\lambda_i}).$$

(c) Show that $|G^{F'}| = \prod_{i=0}^{n-1}(q^n - (-1)^{n-i}q^i)$.

The above order formulas are formally obtained by changing q to $-q$ in the corresponding formulas for $\mathrm{GL}_n(\mathbb{F}_q)$ (and adjusting the total sign); a conceptual explanation for this phenomenon is given by §11.19 of Steinberg (1968).

[*Hint.* (b) Check that $T_{w_\lambda w_0}^{F'} \cong \{t \in T_0 \mid F(t) = n_w^{-1}t^{-1}n_w\}$, where F is the standard Frobenius map on G; then argue as in Example 4.3.9. (c) We have $W^{F'} = \mathfrak{S}_n^{\bar\varphi}$. Now evaluate $\sum_{w \in W^{F'}} q^{l(w)}$ using §10.5.1 of Geck and Pfeiffer (2000). Distinguish the cases where n is even or odd.]

4.7.9 Exercise   Let X be an affine variety defined over $\mathbb{F}_q$, with corresponding Frobenius map $F\colon X \to X$. Let H be a finite group of automorphisms of X such that F commutes with every element of H.

(a) Let X/H be the *affine quotient* of X by H and $\pi\colon X \to X/H$ be the natural map; see Theorem 2.5.10. Show that there is a unique Frobenius map $\bar{F}\colon X/H \to X/H$ such that $\bar{F} \circ \pi = \pi \circ F$.

(b) By Exercise 4.7.4(a), the composition $gF = Fg(g \in H)$ is a Frobenius map for an $\mathbb{F}_q$-rational structure on X. Show that $|(X/H)^{\bar{F}}| = \frac{1}{|H|}\sum_{g \in H}|X^{gF}|$.

[*Hint.* (a) The fact that all elements of H commute with F immediately implies that $\pi \circ F$ is constant on the orbits of H on X. Hence, by the universal property in §2.5.8, there exists a unique morphism $\bar{F}\colon X/H \to X/H$ such that $\bar{F} \circ \pi = \pi \circ F$. It remains to check that $\bar{F}$ is a Frobenius map. Recall from Theorem 2.5.10 that $\pi\colon X \to X/H$ is a finite morphism and that $A[X/H] = \{\bar{f}\colon X/H \to k \mid \bar{f} \circ \pi \in A[X]\}$. Now check the two conditions in Definition 4.1.1.

(b) Consider the set $M := \{(x, g) \in X \times H \mid F(g.x) = x\}$ and compute the cardinality of M in two different ways. Firstly, $|M|$ is the sum of all cardinalities $|X^{gF}|$ where g runs over the elements of H. Secondly, for a given $x \in X$, we have $\{g \in H \mid F(g.x) = x\} \neq \varnothing$ if and only if the H-orbit O_x of x is F-stable. Thus, denoting by $x_1, \ldots, x_n \in X$ representatives for the F-stable H-orbits on H, we have $n = |(X/H)^F|$ and $|M| = \sum_{i=1}^n |O_{x_i}| |\{g \in H \mid F(g.x_i) = x_i\}|$. But we also have $|\{g \in H \mid F(g.x) = x\}| = |\text{Stab}_H(x)|$ for any $x \in X$, and so the above expression equals $n|H|$, as required.]

4.7.10 Exercise   Let X be an affine variety and H be a finite group of automorphisms of X. Let $g\colon X \to X$ be an automorphism of finite order which commutes with every element of H. Show that g induces an automorphism $\bar{g}$ on the affine quotient X/H and that

$$\mathcal{L}(\bar{g}, X/H) = \frac{1}{|H|} \sum_{h \in H} \mathcal{L}(gh, X).$$

[*Hint.* Choose a Frobenius map $F\colon X \to X$ which commutes with g and all elements of H. Then use Theorem 4.4.12 and Exercise 4.7.9.]

4.7.11 Exercise   Let G be a finite group with a split BN-pair, where U is a finite p-group and H is an abelian group of order prime to p. The purpose of this exercise is to construct the Steinberg representation of G, following Steinberg (1957). Let $K[G]$ be the group algebra of G over a field K. We define

$$e := \sum_{b \in B} \sum_{w \in W} (-1)^{l(w)} b\, n_w \in K[G].$$

Let I be the right ideal in $K[G]$ generated by e. Then we have a representation $\rho\colon G \to \text{End}_K(I)$, where G acts on I by right multiplication.

(a) Write $B = UH$. Then show that the set $\{eu \mid u \in U\}$ is a vector space basis for I; in particular, we have $\dim_K I = |U|$. Show that the restriction of ρ to U is the regular representation of U.

(b) Assume that the characteristic of K is either 0 or prime to $[G : B]$. Then show that ρ is an irreducible representation of G.

(c) Assume that $K = \mathbb{C}$. Then show that ρ occurs with multiplicity 1 in the permutation representation of G on the cosets of B.

The character of ρ will be denoted by St_G and called the *Steinberg character* of G. We have $\mathrm{St}_G(1) = |U|$ and $\mathrm{St}_G(u) = 0$ for all non-unit $u \in U$.

[*Hint.* (a), (b) This follows by tricky (but elementary) computations using the axioms of a split BN-pair. See Steinberg (1957) and also Steinberg's own comments Steinberg (1997): pp. 580–6. To prove (c), one can argue as follows. By Frobenius reciprocity, the required multiplicity is given by $\frac{1}{|B|} \sum_{b \in B} \mathrm{St}_G(b)$. Now let $b \in B$ and write $b = vh$ with $v \in U, h \in H$. Then $eh = e$ and so $eub = eh^{-1}uvh$, where $h^{-1}uvh \in U$. Thus, b permutes the elements in the basis of I given in (a). So $\mathrm{St}_G(b)$ is the number of basis elements fixed by b. This already shows that the above sum is positive. On the other hand, by (a), St_G occurs with multiplicity 1 in the permutation representation of G on G/U.]

Bibliography

Atiyah, M. F., and Macdonald, I.G. (1969), *Introduction to commutative algebra*. Addison–Wesley Publishing Company.

Aschbacher, M. (2000), *Finite group theory* (2nd edn). Cambridge Studies in Advanced Mathematics, vol. 10. Cambridge University Press.

Borel, A. (1991), *Linear algebraic groups* (2nd edn, enlarged). Graduate Texts in Mathematics vol. 126. Springer-Verlag, Berlin–Heidelberg–New York.

Bourbaki, N. (1968), *Groupes et algèbres de Lie*, Chapters 4, 5, and 6. Hermann, Paris.

Carter, R. W. (1972), *Simple groups of Lie type* Wiley, New York. Reprinted 1989 as Wiley Classics Library Edition.

Carter, R. W. (1985), *Finite groups of Lie type: conjugacy classes and complex characters*. Wiley, New York. Reprinted 1993 as Wiley Classics Library Edition.

Carter, R. W., and Geck, M. (1998) ed., *Representations of reductive groups*. Publications of the Newton Institute, Cambridge University Press.

Chevalley, C. (1946), *Theory of Lie groups*. Princeton University Press (fifteeth printing, 1999).

Chevalley, C. (1956–8), *Classification des groupes de Lie algébriques*. Séminaire École Normale Supérieure, Mimeographed Notes, Paris.

Cox, D., Little, J., and O'Shea, D. (1992), *Ideals, varieties, and algorithms: an introduction to computational algebraic geometry and commutative algebra* (2nd edn). Undergraduate Texts in Mathematics. Springer Verlag, New York.

Curtis, C. W. (1998), Representations of Hecke algebras. *Astérisque* **168**, 13–60.

Deligne, P. (1977), *Séminaire de Géométrie Algébrique du Bois-Marie SGA* $4\frac{1}{2}$. *Cohomologie étale*. In collaboration with J. F. Boutot, A. Grothendieck, L. Illusie, and J. L. Verdier.

Lecture Notes in Mathematics, vol. 569. Springer-Verlag, Berlin–New York.

Deligne, P., and Lusztig, G. (1976), Representations of reductive groups over finite fields. *Annals of Mathematics* **103**, 103–61.

Dieudonné, J. (1971), *La géométrie des groupes classiques* (3rd edn). Springer-Verlag, Berlin–Heidelberg–New York.

Dieudonné, J. (1974), *Cours de géométrie algébrique* (vol. 2). Précis de Géométrie Algébrique Élémentaire. Presses Universitaires de France.

Digne, F., and Michel, J. (1985), *Fonctions $\mathcal{L}$ des variétés de Deligne–Lusztig et descente de Shintani*. Mémoires de la Société Mathématique de France, vol. 20.

Digne, F., and Michel, J. (1991), *Representations of finite groups of Lie type*. London Mathematical Society Students Texts, vol. 21. Cambridge University Press.

Dwork, B. (1960), On the rationality of the zeta function of an algebraic variety. *American Journal of Mathematics* **82**, 631–48.

Dye, R. H. (1977), A geometric characterization of the special orthogonal groups and the Dickson invariant. *Journal of the London Mathematical Society* **15**, 472–6.

Ennola, V. (1963), On the characters of the finite unitary groups. *Annales Academiae Scientarium Fennicae* **323**, 120–55.

Fischer, G. (1994), *Ebene algebraische Kurven*. Vieweg–Verlag.

Fogarty, J. (1969), *Invariant theory*. Benjamin, New York.

Fulton, W. (1969), *Algebraic curves: an introduction to algebraic geometry*, Addison-Wesley.

Fulton, W., and Harris, J. (1991), *Representation theory, a first course*. Graduate Texts in Mathematics, vol. 129. Springer-Verlag, Berlin–Heidelberg–New York.

Geck, M., Hiss, G, Lübeck, F., Malle G., and Pfeiffer, G. (1996) CHEVIE—A system for computing and processing generic character tables for finite groups of Lie type, Weyl groups and Hecke algebras, *Applicable Algebra in Engineering Communication, and Computing* **7**, 175–210. See also the homepage at `http://www.math.rwth-aachen.de/~CHEVIE`

Geck, M., and Malle, G. (1999), On special pieces in the unipotent variety. *Experimental Mathematics* **8**, 281–90.

Geck, M., and Pfeiffer, G. (2000), *Characters of finite Coxeter groups and Iwahori–Hecke algebras*. London Mathematical Society

Monographs, New Series, vol. 21. Oxford University Press, New York.

Goodman, R., and Wallach, N. R. (1998), *Representations and invariants of the classical groups*. Encyclopedia of Mathematics and its Applications vol. 68. Cambridge University Press.

Gorenstein, D., Lyons, R., and Solomon, R. (1994), *The classification of the finite simple groups*. Mathematical Surveys and Monographs, vol. 40, no. 1. American Mathematical Society, Providence, RI.

Green, J. A. (1955), The characters of the finite general linear groups, *Transactions of the American Mathematical Society* **80**, 402–47.

Greuel, G. M., and Pfister, G. (2002), *A singular introduction to commutative algebra*, with contributions by O. Bachmann, C. Lossen, H. Schönemann. Spinger-Verlag, Berlin-New York.

Grove, L. C. (1982), Dickson's pseudodeterminant without matrices. *Proceedings of the American Mathematical Society* **86**, 695–6.

Grove, L. C. (2002), *Classical groups and geometric algebra*. Graduate Studies in Mathematics, vol. 39. American Mathematical Society, Providence, RI.

Hartshorne, R. (1977), *Algebraic geometry*. Graduate Texts in Mathematics, vol. 25. Springer-Verlag, Berlin–Heidelberg–New York.

Humphreys, J. E. (1972), *Introduction to Lie algebras and representation theory*. Graduate Texts in Mathematics, vol. 9. Springer-Verlag, Berlin–Heidelberg–New York.

Humphreys, J. E. (1991), *Linear algebraic groups* (2nd edn). Graduate Texts in Mathematics, vol. 21. Springer-Verlag, Berlin–Heidelberg–New York.

Huppert, B. (1983), *Endliche Gruppen I*. Grundlehren der mathematischen Wissenschaften, vol. 134. Springer-Verlag, Berlin–Heidelberg–New York.

Koblitz, N. (1984), *p-adic numbers, p-adic analysis, and zeta-functions* (2nd edn), Graduate Texts in Mathematics, vol. 58. Springer-Verlag, Berlin–Heidelberg–New York.

Lang, S. (1984), *Algebra* (2nd edition), Addison–Wesley Publishing Company.

Lang, S. and Weil, A. (1954), Number of points of varieties in finite fields. *American Journal of Mathematics* **76**, 819–27.

Lusztig, G. (1976a), On the finiteness of the number of unipotent classes. *Inventiones Mathematicae* **34**, 201–13.

Lusztig, G. (1976b), Coxeter orbits and eigenspaces of Frobenius. *Inventiones Mathematicae* **38**, 101–59.

Lusztig, G. (1977), *Representations of finite Chevalley groups.* C.B.M.S. Regional Conference Series in Mathematics, vol. 39. American Mathematical Society, Providence, RI.

Lusztig, G. (1984), *Characters of reductive groups over a finite field.* Annals of Mathematical Studies, vol. 107. Princeton University Press.

Lusztig, G. (1988), On the representations of reductive groups with disconnected centre. *Astérique* **168**, 157–66.

Lusztig, G. (1991), Intersection cohomology methods in representation theory. *Proceedings of the International Congress of Mathematics, Kyoto, Japan, 1990.* Springer-Verlag, Berlin–New York, pp. 155–74.

Lusztig, G. (1992), Remarks on computing irreducible characters. *Journal of the American Mathematical Society* **5**, 971–86.

Lusztig, G. (1999), A survey of group representations. *Nieuw Archiev voor Wiskunde* **17**, 483–89.

Macdonald, I. G. (1995), *Symmetric functions and Hall polynomials* (2nd edn). Oxford University Press.

Mumford, D. (1988), *The red book of varieties and schemes.* Lecture Notes in Mathematics, vol. 1358. Springer-Verlag, Berlin–New York.

Reid, M. (1988), *Undergraduate algebraic geometry.* London Mathematical Society Students Texts, vol. 12. Cambridge University Press.

Rossmann, W. (2002), *Lie groups: an introduction through linear groups.* Oxford Graduate Texts in Mathematics, vol. 5. Oxford University Press.

Schmidt, W. K. (1976), *Equations over finite fields: an elementary approach.* Lecture Notes in Mathematics, vol. 536. Springer-Verlag, Berlin–New York.

Serre, J.-P. (1977), *Linear representations of finite groups.* Graduate Texts in Mathematics, vol. 42. Springer-Verlag, Berlin–Heidelberg–New York.

Shafarevich, I. R. (1994), *Basic algebraic geometry.* (vol. 1): *varieties in projective space* (2nd edn). Springer-Verlag, Berlin, Heidelberg.

Shoji, T. (1995), Character sheaves and almost characters of reductive groups. *Advances in Mathematics* **111**, (I) 244–313, (II) 314–354.

Spaltenstein, N. (1982), *Classes unipotentes et sous-groupes de Borel.* Lecture Notes in Mathematics, vol. 946. Springer-Verlag, Berlin–New York.

Spanier, E. H. (1966), *Algebraic topology.* Springer-Verlag, New York.

Springer, T. A. (1998), *Linear algebraic groups* (2nd edn). Progress in Mathematics, vol. 9. Birkhäuser, Boston–Basel–Berlin.

Springer, T. A., and Steinberg, R. (1970), Conjugacy classes. In: Borel, A. et al., *Seminar on algebraic groups and related finite groups.* Lecture Notes in Mathematics vol. 131, Springer-Verlag, Berlin–Heidelberg; see also Steinberg (1997), pp. 293–394.

Srinivasan, B. (1979), *Representations of finite Chevalley groups.* Lecture Notes in Mathematics, vol. 764. Springer-Verlag, Berlin–New York.

Steinberg, R. (1957), Prime power representations of finite linear groups II. *Canadian Journal of Mathematics* **9**, 347–51; see also Steinberg (1997), pp. 35–9.

Steinberg, R. (1967), *Lectures on Chevalley groups.* Mimeographed notes, Department of Mathematics, Yale University.

Steinberg, R. (1968), Endomorphisms of linear algebraic groups. *Memoirs of the American Mathematical Society* **80**, 1–108; see also Steinberg (1997), pp. 229–85.

Steinberg, R. (1974), *Conjugacy classes in algebraic groups.* Lecture Notes in Mathematics, vol. 366. Springer-Verlag, Berlin–New York.

Steinberg, R. (1977), On theorems of Lie–Kolchins, Borel and Lang. In: *Contributions to algebra: a collection of papers dedicated to Ellis Kolchin.* Academic Press; see also Steinberg (1997), pp. 467–72.

Steinberg, R. (1997), *Collected papers* (with a foreword by J.-P. Serre). American Mathematical Society, Providence, RI.

Sturmfels, B. (1993), *Algorithms in invariant theory.* Texts and Monographs in Symbolic Computation. Springer-Verlag, Wien–New York.

Suzuki, M. (1962), On a class of doubly transitive groups. *Annals of Mathematics* **75**, 105–45.

Taylor, D. E. (1992), *The geometry of classical groups.* Sigma Series in Pure Mathematics, vol. 9. Heldermann-Verlag, Berlin.

Tits, J. (1962), Théorème de Bruhat et sous-groupes paraboliques. *Comptes Rendus des Séances de l'Académie des Sciences* (Paris) **254**, 2910–12.

Ward, H. N. (1966), On Ree's series of simple groups, *Transactions of the American Mathematical Society* **121**, 62–89.

Weil, A. (1949), Number of solutions of equations in finite fields. *Bulletin American Mathematical Society* **55**, 497–508.

Index

The manufacturer's authorised representative in the EU for product safety is Oxford
University Press España S.A. of El Parque Empresarial San Fernando de Henares,
Avenida de Castilla, 2 – 28830 Madrid (www.oup.es/en or product.safety@oup.com).
OUP España S.A. also acts as importer into Spain of products made by the manufacturer.

Printed in the USA/Agawam, MA
June 4, 2026

905945.034